T0310248

MICROBIOLOGY OF DRINKING WATER PRODUCTION AND DISTRIBUTION

MICROBIOLOGY OF DRINKING WATER PRODUCTION AND DISTRIBUTION

GABRIEL BITTON

Department of Environmental Engineering Sciences
Engineering School of Sustainable Infrastructure & Environment
University of Florida, Gainesville, Florida, USA

WILEY Blackwell

Library of Congress Cataloging-in-Publication Data has been applied for.

ISBN: 978-1-1187-4392-8 (cloth)

Cover image: Water drops © AndrewJohnson, Bubbles in a water glass © Sergey_Peterman

Printed in Singapore

10 9 8 7 6 5 4 3 2 1

To Benjamin Noah and to all my family across the oceans.
This book is dedicated to all the children around the world who do not have access to safe drinking water.

CONTENTS

PREFACE   xi

1 MICROBIAL CONTAMINANTS IN DRINKING WATER   1
   1.1 Introduction   1
   1.2 Transmission Routes of Pathogens and Parasites   1
   1.3 Major Pathogens and Parasites of Health Concern in Drinking
        Water   6
   Web Resources   27
   Further Reading   27

2 MICROBIOLOGICAL ASPECTS OF DRINKING WATER
   TREATMENT   29
   2.1 Introduction   29
   2.2 Worldwide Concern Over Drinking Water Safety   30
   2.3 Microbiological Quality of Source Water   33
   2.4 Overview of Processes Involved in Drinking Water Treatment
        Plants   35
   2.5 Process Microbiology and Fate of Pathogens and Parasites in
        Water Treatment Plants   36
   2.6 Waste Residuals from Water Treatment Plants   55
   2.7 Drinking Water Quality at the Consumer's Tap   55
   Web Resources   62
   Further Reading   63

3 DRINKING WATER DISINFECTION   65
   3.1 Introduction   65
   3.2 Chlorine   66
   3.3 Chlorine Dioxide   72
   3.4 Ozone   74
   3.5 Ultraviolet Light   76
   3.6 Use of Photocatalists in Water Disinfection   85
   3.7 Physical Removal/Inactivation of Microbial Pathogens   86

Web Resources   89
Further Reading   89

4   DRINKING WATER DISTRIBUTION SYSTEMS: BIOFILM
    MICROBIOLOGY   91
    4.1   Introduction   91
    4.2   Biofilm Development in WDSs   93
    4.3   Growh of Pathogens and Other Microorganisms in WDSs   105
    4.4   Some Advantages and Disadvantages of Biofilms in Drinking
          Water Treatment and Distribution   109
    4.5   Biofilm Control and Prevention   112
    Web Resources   114
    Further Reading   115

5   ESTHETIC AND OTHER CONCERNS ASSOCIATED WITH
    DRINKING WATER TREATMENT AND DISTRIBUTION   117
    5.1   Introduction   117
    5.2   Taste and Odor Problems in Drinking Water Treatment Plants   117
    5.3   Algae and Cyanobacteria   121
    5.4   Fungi   127
    5.5   Actinomycetes   129
    5.6   Protozoa   130
    5.7   Invertebrates   132
    5.8   Endotoxins   134
    5.9   Iron, Manganese, and Sulfur Bacteria   134
    5.10 Nitrifying Bacteria in Water Distribution Systems   135
    Web Resources   137
    Further Reading   138

6   BIOLOGICAL TREATMENT AND BIOSTABILITY OF
    DRINKING WATER   141
    6.1   Introduction   141
    6.2   Biological Treatment of Drinking Water   141
    6.3   Assessment of Biostability of Drinking Water   143
    Web Resources   151
    Further Reading   151

7   BIOTERRORISM AND DRINKING WATER SAFETY   153
    7.1   Introduction   153
    7.2   Early History of Biological Warfare   153
    7.3   BW Microbial Agents and Biotoxins   154
    7.4   Deliberate Contamination of Water Supplies with BW Agents or
          Biotoxins   161
    7.5   Early Warning Systems for Assessing the Contamination of
          Source Waters or Water Distribution Systems   165

7.6   Protection of Drinking Water Supplies   168
7.7   Disinfection of BW-Contaminated Drinking Water   169
Web Resources   170
Further Reading   171

8   WATER TREATMENT TECHNOLOGIES FOR DEVELOPING
     COUNTRIES   173
8.1   Introduction: Water for a Thirsty Planet   173
8.2   Some Statistics of Waterborne Diseases in Developing Countries   174
8.3   Some HWT Methods or Technologies in Use in Developing
       Countries   175
8.4   Personal Portable Water Treatment Systems for Travelers and
       Hikers   189
Web Resources   193
Further Reading   194

9   BOTTLED WATER MICROBIOLOGY   195
9.1   Introduction   195
9.2   Sources and Categories of Bottled Water   197
9.3   Bottled Water Microorganisms   199
9.4   Regulations Concerning Bottled Water   203
Web Resources   204
Further Reading   205

10   INTRODUCTION TO MICROBIAL RISK ASSESSMENT FOR
      DRINKING WATER   207
10.1   Health-Based Targets for Drinking Water   207
10.2   Quantitative Microbial Risk Assessment (QMRA)   208
10.3   Some Examples of Use of Risk Assessment to Assess the Risk
        of Infection or Disease From Exposure to Microbial Pathogens   212
Web Resources   216
Further Reading   216

REFERENCES   217

INDEX   289

PREFACE

In industrial countries, we take safe drinking water for granted. Due to an increase in world population, the microbiological safety of drinking water is becoming a worldwide concern. Following a meeting of drinking water experts in 1995, it was predicted that water safety will be a major concern in the 21st century (Ford and Colwell, 1996). Concerns have been raised over the emergence of antibiotic resistance and chlorine-resistant microorganisms such as *Cryptosporidium*. Consequently, the largest documented waterborne cryptosporidiosis outbreak occurred in 1993 in Milwaukee, Wisconsin, where 403,000 people became ill, resulting in 4400 hospitalized patients and 54 deaths. Furthermore, aging populations in developed countries and increasing use of immunosuppressive drugs have led to decreased immunity to waterborne pathogens and parasites. The problems are more serious in developing countries who suffer from unsafe drinking water and poor sanitation, and where children are, unfortunately, the main victims of infectious and parasitic diseases.

The long-term effects of chemical toxicants in drinking water have received much attention by investigators and, comparatively, less efforts have been devoted to the microbiological safety of this precious resource.

This book is divided into 10 chapters. The first chapter introduces the reader to the topic of microbial pathogens and parasites of concern in drinking water safety. Chapter 2 deals with the microbiology of the treatment processes involved in conventional water treatment plants. Advances in drinking water research followed by the establishment of multiple barriers against microbial pathogens and parasites have significantly increased the safety of the water we drink daily. This multiple-barrier system includes source water protection, reliable water treatment (pretreatment, coagulation, flocculation, sedimentation, filtration, disinfection, water softening, membrane filtration, activated carbon treatment, and the potential use of nanotechnology for water purification). Household water treatment processes via the use of point-of-use (POU) devices are also included in this chapter. Chapter 3 discusses the disinfection step that is an essential and final barrier against human exposure to disease-causing pathogenic microorganisms, including viruses, bacteria, and protozoan parasites. Disinfection of drinking water is probably the most significant preventive measure in human history. Many waterborne outbreaks are attributed to the degradation of water quality in water distribution systems (WDS) through cross-connections, main breaks, back siphonage, or negative pressure events. The main topic covered in Chapter 4 is biofilm microbiology with emphasis on biofilm formation and the factors involved in its development in water distribution pipes. The fate of pathogens (e.g., nontubercular mycobacteria, *Legionella*, protozoan parasites, enteric viruses, and opportunistic pathogens in general) is also covered. Chapter 5 covers other microbiological topics of concern in WDS. These topics include taste and odor problems, cyanotoxins produced by cyanobacteria, fungi, protozoa with emphasis on free living amoebas, microinvertebrates, iron and manganese bacteria, and the occurrence of nitrifying bacteria. Chapter 6 deals with the biotreatment and biostability of drinking water and covers the various methods used to assess biostability of drinking water. Drinking water safety can be compromised by the deliberate or accidental contamination of drinking water resources. Chapter 7 covers the major biowarfare microbial agents and biotoxins that could be used to deliberately contaminate drinking water. An estimated 1.1 billion of the world's population does not have access to safe clean water, and approximately 2.6 billion people lack improved sanitation. As a result, approximately 2.2 million people die each year from waterborne diseases. The World Health Organization estimates that endemic diarrhea accounts for 17% of all deaths among children less than 5 years of age. Chapter 8 covers the major treatment technologies for improving the quality of drinking water in developing countries. Chapter 9 addresses the topic of bottled water microbiology. Although bottled water is a useful resource in emergency situations, its quality may sometimes be contaminated via introduction of microbial pathogens at the source or during bottling. Chapter 10 describes the steps involved in quantitative microbial risk assessment (QMRA) and gives examples of the use of risk assessment to estimate the risk of exposure to bacterial, viral, and protozoan pathogens and predict the burden of waterborne diseases on a given community.

This book can serve as a textbook for courses on drinking water microbiology and would be useful in other courses in environmental engineering programs. Due to the author's extensive review of the literature pertaining to drinking water microbiology, this book can also serve as a reference text for engineers and scientists interested in the interface between public health microbiology and drinking water treatment and distribution.

I am grateful to Nancy, Julie, Thomas, Natalie, Jonathan, Ari-Gabriel, Benjamin Noah, my family across the globe, and my close friends and colleagues for their enthusiastic support for this project.

GABRIEL BITTON
Gainesville, FL
April 18, 2014

MICROBIAL CONTAMINANTS IN DRINKING WATER

1.1 INTRODUCTION

In water treatment plants, the goal is to produce pathogen- and parasite-free drinking water, not necessarily sterile water. There are, however, several sources of contamination in a potable water system (Percival et al., 2000):

- Drinking water source (see Chapter 2).
- Inadequate treatment in the water treatment plant.
- Water distribution system (WDS): Treated water quality may deteriorate in the distribution system. Pathogens and parasites may be introduced into treated water through cracks in the water pipes, back-siphonage or cross-contamination.
- Biofilm development which may alter water quality.

This chapter surveys the major microbial pathogens and parasites which may contaminate drinking water.

1.2 TRANSMISSION ROUTES OF PATHOGENS AND PARASITES

Transmission involves the transport of an infectious agent from the reservoir to a host. It is the most important link in the chain of infection. Pathogens can be transmitted from the reservoir to a susceptible host by various routes. The transmission pathways of water-related pathogens are summarized in Figure 1.1 (WHO, 2011c).

1.2.1 Person-to-Person Transmission

The most common route of transmission of infectious agents is from person to person. Examples of direct contact transmission are the sexually transmitted diseases such

Microbiology of Drinking Water Production and Distribution, First Edition. Gabriel Bitton.
© 2014 John Wiley & Sons, Inc. Published 2014 by John Wiley & Sons, Inc.

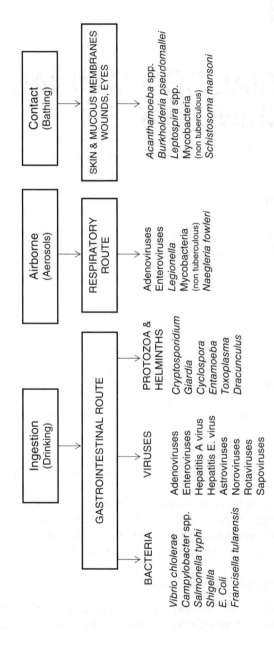

Figure 1.1 Transmission pathways for water-related pathogens and parasites. Adapted from WHO (2011c). *Guidelines for Drinking-Water Quality.* 4th Ed. World Health Organization, Geneva, Switzerland.

as syphilis, gonorrhea, herpes, or acquired immunodeficiency syndrome (AIDS). Coughing and sneezing discharge very small droplets containing pathogens within a few feet of the host (droplet infection).

1.2.2 Waterborne Transmission

Water is essential to life on Planet Earth. Humans generally consume from 2 to 4 L water/day. The composition of drinking water has positive (e.g., iron, fluoride) and, sometimes, negative effects on human health because it may contain microbiological, chemical, and radiological contaminants. When contaminated with human and animal wastes, water contributes to (Mintz et al., 2001)

- *Waterborne diseases*: Infections with microbial pathogens may result in gastrointestinal problems (e.g., diarrhea) or in systemic illnesses (e.g., hepatitis caused by viruses; kidney failure resulting from infection with *Escherichia coli* O157:H7) (Krewski et al., 2002).
- *Water-washed diseases* (e.g., trachoma, lice, tick-borne disease) linked to poor hygiene resulting from poor access to safe water.
- *Water-based diseases* such as schistosomiasis caused by parasitic worms (e.g., *Schistosoma mansoni*) that live inside freshwater snails. The infectious form of the worm, called cercariae, contaminates the water and causes infection following contact with the skin. The parasite causes damages in several organs such as the intestines, skin, liver, and the brain.
- *Water-related diseases* caused by insect vectors that breed in water (e.g., malaria, Dengue fever, yellow fever).

In 1854, John Snow, an English physician and epidemiologist, first established the waterborne transmission of pathogens. Following a cholera epidemics that resulted in 500 deaths, he noted a relationship between the cholera epidemics and consumption of water from the Broad Street well in London. In 1893, Robert Koch, isolated the bacterium that causes cholera, establishing the link between drinking water contamination and the onset of cholera. At the beginning of the twentieth century, cholera and typhoid fever were the major waterborne diseases of concern to public health officials (more details are given by Medena et al., 2003). Other pathogens and parasites have emerged since and include, among others, *Cryptosporidium, Giardia, E. coli* O157:H7, and enteric viruses.

In 1996, diarrheal diseases contracted worldwide mainly via contaminated water or food, killed 3.1 million people, most of them were children (WHO, 1996b). In the United States, waterborne disease outbreaks are generally reported to the US Environmental Protection Agency (U.S. EPA) and the Centers for Disease Control and Prevention (CDC) by local epidemiologists and health authorities (Craun, 1986, 1988). Many of the outbreaks are caused by untreated or inadequately treated groundwater and surface waters. Gastrointestinal illnesses of unidentified etiology and giardiasis are the most common waterborne diseases of groundwater and surface water systems (Craun, 1988).

TABLE 1.1   Estimated number of cases per year caused by infection with selected bacterial pathogens in the United States

Foodborne Pathogen	Estimated Cases ($\times 10^3$)	Commonly Implicated Foods
Campylobacter jejuni	4000	Poultry, raw milk, untreated water
Salmonella (nontyphoid)	2000	Eggs, poultry, meat, fresh produce, other raw foods
Escherichia coli O157:H7	25	Ground beef, raw milk, lettuce, untreated water, unpasteurized cider/apple juice
Listeria monocytogenes	1.5	Ready-to-eat foods (e.g., soft cheese, deli foods, pate)
Vibrio sp. (e.g., *V. vulnificus*)	10	Seafood (raw, undercooked, cross-contaminated)

Source: Adapted from Altekruse et al. (1997). Emerging Infect. Dis. 3: 285–293.

1.2.3 Foodborne Transmission

Unsanitary practices during food production or preparation are the cause of the transmission of several pathogens and parasites. Food affects millions of consumers every year, resulting in thousands of deaths. Emerging foodborne pathogens reported in the United States are shown in Table 1.1 (Altekruse et al., 1997). Raw vegetables and fruits may become contaminated following irrigation with fecally contaminated water or as a result of being handled by an infected person during processing, storage, distribution, or final preparation (Seymour and Appleton, 2001).

Shellfish (e.g., oysters, clams, mussels) consumption is significant in the transmission of bacterial and viral pathogens and protozoan cysts and oocysts. Shellfish often live in polluted estuarine water. As filter feeders, they process large quantities of water (4–20 L/h) and accumulate pathogens and parasites mainly in their digestive tissues. They are of health concern because they are often eaten raw or insufficiently cooked. The development of molecular techniques has helped in the detection of pathogens and parasites in shellfish. Various types of pathogenic viruses were detected in shellfish samples around the world. These include enteroviruses, noroviruses, hepatitis A virus, rotaviruses, astroviruses, and adenoviruses (Beuret et al., 2003; Costantini et al., 2006; Gentry et al., 2009; Hernroth et al., 2002; Le Guyader et al., 1996, 2000, 2009; Nenonen et al., 2008; Pintó et al., 2009; Wang et al., 2008). Infectious oocysts of *Cryptosporidium* were also detected in mussels and cockles in Spain (Gomez-Bautista et al., 2000).

1.2.4 Airborne Transmission

Some diseases (e.g., Q fever, some fungal diseases) can be spread by airborne transmission. This route is important in the transmission of biological aerosols generated by wastewater treatment plants or spray irrigation with wastewater effluents. As regard drinking water, *Legionella* and *Mycobacterium* can be transmitted by the airborne route.

1.2.5 Vector-Borne Transmission

The most common vectors for disease transmission are arthropods (e.g., fleas, insects) or vertebrates (e.g., rodents, dogs and cats). The pathogens may or may not multiply inside the arthropod vector. Some vector-borne diseases are malaria (caused by *Plasmodium*), yellow fever or encephalitis (both due to arboviruses), and rabies (from virus transmitted by the bite of rabid dogs or cats).

1.2.6 Fomites

Some pathogens may be transmitted by nonliving objects or *fomites* (e.g., clothes, utensils, toys, stationary and cell phones, computer keyboards, and other environmental surfaces). Fomites frequently become contaminated with pathogens and parasites from several sources (e.g., contaminated hands, bodily secretions, airborne aerosols) and are responsible for the transmission of enteric (e.g., rotaviruses, noroviruses) and airborne viruses (e.g., rhinoviruses, influenza viruses). Table 1.2 shows examples of virus-contaminated environmental surfaces (Boone and Gerba, 2007). Following contamination of the surface, viruses can survive for relatively long periods, depending on the fomite properties, environmental factors (temperature, relative humidity, pH, light, other microbes present) and virus type and strain. Fomite-to-finger

TABLE 1.2 **Buildings or surfaces where viruses have been detected or survived**

	Location of Virus	
Virus	Buildings	Surfaces
Rhinovirus	Not found	Skin, hands, door knobs, faucet
Influenza virus	Day care centers, homes Nursing homes	Towels, medical cart items
Coronavirus	Hospitals, apartment	Phones, doorknobs, computer gloves, sponges
Norovirus	Nursing homes, hotels hospital wards, cruise ships, recreational camps	Carpets, curtains, lockers, bed covers, bed rails, drinking cup, water jug handle, lampshade
Rotavirus	Day care centers, pediatric ward	Toys, phone, toilet handles, sinks, water fountains, door handles, refrigerator handles, water play tables, thermometers, play mats, china, paper, cotton cloth, glazed tile, polystyrene
Hepatitis A virus	Hospitals, schools, institutions for mentally handicapped, animal care facilities, bars	Drinking glasses, paper, china cotton cloth, latex gloves, glazed tile, polystyrene
Adenovirus	Schools, pediatric wards, nursing homes	Paper, china

Source: Adapted from Boone and Gerba (2007). Appl. Environ. Microbiol. 73: 1687–1696.

transfer efficiency for *E. coli, Staphylococcus aureus, Bacillus thuringiensis,* and MS2 phage is higher for nonporous surfaces (e.g., stainless steel, glass, ceramic tile) than for porous surfaces (e.g., polyester, paper currency), and is also higher under high relative humidity than under low relative humidity (Lopez et al., 2013). Pathogen transmission by fomites can be controlled via thorough and frequent hand washing and proper disinfection of the contaminated surfaces.

1.3 MAJOR PATHOGENS AND PARASITES OF HEALTH CONCERN IN DRINKING WATER

It is estimated that there are 1407 species of pathogens infecting humans, and 177 of the species are considered emerging or reemerging pathogens. These pathogens include bacteria (538 species), viruses (208 types), fungi (317 species), parasitic protozoa (57 species), and helminths (287 species) (Woolhouse, 2006). Infection of susceptible hosts may sometimes lead to overt disease. The development of the disease depends on various factors, including infectious dose, pathogenicity, host susceptibility, and environmental factors. One important factor is the *minimal infective dose* (MID) which varies widely (a few viruses or oocysts to 10^7 bacterial cells) with the type of pathogen or parasite (Table 1.3; Bitton, 2011). Some organisms, however, are *opportunistic pathogens* that cause disease mostly in immunocompromised individuals (e.g., infants, the elderly, patients on antimicrobial chemotherapy, HIV, or cancer patients) (see review of Mota and Edberg, 2002). Domestic and wild animals may serve as reservoirs for several diseases called zoonoses that are transmitted from animals to humans. Some examples of zoonoses are rabies, brucellosis, tuberculosis, anthrax, leptospirosis, and toxoplasmosis. Water, wastewater,

TABLE 1.3   Minimal infective doses for some pathogens and parasites

Organism	Minimum Infective dose (MID)
Salmonella spp.	10^4-10^7
Shigella spp.	10^1-10^2
Escherichia coli	10^6-10^8
Escherichia coli O157:H7 < 100	
Vibrio cholerae	10^3
Campylobacter jejuni	about 500
Mycobacterium avium 10^4-10^7 (for mice)	
Yersinia enterocolitica 10^6	
Giardia lamblia	10^1-10^2 cysts
Cryptosporidium	10^1 cysts
Entamoeba coli	10^1 cysts
Ascaris	1–10 eggs
Hepatitis A virus	1–10 PFU

Source: Compiled by Bitton (2011).

TABLE 1.4   U.S. EPA Contaminant Candidate List 3 (CCL3)

Microbial Contaminant	Illness
Bacteria	
Campylobacter jejuni	Causes mild, self-limiting GI illness
Escherichia coli	GI illness and kidney failure
Helicobacter pylori	Causes stomach ulcers and cancer
Legionella pneumophila	Naturally found in aquatic environments. Causes pneumonia
Mycobacterium avium	Opportunistic bacterium. Causes lung disease
Salmonella enterica	Causes mild self-limiting GI illness
Shigella sonnei	Causes mild self-limiting GI illness and bloody diarrhea
Viruses	
Adenoviruses	Cause respiratory and GI illnesses
Caliciviruses	They include norovirus: mild GI illness
Enteroviruses	Wide range of illnesses
Hepatitis A virus	Causes liver disease and jaundice
Protozoa	
Naegleria fowleri	Protozoan parasite found in warm water and causing primary amoebic meningoencephalitis (PAME)

Source: Adapted from U.S. EPA (http://www.epa.gov/safewater/ccl/ccl3.html).

food, or soils can also harbor infectious agents and serve as reservoirs (Berger and Oshiro, 2002; Woolhouse, 2006).

Several pathogenic microorganisms and parasites as well as indigenous microorganisms are commonly found in water supplies and may end up in drinking water. The three categories of pathogens generally encountered in aquatic environments are (Leclerc et al., 2002):

- **Bacterial Pathogens**: Some of these pathogens (e.g., *Salmonella*, *Shigella*, *Vibrio cholerae*) are enteric bacteria. Others (e.g., *Legionella*, *Mycobacterium avium*, *Aeromonas*) are indigenous aquatic bacteria.
- **Viral Pathogens**: They are also released into aquatic environments but are unable to multiply outside their host cells. Their MIDs are generally lower than for bacterial pathogens.
- **Protozoan Parasites**: They are released into aquatic environments as cysts or oocysts which are quite resistant to environmental stress and to disinfection in water and wastewater treatment plants.

Table 1.4 shows the microbial contaminants placed in 2008 on the U.S. EPA Contaminant Candidate List 3 (CCL3) (http://www.epa.gov/safewater/ccl/ccl3.html).

1.3.1 Bacterial Pathogens

Human and animal fecal matter contains billions of bacteria (up to 10^{12} per gram of feces) many of which belong to bacterial genera such as *Bacteroides, Bifidobacterium, Clostridium*, and *Prevotella* (Matsuki et al., 2002). However, some of the bacteria shed in feces are pathogenic and can be transmitted by the waterborne or other routes to humans, leading to enteric infections such as typhoid fever or cholera (Theron and Cloete, 2002).

We will now briefly discuss the main bacterial infectious agents that may be encountered in drinking water.

1.3.1.1 Vibrio cholerae. *Vibrio cholerae* is a gram-negative, coma-shaped with a single polar flagellum, facultatively anaerobic bacterium which is a member of the aquatic microbial community and causes cholera, an endemic disease prevailing in Asia (e.g., Bangladesh) and other parts of the globe. This pathogen releases an enterotoxin that causes a mild to profuse diarrhea, vomiting, and a very rapid loss of fluids and electrolytes, resulting in the patient death relatively in a short period of time if not properly treated. Cholera cases are estimated at 3–5 million, resulting in 100,000–120,000 deaths annually mostly in developing countries like Asia, Africa, and Latin America (WHO, 2011a). There are approximately 200 known serogroups of *Vibrio cholerae*, and only serogroups O1 and O139 are known to cause disease outbreaks (Huk et al., 2002). This pathogen is transmitted mainly via contaminated food and water. *Vibrio cholerae* was detected in contaminated food and drinking water, using immunological or molecular techniques. In aquatic environments, this pathogen generally attaches to solids and plankton such as zooplankton, cyanobacteria, and algae and persists under the viable but nonculturable (VBNC) state (Brayton et al., 1987; Huk et al., 1990).

1.3.1.2 Salmonella. More than 2000 serotypes of Salmonellae are encountered in the environment. An estimated 2–4 million human *Salmonella* infections occur each year in the United States (Feachem et al., 1983). *Salmonella* species are the cause of typhoid and paratyphoid fevers and gastroenteritis. *Salmonella typhi*, the etiological agent of typhoid fever, produces an endotoxin that causes fever, nausea, and diarrhea lasting 3–5 days, and, in severe cases, may be fatal if not properly treated with antibiotics. *Salmonella* is transmitted via contaminated food (e.g., chicken, milk, and eggs) or drinking water. Typhoidal *Salmonella* species (e.g., *S. typhi*, *S. paratyphi*) are associated with waterborne transmission whereas the nontyphoidal species (e.g., *S. typhimurium*) are associated with person-to-person contact and foodborne transmission (WHO, 2011c).

1.3.1.3 Shigella. *Shigella* is a gram-negative, nonmotile rod and is a member of the Enterobacteriaceae. It is the etiologic agent of bacillary dysentery or shigellosis, which affects the large intestine, leading to cramps, diarrhea, fever, and bloody stools resulting from inflammation and ulceration of the intestinal mucosa. Each year, *Shigella* causes over 2 million infections and approximately 600,000 deaths worldwide, especially in developing countries (WHO, 2011c). There are four pathogenic species of *Shigella*: *S. dysenteriae*, *S. flexneri*, *S. boydii*, and *S. sonnei*. *Shigella dysenteriae*, serotype 1, causes severe diarrhea and produces a potent toxin called the Shiga toxin which may lead to hemolytic uremic syndrome (HUS) which results in kidney failure (Torres, 2002).

Shigella is transmitted via person-to-person contact and via the foodborne (e.g., salads or raw vegetables) and waterborne routes. As shown in Table 1.3, its infectious dose is relatively low and can be as low as 10 organisms. Cases of shigellosis have been linked to drinking water and have been documented in homes, schools, and cruising ships. This pathogen has a relatively low persistence in the environment and, thus, little is known about its fate in water treatment plants.

1.3.1.4 *Escherichia coli.*

Escherichia coli, a gram-negative facultative anaerobe, colonizes the gastrointestinal tract of humans and warm-blooded animals. While many *E. coli* strains are harmless, some bear virulence factors and cause diarrhea and other health problems. These diarrheagenic *E. coli* serotypes have been classified as enterotoxigenic (ETEC), enteropathogenic (EPEC), enterohemorrhagic (EHEC), enteroinvasive (EIEC), and enteroaggregative (EAggEC) (Guerrant and Thielman, 1995; Nataro and Kaper, 1998). Molecular techniques (nucleic acid probes and PCR) help distinguish between diarrheagenic strains of *E. coli* and nonpathogenic strains.

Enterohemorrhagic *E. coli* O157:H7 produces shiga-like toxins 1 and 2, encoded by stx_1 and stx_2 genes, respectively. It has a relatively low infectious dose (<100 organisms) and causes bloody diarrhea, particularly, among the very young and very old members of the community (Herwaldt et al., 1992). Infections, if left untreated, may lead to HUS, resulting in kidney damage or failure in children, with 3–5% deaths of patients (Boyce et al., 1995). *Escherichia coli* O157:H7 is the infectious agent in several waterborne outbreaks, some of which are shown in Table 1.5 (Muniesa et al., 2006). The most serious outbreaks associated with drinking water occurred in Cabool, Missouri (243 infected; 4 deaths); Walkerton, Canada (2300 infected; 7 deaths); and New York (775 infected, 2 deaths) (Muniesa et al., 2006; Swerdlow et al., 1992).

Escherichia coli O157:H7 is detected in environmental samples, using culture techniques, immunological methods (Tsai et al., 2000), or molecular methods such as multiplex PCR protocols based on the detection of six major virulence genes (*fliC, stx1, stx2, eae, rfbE, and hlyA*) (Bai et al., 2010).

Another shiga toxin–producing *E. coli* (STEC) is strain O104:H4. It is the causative agent of a 2011 epidemic in Germany traced to the consumption of uncooked sprouts (Muniesa et al., 2012).

1.3.1.5 *Yersinia.*

Yersinia enterocolitica is a gram-negative rod responsible for acute gastroenteritis with invasion and ulceration of the terminal ileum. Foodborne (e.g., milk, tofu, meat, and meat products) outbreaks of yersiniosis have been documented. The role of water is uncertain, but there are instances in which this pathogen was suspected to be the cause of waterborne transmission of gastroenteritis (Schiemann, 1990). This pathogen has been isolated from wastewater effluents, river water, and from drinking water (Bartley et al., 1982; Meadows and Snudden, 1982; Stathopoulos et al., 1990). However, since most of the *Yersinia* spp. detected in drinking water are nonpathogenic, their transmission is likely via human and animal wastes (WHO, 2011c).

TABLE 1.5   Reported waterborne outbreaks caused by EHEC *Escherichia coli* strains

Outbreak	Location	No. Cases (deaths)	Transmission Route
1989	Missouri (USA)	243 (4)	Unchlorinated water supply
1990	Saitama (Japan)	174	Drinking water
1990	Shimane (Japan)	12	Natural water
1991	Oregon (USA)	21	Lake water
1992	Swaziland & South Africa	>100[a]	Untreated water
1992	Scotland (UK)	6	Padding pool
1993	Rotterdam (Holland)	4	Swimming water
1993	Surrey (UK)	6 (1)	Padding pool
1994	New York (USA)	12	Lake water
1995	Illinois (USA)	12	Swimming water (lake)
1996	Scotland (UK)	6	Drinking water
1997	Seinajoki (Finland)	14	Lake water
1998	Wyoming (USA)	157	Drinking water
1999	New York (USA)	775(2)	Drinking water
1999	Connecticut (USA)	11	Lake water
1999	Scotland (UK)	6	Unchlorinated water supply
1999	South Devon (UK)	14 (1)	Beach water
1999	Washington (USA)	37	Lake water
2000	Walkerton (Canada)	2300 (7)	Drinking water
2001	Minnesota (USA)	20	Public beach
2002	Kentucky (USA)	2	Untreated groundwater
2004	Cornwall (UK)	7	Freshwater stream
2004	British Columbia (Canada)	10	Water spray (Rec. park)

Source: Adapted from Muniesa, M., J. Jofre, C. Garcia-Aljaro, and A.R. Blanch (2006). Environ. Sci. Technol. 40: 7141–7149.
[a]Undetermined number of cases.

1.3.1.6   Campylobacter.

Campylobacter (e.g., *C. jejuni*, *C. fetus*) is a thermotolerant (does no grow below 30°C), gram-negative bacterial pathogen that infects both humans and animals. It is highly infective and has a relatively low infectious dose of approximately 500–1000 organisms. Important animal reservoirs of this pathogen are poultry, wild birds, and cattle. It is transmitted to humans by consumption of raw milk, uncooked poultry, or drinking water (Hänninen et al., 2003; Palmer et al., 1983; Taylor et al., 1983, WHO, 2011c). It is the leading cause of foodborne infections and gastroenteritis (fever, nausea, abdominal pains, diarrhea, vomiting) in developed countries (Frost et al., 2002). Some infected individuals may contract the Guillain–Barré syndrome, an acute paralytic disease or the Reiter's syndrome, an arthritic disease (Altekruse et al., 1997). *Campylobacter* shows no or little correlation with traditional bacterial indicators (St-Pierre et al., 2009). It does not grow in the environment but may survive in a VBNC state in aquatic environments (Park, 2002; Rollins and Colwell, 1986).

Arcobacter, mainly *A. butzleri*, is a *Campylobacter*-like bacterial pathogen commonly involved in gastroenteritis in humans and animals. There are 17 species of *Arcobacter*, some of which (mainly *A. butzleri*, *A. cryaerophilus*, *A. skirrowii*) have been associated with diarrhea in humans (Lecican et al., 2013). *Arcobacter* displays several microbiological and clinical characteristics like *Campylobacter jejuni*. However, it is more associated with persistent and watery diarrhea but causes less bloody diarrhea (Vandenberg et al., 2004). *Arcobacter* is a waterborne and foodborne pathogen transmitted mainly via contaminated meat. It is detected via enrichment followed by plating on selective media or via PCR assays (Brightwell et al., 2007).

1.3.1.7 Legionella pneumophila. This bacterial pathogen is the cause of Legionnaires' disease, a type of acute pneumonia, due to the ability of *Legionella* to multiply within the alveolar macrophages resulting in a high fatality rate. The disease was first reported in 1976 in Philadelphia, PA. Thousands of cases are reported each year in the United States (Marston et al., 1997) and other parts of the world. For example, in France, 1044–1527 cases of legionellosis were reported per year between 2003 and 2007 (http://www.invs.sante.fr/surveillance/legionellose, July, 2010; cited by Touron-Bodilis et al., 2011). The disease is a severe form of pneumonia with symptoms including myalgia (muscle pain), fever, chills, coughing, chest and abdominal pain, and diarrhea (Percival et al., 2000). Many nosocomial (i.e., hospital-acquired) legionellosis outbreaks occur in hospital setting where the sources of the pathogens are the potable water distribution system or, sometimes, the hospital cooling towers (Garcia-Nunez et al., 2008; Harb and Kwaik, 2002). However, hyperchlorination (2–10 mg/L free residual chlorine) was able to control this pathogen (Shands et al., 1985).

The reservoirs for this pathogen are the aquatic environment and soils (e.g., potting soils) (Duchin et al., 2000; Koide et al., 1999). The major factors affecting the survival and growth of *Legionella* in natural and engineered (e.g., hot-water tanks) aquatic environments are temperature, disinfectant type and concentration, presence of certain types of nutrients, and the beneficial protective role of protozoan hosts (Buse et al., 2012).

This organism is transmitted mainly by aerosolization of contaminated water and contaminated floating biofilms or soils (Declerck, 2010; Harb and Kwaik, 2002). Outbreaks may occur following exposure to *L. pneumophila* aerosols from cooling towers, humidifiers, shower heads, spas, hot-water faucets, air conditioners, dental equipment, whirlpools, mist machines in produce departments of grocery stores, or mechanical aerosolization of soil particles during gardening (Armstrong and Haas, 2007; Bauer et al., 2008; Bollin et al., 1985; Leoni et al., 2001; Muraca et al., 1988; Ragull et al., 2007). A conceptual exposure model describing the inhalation of *Legionella* aerosols generated by shower heads is described in Figure 1.2 (Schoen and Ashbolt, 2011). In the United States, between 2001 and 2006, *Legionella* was found to be the cause of 29% of outbreaks (Craun et al., 2010). Although *L. pneumophila* has been detected in drinking water systems (Hsu et al., 1984; Tobin et al., 1986), no outbreak has been attributed to the consumption of contaminated drinking water.

Legionella pneumophila growth in aquatic environments and in engineered water systems benefits from its association with other bacteria (Gião et al., 2011; Stout et al., 1985), amoeba (e.g., *Acanthamoeba, Naegleria. Hartmannella, Echinamoeba, Vahlkampfia*) (Barbaree et al., 1986; Barker et al., 1992; Declerck, 2010; Lau and Ashbolt, 2009; Neumeister et al., 1997; Thomas et al., 2008), and ciliates (e.g., *Tetrahymena*) (Fields et al., 1984; Smith–Somerville et al., 1991). As an example, ingested *Legionella* was shown to form pellets inside *Tetrahymena*, a phenomenon

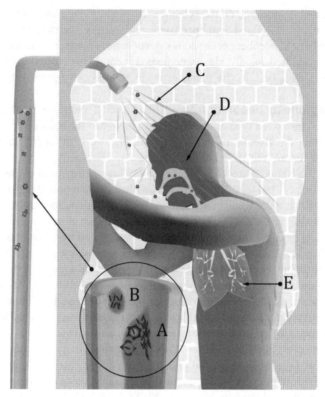

Figure 1.2   Conceptual model for Legionella exposure from inhalation of shower aerosols containing Legionella derived from the in-premise plumbing of buildings. First, Legionella multiply within the premise plumbing biofilm, potentially by colonizing the biofilm or within a protozoan host (A). The biofilm-associated Legionella detaches from the biofilm during a showering event (B), is transported to the shower head, and is aerosolized (C). Finally, the aerosolized Legionella is inhaled (D) and a fraction of that inhaled dose is deposited in the alveolar region of the lungs (E). *Source:* Schoen and Ashbolt (2011). Water Res. 45: 5826–5836. (For a color version, see the color plate section.)

that helps in the survival and transmission of the bacterial pathogen (Berk et al., 2008). This intracellular growth of *Legionella* inside prototozoa provides protection against disinfection and other factors such as low pH and high temperatures (Declerck, 2010; States et al., 1990). Thus, one should consider the role played by protozoa in any control strategy concerning *L. pneumophila*. The control measures for *Legionella* in water distribution systems include heat treatment at 70°C for 30 minutes (however, *Legionella* may become heat-resistant after repeated heat treatments; Allegra et al., 2011), copper and silver treatment, and hyperchlorination up to 50 mg/L (Harb and Kwaik, 2002), monochloramine (Kool et al., 2000; Moore et al., 2006), and chlorine dioxide (Zhang et al., 2009d).

1.3.1.8 Opportunistic Bacterial Pathogens. Opportunistic bacterial pathogens include bacterial heterotrophs such as *Pseudomonas, Aeromonas, Klebsiella, Flavobacterium, Enterobacter, Citrobacter, Serratia, Acinetobacter, Proteus, Providencia, Moraxella, Micrococcus, Bacillus*, and nontubercular mycobacteria (NTM) (Sobsey and Olson, 1983). Of particular concern is *Acinetobacter baumannii* which accounts for about 80% of healthcare-associated infections. This opportunistic pathogen is contracted via contact with contaminated surfaces in hospital setting, and is resistant to all but one (colistin also known as polymyxin E) antibiotic (CDC, 2010). An outbreak of this pathogen was recently reported in the burn unit of Shands hospital in Gainesville, FL. Opportunistic bacterial pathogens are found in water distribution pipes and finished drinking water, and are of health concern mostly to the elderly, newborn babies, sick and immunocompromised people. Figure 1.3 reports on the occurrence of opportunistic bacterial pathogens such as *L. pneumophila* and *Pseudomonas aeruginosa* in unchlorinated drinking water samples from eight water treatment plants in the Netherlands (van der Wielen and van der Kooij, 2013).

Among the bacterial opportunistic pathogens, the NTM are an important subgroup because they cause lung disease globally. It is estimated that their prevalence in the United States is 30–60 cases per 100,000 people (Prevots et al., 2010). Frequently isolated NTM are *Mycobacterium avium* complex (MAC) (i.e., *M. avium* and *M. intracellulare*) which causes about 80% of lung disease cases, and *M. avium subspecies paraturberculosis* (MAP) which infect both humans and animals (Falkinham, 2002; Falkinham et al., 2001; Parrish et al., 2008).

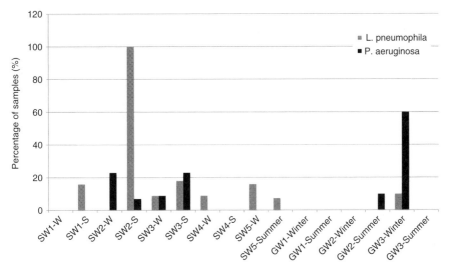

Figure 1.3 Percentages of drinking water samples positive for *L. pneumophila*, and *P. aeruginosa*. Drinking water was sampled from the distribution systems of eight different treatment plants in the winter (W) and the summer (S) of 2010. Adapted from van der Wielen and van der Kooij (2013). Appl. Environ. Microbiol. 79: 825–834.

NTM have been isolated from drinking water and biofilms in distribution lines (see Chapter 4). They may also contaminate hot water tanks, shower heads (e.g., detection of *Mycobacterium mucogenicum*), medical devices, dental units waterlines, hot tubs, indoor swimming pools, soils, and bioaerosols (Covert et al., 1999; Falkinham, 2003; Falkinham et al., 2008; Hilborn et al., 2006; Le Dantec et al., 2002b; Lehtola et al., 2006; Perkins et al., 2009; Rowland and Voorheesville, 2003). In a recent study of a hospital drinking water system, mycobacteria (*Mycobacterium gordonae, M. peregrinum, M. chelonae, M. mucogenicum, M. avium*) were isolated in both water and biofilms in several locations (e.g., shower heads and drains, sink water and drains, water fountain faucet) of the hospital water distribution system (Ovrutsky et al., 2013).

Due to the hydrophobic nature of the waxy and hydrophobic cell wall of mycobacteria, a major route for their transmission is via aerosolization. Shower heads have been implicated in the dissemination of aerosolized microorganisms (e.g., *Legionella, Mycobacterium*). Exposure to these aerosols is responsible for community acquired pneumonia. The hydrophobic cell wall also explains the resistance of mycobacteria to disinfectants and antibiotics (Taylor et al., 2000). Resistance to chemical insult in water distribution systems is also provided by their ability to grow inside free-living amoebas (FLA) which serve as reservoirs and transmission routes for mycobacteria (Corsaro et al., 2010; Miltner and Bermudez, 2000; Taylor et al., 2003; Thomas et al., 2006). NTM have been placed on the on the U.S. EPA drinking water contaminant list.

1.3.1.9 Helicobacter pylori.
Helicobacter pylori, a bacterial agent on the U.S. EPA drinking water contaminant list (CCL), is a gram-negative, spiral shaped bacterium. It causes peptic ulcers, stomach cancer, lymphoma and adenocarcinoma. It is also associated with increased occurrence of iron-deficiency anemia in *Helicobacter*-infected patients (Cardenas et al., 2006; Muhsen and Cohen, 2013). Millions of peptic ulcer cases are reported every year in the United States, with some leading to patient death following surgery (Levin et al., 1998; Sonnenberg and Everhart, 1997). Antibiotics such as metronidazole, tetracycline, amoxicillin, clarithromycin, and azithromycin are used for the treatment of *H. pylori* infections. *Helicobacter pylori* may infect both humans and animals and is transmitted mostly via person-to-person route, and potentially via the waterborne and foodborne routes (Brown, 2000; Hopkins et al., 1993; Klein et al., 1991). *Helicobacter pylori* was detected in surface waters and groundwater samples, using culture-based methods as well as immunological, autoradiography, and molecular-based methods. Its presence was reported in drinking water and biofilms in municipal water distribution systems (Braganca et al., 2007; Gião et al., 2008; Handwerker et al., 1995; Park et al., 2001) where it may persist in the VBNC state (Gião et al., 2008; Moreno et al., 2007). More work is needed to understand the waterborne transmission of this pathogen.

1.3.1.10 Leptospira.
Leptospira is a small aerobic spirochete that can gain access to the host through abrasions of the skin or through mucous membranes of the nose, mouth, and eyes. It causes leptospirosis which is characterized by the dissemination of the pathogen in the patient's blood and the subsequent infection of the kidneys and the

TABLE 1.6 Major waterborne bacterial diseases

Bacterial Agent	Major Disease	Major Reservoir	Main Affected Site
Salmonella typhi	Typhoid fever	Human feces	GI tract
Vibrio cholerae	Cholera	Human feces	GI tract
Shigella	Bacillary dysentery	Human feces	GI tract
Pathogenic *Escherichia Coli*	Gastroenteritis, hemolytic uremic syndrome	Human feces	GI tract
Campylobacter jejuni	Gastroenteritis	Human/animal feces	GI tract
Yersinia enterocolitica	Gastroenteritis	Human/animal feces	GI tract
Legionella pneumophila	Acute respiratory illness (Legionnaire's disease)	Warm to hot contaminated water	Lungs
Mycobacterium tuberculosis	Tuberculosis	Human respiratory exudates	Lungs
Leptospira	Leptospirosis (Weil's disease)	Animal feces and urine	Generalized
Opportunistic pathogens	Various	Natural waters	Mainly GI tract and lungs

Source: Adapted from Sobsey and Olson (1983). Microbial agents of waterborne disease. In: Assessment of Microbiology and Turbidity Standards For Drinking Water, P.S. Berger and Y. Argaman, Eds. Report No. EPA- 570 9 83 001, U.S. Environmental Protection Agency, Office of Drinking Water, Washington, D.C.

central nervous system. Leptospirosis can also result in pulmonary bleeding which could be fatal to infected patients (WHO, 2011c). The disease can be transmitted through direct contact with infected animals (rodents, domestic pets, and wildlife) or though contact with urine-contaminated waters (e.g., bathing). This *zoonotic* disease may strike sewage workers.

A summary of major waterborne bacterial diseases is shown in Table 1.6 (Sobsey and Olson, 1983).

1.3.2 Viral Pathogens

Approximately some 140 types of enteric viruses released in large numbers in the feces of infected individuals can potentially contaminate wastewater and drinking water. Table 1.7 lists the various enteric viruses of public health concern (Bitton, 2011; Schwartzbrod, 1991). They cause a broad spectrum of diseases ranging from gastroenteritis to aseptic meningitis and paralysis. Since enteric viruses do not replicate in the environment outside their host, they occur in relatively small numbers, particularly in drinking water. Their detection in drinking water thus requires a concentration method with the following criteria: applicability to a wide range of viruses, processing of large sample volumes, high recovery rates, reproducibility, rapidity, and low cost. A wide range of concentration methods have been developed to accomplish this task (Gerba, 1987).

TABLE 1.7   Some waterborne human enteric viruses

Virus Group	Serotypes	Some Diseases
Enteroviruses		
Poliovirus	3	Paralysis, aseptic meningitis
Coxsackievirus		
A	23	Herpangia, aseptic meningitis respiratory illness, paralysis, fever
B	6	Pleurodynia, aseptic meningitis, pericarditis, myocarditis, congenital heart anomalies, nephritis, fever
Echovirus	34	Respiratory infection, aseptic meningitis, diarrhea, pericarditis, myocarditis, fever, rash
Enteroviruses (78-71)	4	Meningitis, respiratory illness
Hepatitis A virus (HAV)		Infectious hepatitis
Hepatitis E virus (HEV)		Hepatitis
Reoviruses	3	Respiratory disease
Rotaviruses	4	Gastroenteritis
Adenoviruses	41	Respiratory disease, acute conjunctivitis, gastroenteritis
Norwalk agent (calicivirus)	1	Gastroenteritis
Astroviruses	5	Gastroenteritis

Source: Adapted from Bitton (2011). Wastewater Microbiology, 4th. Edition, Wiley-Blackwell Pub., Hoboken, N.J., 781 pp.
Schwartzbrod, L. Ed. (1991). Virologie des milieux hydriques. TEC & DOC Lavoisier, Paris, France, 304 pp.

Enteric viruses are mainly transmitted via person-to-person contacts. However, they may also be communicated via waterborne transmission either directly (drinking water, swimming, aerosols) or indirectly via contaminated food (e.g., shellfish, vegetables). They may also be communicated via contact with contaminated environmental surfaces found in day-care centers or in hospitals (see Table 1.2). As compared to bacterial pathogens, viruses display a relatively low MID. Epidemiological investigations have shown that enteric viruses are responsible for 4.7–11.8% of waterborne epidemics (Cliver, 1984; Craun, 1988).

We will now cover some major enteric viruses which may contaminate drinking water:

1.3.2.1 Hepatitis Viruses.
Some hepatitis viruses, such as hepatitis B virus (HBV) and hepatitis C virus (HCV), are transmitted via contact with contaminated blood or via sexual contact and will not be discussed here. However, hepatitis A virus (HAV) and hepatitis E virus (HEV) can be transmitted via contaminated drinking water or food (e.g., shellfish).

Hepatitis A virus (HAV) is a 27-nm RNA hepatovirus belonging to the family picornaviridae with a relatively short incubation period (2–6 weeks) and displaying a fecal–oral transmission route, either by person-to-person direct contact, waterborne, or foodborne transmission (Myint et al., 2002). It is the etiological agent of infectious hepatitis which causes liver damage (liver necrosis and inflammation), with jaundice being the typical symptom. Direct contact transmission has been documented mainly in nurseries (especially among infants wearing diapers), mental institutions, prisons, or military camps. Waterborne and foodborne transmission of infectious hepatitis have been documented worldwide, as a result of consuming improperly treated water or contaminated well water, swimming in contaminated lakes and public pools and consuming shellfish grown in wastewater-contaminated waters (De Serres et al., 1999; Mahoney et al., 1992; Schwartzbrod, 1991). An example is the detection of enteric viruses (HAV, noroviruses, and rotaviruses) in up to 20% of market oyster samples in China (Wang et al., 2008).

Due to some difficulties in detecting HAV via tissue culture assays, molecular and immunological techniques are now used for its detection in environmental samples (Altmar et al., 1995; Yeh et al., 2008).

Hepatitis E virus (HEV) is a single-stranded RNA virus classified in the *Hepeviridae* family (Emerson and Purcell, 2006). This virus is not well characterized due to the lack of a tissue culture cell line for its assay but is now detected using molecular techniques (e.g., reverse transcriptase-polymerase chain reaction (RT-PCR)) which show that HEV sequences are classified as four genotypes. Unlike HAV, it is mainly transmitted via fecally contaminated water while the person-to-person transmission is very low (Myint et al., 2002). It is also acquired through consumption of raw or undercooked pork products (zoonotic route) and is inactivated by heating the contaminated food product at an internal temperature of 71°C for 20 minutes (Barnaud et al., 2012). HEV is endemic in several regions in Africa, Asia, and in Mexico. It is also found in urban wastewater in nonendemic areas. HEV attacks mostly young adults and pregnant women and has clinical symptoms similar to those of HAV (Moe, 1997). The fatality rates are 2–3% in the general population and as high as 32% for pregnant women in their third trimester (Bader 1995; Haas et al., 1999).

1.3.2.2 *Viral Gastroenteritis.* Gastroenteritis is caused by protozoan parasites, bacterial, and viral pathogens such as rotaviruses, caliciviruses, enteric adenoviruses, and astroviruses.

Rotaviruses. They are 70-nm particles containing double-stranded RNA surrounded with a double-shelled capsid and belong to the Reoviridae family (Figure 1.4). They are the major cause of infantile acute gastroenteritis in children younger than 2 years of age and contribute significantly to childhood mortality in developing countries. The virus is spread mainly by the fecal–oral route, but a respiratory route has also been suggested (Flewett, 1982). Rotaviruses have been implicated in several waterborne outbreaks around the world (Gerba et al., 1985; Williams and Akin, 1986). Since rotavirus A infection is the most important cause of infant mortality worldwide, two vaccines, Rotarix and RotaTeq, are now available worldwide (Jiang et al., 2010).

Methodology for detecting rotaviruses in environmental samples includes tissue cultures (using MA-104, a cell line derived from fetal rhesus monkey kidney), commercial ELISA kits, or RT-PCR (Gajardo et al., 1995). Using RT-PCR, rotavirus RNA was detected in drinking water at homes with children suffering from rotaviral acute gastroenteritis (Gratacap-Cavallier et al., 2000). Group A rotaviruses were also

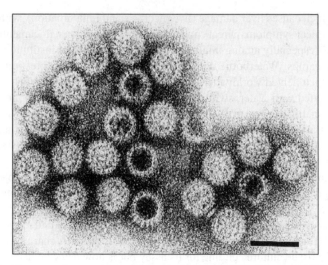

Figure 1.4 Electron micrograph of rotaviruses (bar = 100 nm). Courtesy of F.M. Williams, U.S EPA.

detected in 1.7% of treated drinking water samples in South Africa (van Zyl et al., 2006).

Human caliciviruses. Gastroenteritis is also caused by caliciviruses which are small round viruses genetically related to the 27-nm Norwalk-like virus (NLV) (Figure 1.5). About half of the gastroenteritis outbreaks are due to noroviruses (Atmar and Estes, 2006; Le Guyader et al., 1996). Transmission of these viruses is through direct contact, food (e.g., via shellfish consumption, contaminated fruits and vegetables, food handlers), and water (e.g., drinking water, ice, well water, bottled water) (Hewitt et al., 2007; Lopman et al., 2003; Parshionikar et al., 2003), and outbreaks have been documented around the world, particularly on cruise ships (Table 1.8; Boccia et al., 2002; Gerba et al., 1985; Kukkula et al., 1999; Williams and Akin, 1986). Among the five norovirus genogroups, groups I, II, and IV cause gastroenteritis in humans. The virus causes diarrhea and vomiting and appears to attack the proximal small intestine. Noroviruses and enteroviruses have been placed on the U.S. EPA Contaminant Candidate List 3 (CCL3; Table 1.4) along with other viral and bacterial pathogens and protozoan parasites (http://www.epa.gov/safewater/ccl/ccl3.html). These viruses are stable under environmental conditions and appear to have a relatively low infectious dose estimated at 10–100 virus particles. There are some difficulties in propagating human noroviruses in tissue cultures, but recent studies on the development of *in vitro* infectivity assays for these viruses have shown some success with Caco-2 and C2BBe1 (a clone of Caco-2 cells) cells (Straub et al., 2011, 2013). Environmental monitoring for these viruses necessitates sensitive methods such as immunomagnetic separation (IMS) followed by RT-PCR or RT-PCR-DNA enzyme immunoassay for their detection in stools, shellfish, and aquatic samples (Meschke and Sobsey, 2002; Schwab et al., 2001; Wolf et al., 2007). They are also detected in environmental samples, using U.S. EPA method 1615 which includes a concentration step (using 1MDS

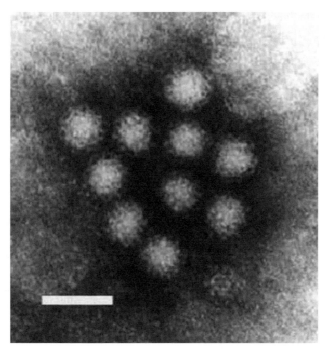

Figure 1.5   Electron micrograph of Norwalk virus (norovirus) (bar = 50 nm). *Source:* Courtesy of F.M. Williams, U.S. EPA.

TABLE 1.8   Some examples of waterborne outbreaks of Norwalk-like virus

Location	Number of Ill Persons	Remarks
Pennsylvania	350	Insufficiently chlorinated drinking water (attack rate = 17–73%)
Tacoma, WA	Approx. 600	Well of 41.4 m depth (attack rate = 72%)
Arcata, CA	30	Sprinkler irrigation system not suitable for human consumption
Maryland	126	Well 95 ft deep (attack rate = 64%)
Rome, GA	Approx. 1500	Spring water (attack rate = 72%)
Tate, GA	Approx. 500	Springs and wells
Finland	1500–3000	Contaminated drinking water. Well contaminated by river water. Finding of NLV genotype 2 in tap water
Italy	344	Outbreak of gastroenteritis at a tourist resort. NLV found in 22 out of 28 stool specimens. Contaminated water and ice

Source: Adapted from Boccia et al. (2002); Gerba et al. (1985); Kukkula et al. (1999); Williams and Akin (1986).

or Nano-Ceram filters) followed by elution with beef extract–glycine and detection via cell culture (feasible for murine norovirus) and reverse transcriptase quantitative PCR (Cashdollar et al., 2013; Fout et al., 2010).

Enteric adenoviruses. They are nonenveloped double-stranded DNA viruses, with a 80–120 nm diameter. They are emerging pathogens which cause infections leading to diarrhea and vomiting. This group comprises two important serotypes (types 40 and 41) among the 51 known human adenoviruses (Herrmann and Blacklow, 1995; Yates et al., 2006). They have been detected in wastewater effluents, surface waters (seawater, river water), swimming pool water, and drinking water (Albinana-Gimenez et al., 2006; Enriquez et al., 1995; van Heerden et al., 2005; Jiang, 2006) and have been placed by the U.S. EPA on the drinking water Contaminant Candidate List (CCL3).

Enteric adenoviruses are detected, using cell infectivity assays, a commercial monoclonal ELISA method, or PCR techniques (Allard et al., 1992; Wood et al., 1989)

Astroviruses. They are 27- to 34-nm single-stranded RNA viruses with a characteristic star-like appearance. They spread via person-to-person contacts and via contaminated food or drinking water. Patients have mild, watery diarrhea lasting 3–4 days or more for immunocompromised patients (Matsui, 1995). Since astroviruses affect mostly children and immunocompromised adults, they have been associated with gastroenteritis outbreaks in day- care centers and homes for the elderly (Marshall et al., 2007; Mitchell et al., 1999).

Astroviruses can be detected with molecular probes (e.g., RT-PCR, RNA probes), monoclonal antibodies, or in cell cultures (e.g., Caco cell cultures) treated with trypsin (Jonassen et al., 1993; Nadan et al., 2003; Pintó et al., 1996).

1.3.3 Protozoan Parasites

We will now cover the major waterborne pathogenic protozoa which may potentially affect humans through consumption of contaminated drinking water (Table 1.9; Bitton, 2011). Protozoan parasites were the etiological agents in about 200 waterborne outbreaks worldwide between 2004 and 2010, and *Cryptosporidium* spp. and *Giardia lamblia* were involved in 95.5% of the outbreaks (Baldursson and Karanis, 2011).

1.3.3.1 Giardia. *Giardia* is a protozoan parasite which persists as ovoid cysts (8–12 μm long and 7–10 μm large) in the environment. Upon ingestion, the cysts evolve into trophozoites which attach to the epithelial cells of the upper small intestine where they multiply and interfere with the absorption of fats and other nutrients. They then encyst as they reach the colon (Rochelle, 2002). Infection occurs upon ingestion of the cysts. *Giardia* infectious dose is relatively low and may be as low as 10 cysts in humans (Rendtorff, 1979). *Giardia* has an incubation period of 1–8 weeks and causes diarrhea, abdominal pains, nausea, fatigue, and weight loss.

The first major documented waterborne outbreak of giardiasis in the United States occurred in 1974 in Rome, NY, and affected some 5000 people (10% of Rome

TABLE 1.9 Major waterborne diseases caused by protozoa

Organism	Disease (Site Affected)	Major Reservoir
Giardia lamblia	Giardiasis (GI tract)	Human and animal feces
Cryptosporidium	Profuse and watery diarrhea; weight loss, nausea, low grade fever (GI tract) tract)	Human and animal feces
Cyclospora	Watery diarrhea alternating with constipation (GI tract)	Feces, contaminated fruits and vegetables
Entamoeba histolytica	Amoebic dysentery (GI tract)	Human feces
Acanthamoeba castellani	Amoebic meningoencephalitis (central nervous system)	Soil and water
Naegleria gruberi	Amoebic meningoencephalitis (central nervous system)	Soil and water
Balantidium coli	Dysentery/intestinal ulcers (GI tract)	Human feces
Microsporidia	Chronic diarrhea, dehydration, weight loss (GI tract)	Feces

population). This was due to the consumption of drinking water that was chlorinated but did not go through filtration. Several other waterborne outbreaks were reported since in other parts of the United States and worldwide. *Giardia* was recognized as one of the most important etiological agents in waterborne disease outbreaks (Craun, 1984). Hence, in the United States, *Giardia* was the etiologic agent responsible for 27% of the outbreaks reported in 1989–1990, and for 16.6% of the outbreaks reported in 1993–1994 (Herwaldt et al., 1992; Kramer et al., 1996). It is estimated that as many as 2.5 million cases of giardiasis occur annually in the United States (Furness et al., 2000) and that *Giardia* is responsible for 100 million mild cases and one million severe cases per year, worldwide (Smith, 1996). Most of these outbreaks are generally associated with the consumption of untreated or inadequately treated water (e.g., water chlorinated but not filtered or faulty filter design). This led the U.S. EPA to amend the Safe Drinking Water Act (Surface Water Treatment Rule) and now requires filtration and disinfection for all surface waters and groundwater under the direct influence of surface water to control the transmission of *Giardia* spp. and enteric viruses (U.S. EPA, 1989). A good correlation was found between the removal of *Giardia* cysts and water turbidity (LeChevallier et al., 1991b; Xagoraraki et al., 2004).

Giardia cysts were monitored in several wastewater treatment plants and their level in raw wastewater may be as high as 10^5 cysts/L (Cacciò et al., 2003; Casson et al., 1990; Jakubowski and Eriksen, 1979). However, cyst concentration decreases significantly after wastewater treatment (Correa et al., 1989; Rose et al., 1989). In potable water supplies (lakes, reservoirs, rivers, springs, groundwater), the presence of *Giardia* cysts was detected in 16% to 81% of the samples (LeChevallier et al., 1991b; Rose et al., 1991b). In water treatment plants in Canada, 80% of which use chlorination without filtration, the presence of cysts in treated water was reported in

18% of the samples but only 3% of the samples contained viable cysts (Wallis et al., 1996). In Japan, *Giardia* was detected in 12% of drinking water samples with a mean concentration of 0.8 cysts/1000 L (Hashimoto et al., 2002).

Giardia cysts occur at relatively low concentrations in aquatic environments, particularly drinking water. Thus, large volumes of the aquatic sample must be concentrated in order to detect this parasite in drinking water. Various concentration methods, including ultrafiltration, vortex flow filtration, or adsorption to polypropylene and yarn wound cartridge filters, have been proposed. This is followed by a method to detect the cysts in the concentrates. Ideally, the detection method should give information about the viability and infectivity of the cysts. This is the case for *in vivo* infectivity assays using gerbils as hosts. However, more rapid alternative methods have been developed and include the use of cell cultures (e.g., caco-2 cells), *in vitro* excystation, and fluorogenic dyes such as propidium iodide (PI) in combination with fluorescein diacetate (FDA) or DAPI (Labatiuk et al., 1991; Sauch et al., 1991; Schupp and Erlandsen, 1987). Other proposed fluorogenic vital stains are SYTO-9 and SYTO-59 (Bukhari et al., 2000).

U.S. EPA methods 1622 and 1623 (McCuin and Clancy, 2003; U.S. EPA, 2001) consist of filtration of the sample through a pleated membrane capsule (1 µm) or a compressed foam filter, followed by elution, purification by IMS of the cysts, and observation under a florescence microscope following staining with fluorescein isothiocyanate (FITC) conjugated monoclonal antibody (FAb) and counterstaining with DAPI. Cyst recovery is significantly improved by heating at 80°C for 10 minutes prior to DAPI staining (Ware et al., 2003). A novel concentration method consists of filtration, resuspension, IMS, and detection via flow cytometry. This method recovers 97% of *Giardia* cysts from tap water and takes less than 2 hours to complete (Keserue et al., 2011). Molecular-based techniques can also be used to detect *Giardia* cysts in concentrates. Some of these methods (e.g., measuring the amount of RNA before and after excystation, PCR amplification of heat shock-induced mRNA) can also provide information on cyst viability (Abbaszadegan et al., 1997; Mahbubani et al., 1991).

1.3.3.2 *Cryptosporidium.*

Cryptosporidium parvum (genotype 2) is a coccidian protozoan parasite that infects both humans and animals while *C. hominis* (genotype 1) is specific to humans. This parasite persists in the environment as a thick-walled oocyst (5–6 µm in size; Figure 1.6) which, following ingestion, undergoes excystation to release infective *sporozoites* in the gastrointestinal tract. The infective dose of oocysts is relatively low, as 1–10 oocysts can initiate infection. The infection may result in a profuse and watery diarrhea which typically lasts for 10–14 days and is often associated with weight loss and sometimes nausea, vomiting, and low grade fever (Adal et al., 1995; Current, 1987; Rose, 1990). The duration of the symptoms and the outcome depend on the immunological status of the patient. A study on the occurrence of *Cryptosporidium* in drinking water in China showed that the infection rate in immunodeficient patients (e.g., AIDS patients) was 18 times higher than in immunocompetent ones (Xiao et al., 2012). This parasite is transmitted via person to person, waterborne, foodborne, and zoonotic (i.e., transmission of the pathogen from infected animals to humans) routes. Several drinking water-associated outbreaks were

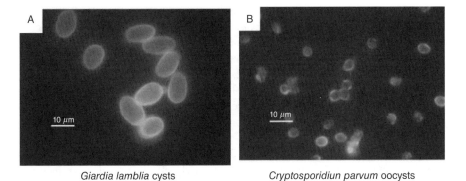

Giardia lamblia cysts   Cryptosporidiun parvum oocysts

Figure 1.6 Immunofluorescence images of *Giardia lamblia* cysts (A) and *Cryptosporidium parvum* oocysts (B). Courtesy of H.D.A. Lindquist, U.S. EPA. (For a color version, see the color plate section.)

documented around the globe and two notorious ones occurred in the United States. An outbreak in Carrollton, Georgia, affected approximately 13,000 people and was associated with the consumption of drinking water from a water treatment plant with ineffective flocculation and faulty rapid sand filtration (Hayes et al., 1989). The largest documented waterborne cryptosporidiosis outbreak occurred in 1993 in Milwaukee, Wisconsin, where 403,000 people became ill, resulting in 4400 hospitalized patients and 54 deaths (Kaminski, 1994; MacKenzie et al., 1994).

Cryptosporidium oocyst detection in drinking water involves a concentration method followed by detection of oocysts in the concentrates, using various approaches. Concentration techniques are based on the retention of oocysts on polycarbonate filters, polypropylene cartridge filters, vortex flow filtration, or hollow fiber ultrafilters (Kuhn and Oshima, 2001; Mayer and Palmer, 1996; Schaefer, 1997). The oocysts are generally detected in the concentrates using polyclonal or monoclonal antibodies, flow cytometry, molecular techniques (e.g., PCR, FISH assays), or electronic imaging of the fluorescent oocysts. A two-color FISH assay was developed to distinguish between *C. parvum* and *C. hominis* (Alagappan et al., 2009). The U.S. EPA developed a method (Method 1623; U.S. EPA, 2001) based on IMS followed by a fluorescent antibody detection technique. This method displayed oocyst recovery efficiencies similar to cell culture-PCR method. However, the recovery efficiency of method 1623 can be reduced by the presence of suspended solids in water (Krometis et al., 2009). Oocyst viability and infectivity are determined by *in vitro* excystation, mouse infectivity assays, *in vitro* cell culture infectivity assays, or staining with fluorogenic vital dyes such as DAPI (4,6-diamino-2-phenylindole), propidium iodide, SYTO-9 or SYTO-59 (Bukhari et al., 2000; Campbell et al., 1992; Shin et al., 2001). A combination of IMS and RT-PCR targeting the *hsp70* heat shock-induced mRNA also provides information about oocyst viability (Garcés-Sanchez et al., 2009; Hallier-Soulier and Guillot, 2003). Treatment of samples with propidium monoazide (PMA penetrates only dead oocysts where it prevents DNA amplification) prior to PCR analysis also helps detect viable *Cryptosporidium* oocysts in water (Brescia et al., 2009). The fluorescent antibody–microscopy assay was found to be the most sensitive method for detecting *C. parvum* and *C. hominis* in drinking water (Johnson et al., 2012).

These methods were used in several surveys which showed the presence of oocysts in source waters (Hashimoto et al., 2002; Johnson et al., 1995; LeChevallier et al., 1991b; Ono et al., 2001; Rose et al., 1991b). A recent study showed a mean concentration of 0.7 oocysts/10 L in source waters for 66 drinking water treatment plants across China. Oocysts concentrations were 1.6×10^{-3} per liter following conventional treatment, 5.7×10^{-5} per liter following microfiltration, and 9.7×10^{-5} per liter following ozonation (Xiao et al., 2012). The methods have shown that traditional drinking water treatment processes, particularly chlorination, do not effectively remove/inactivate *Cryptosporidium* oocysts, as surveys revealed that this parasite is detected in finished drinking water (Aboytes et al., 2004; Castro-Hermida et al., 2008). Some 27% of 82 water treatment plants in the United States released *Cryptosporidium* oocysts (Aboytes et al., 2004). In Shanghai, China, almost one-third of 50 water samples from the Huangpu River were positive for this parasite which was detected in finished drinking water (Feng et al., 2011). A 1-year monitoring study in Scotland revealed several *Cryptosporidium* species and genotypes in drinking water. The most frequently encountered species was *C. ubiquitum* followed by *C. andersoni* and *C. parvum* (Nichols et al., 2010).

1.3.3.3 Cyclospora.   *Cyclospora cayetanensis* is an 8–10 µm diameter coccidian parasite which was first reported in the 1980s, and has been often mentioned in the literature as a "cyanobacterium-like body" (Adal et al., 1995; Soave, 1996). It infects the duodenum and jejunum in the gastrointestinal tract. This parasite has an incubation period of approximately 1 week but may be longer for immunocompromised people such as AIDS patients. The clinical symptoms of cyclosporiasis include long lasting watery diarrhea sometimes alternating with constipation, abdominal cramps, nausea, weight loss, sometimes vomiting, anorexia, and fatigue (Ortega, 2002).

Cyclosporiasis is endemic in countries such as Nepal, Haiti, Indonesia, Guatemala, and Peru. In the United States, some outbreaks appear to be associated with drinking water (CDC, 1991; Huang et al., 1996; Rabold et al., 1994) but most infections are associated with the consumption of contaminated fruits and vegetables such as imported raspberries or basil (Herwaldt, 2007; Herwaldt et al., 1997; Lopez et al., 2001). In 2013, CDC investigation of an outbreak of cyclosporiasis in several states in the United States revealed 250 cases of *Cyclospora* infection (CDC, 2013).

Detection of *Cyclospora* in water involves a concentration step, using methods already described for *Giardia* and *Cryptosporidium*. The natural autofluorescence of *Cyclospora* oocysts allows their detection in sample concentrates as blue circles under a fluorescence microscope. Molecular techniques such as fluorogenic probes and PCR can also be used to detect the oocysts (Steele et al., 2003; Varma et al., 2003) and some methods can distinguish *Cyclospora cayetanensis* oocysts from other *Cyclospora* species (Shields and Olson, 2003). The viability/infectivity of *Cyclospora* oocysts cannot be assessed due to the lack of *in vivo* or *in vitro* culture assays for this parasite.

1.3.3.4 Microsporidia. Microsporidia are obligate intracellular protozoan parasites which infect humans and comprise approximately 14 species (e.g., *Encephalitozoon intestinalis, Encephalitozoon cuniculi, Enterocytozoon bienusi*) (Didier et al., 2004). Individuals ingesting the small (1–5 μm) spores experience chronic diarrhea, dehydration, and significant weight loss. Microsporidia are transmitted via person to person, zoonotic, airborne, and water routes (Didier et al., 2004; Schaefer, 1997) and have been included in the U.S. EPA Contaminant Candidate List (U.S EPA, 1998). Microsporidia have been detected in wastewater, drinking water, and recreational waters. A multiplex FISH assay revealed that in recreational waters, the spore numbers increased with bathers' density (Graczyk et al., 2007). They were detected via PCR in 2 out of 16 samples of drinking water (Izquierdo et al., 2011).

Microsporidia are relatively well removed by conventional water treatment processes such as coagulation, sedimentation, and filtration (Gerba et al., 2003).

1.3.3.5 Entamoeba histolytica. Cysts (10–15 μm in diameter) of this protozoan parasite are generally transmitted via contaminated water and food in developing countries. The ingested cysts evolve into trophozoites (10–60 μm) in the gastrointestinal tract and cause amoebiasis or amoebic dysentery, resulting in diarrhea alternating with constipation. This parasite may also invade other organs such as the liver or the lungs. It is transmitted via person-to-person contact and contaminated food and drinking water. Waterborne transmission of this parasite is relatively rare in the United States but continues to prevail in developing countries (Bitton, 2011; WHO, 2011c).

1.3.3.6 Naegleria. *Naegleria* is a free-living protozoan that is generally found in thermal spring waters, thermally polluted effluents, and drinking water supply wells with about 10% of the wells being positive for *N. fowleri* (Marciano-Cabral, 1988). *Naegleria fowleri* is the causative agent for primary amoebic meningoencephalitis (PAME), a fatal disease first reported in the 1960s. *Naegleria* enters in the nasal cavity and migrates to the central nervous system. This disease can be contracted following diving in warm lakes and disturbance of the sediments.

There is a need for rapid detection of *Naegleria* since the disease is fatal after 4–5 days. Detection methods for *Naegleria* include cytometry, API ZYM system based on detection of enzyme activity, monoclonal antibodies, DNA probes, or PCR techniques (Behets et al., 2003; Kilvington and Beeching, 1995a, 1995b; Kilvington and White, 1985; Pougnard et al., 2002; Robinson et al., 2006; Visvesvara et al., 1987). *Naegleria fowleri* can be specifically and rapidly (less than 6 hours) detected in biofilms and drinking water samples, without preculturing, via total DNA extraction followed by RT-PCR (Puzon et al., 2009).

1.3.3.7 Toxoplasma gondii. *Toxoplasma gondii* is a coccidian parasite that uses cats as hosts and causes parasitic infections in humans worldwide (Dubey, 2002). Many infections are congenitally acquired and can cause ocular disease in children. Others are postnatally acquired and lead to enlargement of the lymph nodes. This parasite causes most damage among AIDS patients and other immunosuppressed individuals. In infected pregnant women, this parasite can cause spontaneous abortion and damage to the fetus. Humans become infected through contact with infected cats, or following ingestion of uncooked or undercooked meat, drinking water, or ice contaminated with *T. gondii* oocysts. An outbreak of toxoplasmosis was linked to the consumption of drinking water from a reservoir in Canada (Isaac-Renton et al., 1998). Preventive measures are directed mostly toward pregnant women who should avoid contact with cats as well as contaminated meat (Dubey, 2002).

More information about the role of protozoa in water distribution systems is available in Chapter 5.

Table 1.10 summarizes the major pathogens and parasites transmitted via drinking water (WHO, 2011c).

TABLE 1.10   Pathogens and parasites which can be transmitted via drinking water

Pathogen	Persistence in Water Supplies	Resistance to Chlorine	Relative Infectivity
Bacteria			
Burlholderia pseudomallei	May multiply	Low	Low
Campylobacter jejuni	Moderate	Low	Moderate
Escherichia coli—Pathogenic	Moderate	Low	Low
E. coli—Enterohaemorrhagic	Moderate	Low	High
Francisella tularensis	Long	Moderate	High
Legionella spp.	May multiply	Low	Moderate
Leptospira	Long	Low	High
Mycobacteria (nontuberculous)	May multiply	Moderate	Low
Salmonella typhi	Moderate	Low	Low
Other salmonellae	May multiply	Low	Low
Shigella spp.	Short	Low	High
Vibrio cholerae	Short to long	Low	Low
Viruses			
Adenoviruses	Long	Moderate	High
Astroviruses	Long	Moderate	High
Enteroviruses	Long	Moderate	High
Hepatitis A virus	Long	Moderate	High
Hepatitis E virus	Long	Moderate	High
Noroviruses	Long	Moderate	High
Rotaviruses	Long	Moderate	High
Sapoviruses	Long	Moderate	High
Protozoa			
Acanthamoeba spp.	May multiply	High	High
Cryptosporidium hominis/parvum	Long	High	High
Cyclospora cayetanensis	Long	High	High
Entamoeba histolytica	Moderate	High	High
Giardia intestinalis	Moderate	High	High
Naegleria fowleri	May multiply	Moderate	Moderate
Helminths			
Dracunculus medinensis	Moderate	Moderate	High
Schistosoma spp.	Short	Moderate	High

Source: Adapted from WHO (2011c). Guidelines for Drinking-Water Quality. 4th Ed. World Health Organization, Geneva, Switzerland.

WEB RESOURCES (AS OF MAY 23, 2013)

Viral Pathogens

http://www.ncbi.nlm.nih.gov/ICTVdb/Images/index.htm
(excellent EM pictures and diagrams of viruses and comprehensive description of phylogeny from the Universal Virus Database of the International Committee on Taxonomy of Viruses)

http://www.virology.net/Big_Virology/BVDiseaseList.html

Protozoan Parasites

http://www.epa.gov/microbes/
(microbiological exposure to pathogens)

http://www.medicine.cmu.ac.th/dept/parasite/framepro.htm
(protozoan parasites pictures)

Bacterial Pathogens

http://www.nmpdr.org/FIG/wiki/view.cgi/Main/CellsReagentsDatabasesImages
(national microbial pathogens data resource)

FURTHER READING

Adal, K.A., C.R. Sterling, and R.L. Guerrant. 1995. *Cryptosporidium* and related species. In *Infections of the Gastrointestinal Tract*, M.J. Blaser, P.D. Smith, J.I. Ravdin, H.B. Greenberg, and R.L. Guerrant (Eds) Raven Press, New York, pp. 1107–1128.

Baldursson, S. and P. Karanis. 2011. Waterborne transmission of protozoan parasites: review of worldwide outbreaks – An update 2004–2010. *Water Res.* 45:6603–6614.

Bitton, G. 2011. *Wastewater Microbiology*, 4th edition. Wiley-Blackwell, Hoboken, NJ, 781 pp.

Cliver, D.O. 1984. Significance of water and environment in the transmission of virus disease. *Monog. Virol.* 15:30–42.

Craun, G.F., J.M. Brunkard, J.S. Yoder, V.A. Roberts, J. Carpenter, T. Wade, R.L. Calderon, J.M. Roberts, M.J. Beach, and S.L. Roy. 2010. Causes of outbreaks associated with drinking water in the United States from 1971 to 2006. *Clin. Microbiol. Rev.* 23(3):507–528.

Dufour, A., M. Snozzi, W. Koster, J. Bartram, E. Ronchi, and L. Fewtrell (Eds). 2003. *Microbial Safety of Drinking Water: Improving Approaches and Methods*. ISBN 92 4 154630 1 (WHO) and 1 84339 036 1 (IWA Publishing).

Gerba, C.P., S.N. Singh, and J.B. Rose. 1985. Waterborne gastroenteritis and viral hepatitis. *CRC Crit. Rev. Environ. Contam.* 15:213–236.

LeChevallier, M.W. and K.K. Au. 2004. *Water Treatment and Pathogen Control: Process Efficiency in Achieving Safe Drinking Water*. IWA Publishing, London, UK. ISBN: 1 84339 069 8.

Leclerc, H., L. Schwartzbrod, and E. Dei-Cas. 2002. Microbial agents associated with waterborne disease. *Crit. Rev. Microbiol.* 28:371–409.

Levin, R.B., P.R. Epsstein, T.E. Ford, W. Harrington, E. Olson, and E.G. Reichard. 2002. U.S. drinking water challenges in the twenty-first century. *Environ. Health Perspect.* 110 (suppl. 1):43–52.

Matsui, S.M. 1995. Astroviruses. In *Infections of the Gastrointestinal Tract*, M.J. Blaser, P.D. Smith, J.I. Ravdin, H.B. Greenberg, and R.L. Guerrant (Eds). Raven Press, New York, pp. 1035–1045.

McFeters, G.A. (Ed.) *Drinking Water Microbiology*. Springer, New York, 502 p.

Meschke, J.S. and M.D. Sobsey. 2002. Norwalk-like viruses: detection methodologies and environmental fate. In *Encyclopedia of Environmental Microbiology*, G. Bitton (editor-in-chief). Wiley-Interscience, New York, pp. 2221–2235.

Moe, C.L. 1997. Waterborne transmission of infectious agents. In *Manual of Environmental Microbiology*, C.J. Hurst, G.R. Knudsen, M.J. McInerney, L.D. Stetzenbach, and M.V. Walter (Eds). ASM Press, Washington, DC, pp. 136–152.

Muniesa, M., J. Jofre, C. Garcia-Aljaro, and A.R. Blanch. 2006. Occurrence of *Escherichia coli* O157:H7 and other enterohemorrhagic *Escherichia coli* in the environment. *Environ. Sci. Technol.* 40:7141–7149.

Myint, K.S.A., J.R. Campbell, and A.L. Corwin. 2002. Hepatitis viruses (HAV-HEV). In *Encyclopedia of Environmental Microbiology*, G. Bitton (editor-in-chief). Wiley-Interscience, New York, pp. 1530–1540.

Percival, S.L., J.T. Walker, and P.R. Hunter. 2000. *Microbiological Aspects of Biofilms and Drinking Water*. CRC Press, Boca Raton, FL, 229 p.

Schwartzbrod, L. (Ed.) 1991. *Virologie des Milieux Hydriques*. TEC & DOC Lavoisier, Paris, France, 304 p.

Theron, J. and T.E. Cloete. 2002. Emerging waterborne infections: contributing factors, agents, and detection tools. *Crit. Rev. Microbiol.* 28:1–26.

WHO. 2011. *Guidelines for Drinking-Water Quality*, 4th editon. World Health Organization, Geneva, Switzerland. http://www.who.int/water_sanitation_health/publications/2011/dwq_guidelines/en/.

Wyn-Jones, A.P. and J. Sellwood. 2001. Enteric viruses in the aquatic environment. *J. Appl. Microbiol.* 91:945–962.

2

MICROBIOLOGICAL ASPECTS OF DRINKING WATER TREATMENT

2.1 INTRODUCTION

Water is essential to life on Planet Earth. It is essential for food production, industrial growth, and environmental sustainability. The total quantity of water withdrawn by our thirsty world was estimated at 3906 km^3 in 1995. Projections for 2025 predict a 50% increase for domestic, livestock, and industrial uses, leaving less water for irrigation and therefore endangering food production. Figure 2.1 (Rosegrant et al., 2002) shows the total water withdrawal by regions around the globe and the projections for 2025.

Early on, the ancient civilizations (Egyptians, Greeks, and the Romains) were already preoccupied with water hygiene and sanitation (see review of the topic by Bond et al., 2013). A major breakthrough was due to Antonie van Leeuwenhoek who, in 1684, described "animalcules" in water, and later on, in the nineteenth century, Robert Koch and Joseph Lister, showed that the "animalcules" were indeed microorganisms (Batterman et al., 2009; Bachmann and Edyvean, 2006). Another major milestone is when John Snow, a British physician, established a link between microorganisms and water-related diseases like cholera.

The quest for safe drinking water dates back to at least 4000 years ago when a medical philosopher advocated the use of boiling, sunlight, charcoal filter, and copper vessels to treat water (Symons, 2006). Sir Francis Bacon in 1627 mentioned the use of sedimentation, coagulation, filtration, boiling, and distillation for water treatment. Some other milestones in the history of water treatment technologies are shown in Table 2.1; Bachmann and Edyvean, 2006; Baker and Taras, 1981; Lorch, 1987). Drinking water treatment has been suggested as one of the 10 great achievements of the twentieth century (CDC, 1999).

Microbiology of Drinking Water Production and Distribution, First Edition. Gabriel Bitton.
© 2014 John Wiley & Sons, Inc. Published 2014 by John Wiley & Sons, Inc.

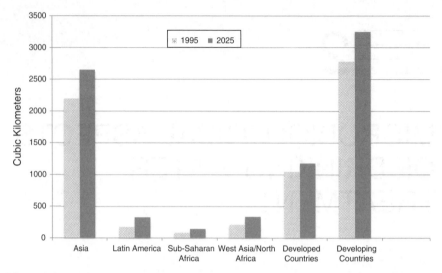

Figure 2.1 Total water withdrawal by regions around the globe for 1995 and the projections for 2025. Adapted from Rosegrant et al. (2002). *Global Water Outlook to 2025: Averting an Impending Crisis.* Food Policy Report. Inter. Food Policy Res. Inst. (IFPRI) and Intern. Water Management Inst. (IWMI), 28pp.

2.2 WORLDWIDE CONCERN OVER DRINKING WATER SAFETY

Drinking water safety is a worldwide concern. Contaminated drinking water has the greatest impact on human health worldwide, especially in developing countries. It is estimated that more than one billion of the world's population does not have access to safe clean water and approximately 2.6 billion lack improved sanitation (WHO/UNICEF, 2000, 2004; WHO, 2003). The World Health Organization (WHO) reported that 80% of diseases and one-third of deaths in developing countries are due to consumption of contaminated water (WHO, 1996a). Nontreated or improperly treated water is a major cause of illness in developing countries. WHO estimates that water-related diseases are responsible for more than 5 million deaths per year (WHO, 1996b). Twenty-five percent of hospital beds were occupied by people who became ill after consuming contaminated water (WHO, 1979) and this number jumps to 50% if we add malaria, dengue fever, and other vector-borne diseases which require water for insect propagation. Diarrheal diseases are third largest cause of morbidity (1 billion episodes/year) and the fifth largest cause of mortality (2.2 million deaths/year) worldwide (Montgomery and Elimelech, 2008; Pond et al., 2004).

Children are unfortunately the main victims of diarrheal diseases. Drinking contaminated water is responsible for 200 deaths/hour among children in the developing world (Gadgil, 2008). As regard children younger than the age of 5, diarrheal diseases are responsible for 16% of deaths worldwide and are the second leading cause of death after pneumonia (Figure 2.2; WHO/UNICEF, 2009). The problem is most acute in Sub-Saharan Africa where 42% of the population lacks safe water and 64%

TABLE 2.1   Some milestones in the history of drinking water treatment technologies

Timeline	Technology	Goal
2000 B.C.	Sanskrit documents (*Sus'ruta Samhita*) stated that *"impure water should be purified by being boiled over a fire, or being heated in the sun or by dipping a heated iron into it, or it may be filtered through sand and coarse gravel and then allowed to cool"*	Water purification
Around 1500 B.C.	Biblical times: Concerning the bitter waters of Marah: *"... and he (Moses) cried onto the Lord and the Lord showed him a tree and he cast it into the waters and the waters were made sweet"* (Exodus 15: 22-25)	Taste removal from water
400 B.C.	Hippocrates' sleeve	Particle removal from rainwater (filtration)
300 B.C.	Storage and settling cisterns (Greece, Roman empire)	Particle removal from water (sedimentation)
100 B.C.	Addition of macerated laurel to rainwater (Diophanes, Greece)	Taste and odor control in drinking water
100 A.D.	Immersion of bruised coral or pounded barley in water (Paxamus, Italy)	Taste (and odor?) control in drinking water
500–1500	Stagnation in technology ("Dark Ages" in Europe)	
1680	Sterilization (Antonie van Leeuwenhoek, in the Netherlands)	Killing of "animalcules" (by adding vinegar or alcohol)
1685	Multiple sand filtration to protect soldiers' health in military installations by Lucas Antonius Portius, a physician from Italy	Particles and taste and odor removal from drinking water
1827	Slow sand filtration (Robert Thom and James Simpson, UK)	Particles and taste and odor removal from drinking water
1847	Chlorine first used as a disinfectant in Vienna, Austria	To prevent the spread of "child bed fever"
1854	John Snow linked cholera outbreak with water pump in Broad street, London	
1880	Rapid sand filtration (USA)	Particles and taste and odor removal from drinking water
1893	Slow sand filter installed in Lawrence, MA, USA	Removal of *Salmonella typhi* in drinking water.
1893	Ozone first used as a disinfectant in the Netherlands	
1906	Chlorine and ozone disinfection. Water treatment using ozone in Nice, France	Destruction of pathogens in drinking water
1908	First use of chlorination of drinking water in Jersey City, NJ	Destruction of pathogens in drinking water

(*continued*)

TABLE 2.1 *(Continued)*

Timeline	Technology	Goal
1910	UV first used in the water supply of Marseille, France	Destruction of pathogens in drinking water
1916	First use of UV radiation in a plant in Henderson, KY	
1940	First installation of ozone treatment in the USA	Taste and odor control
1975	First documented waterborne outbreaks caused by enteropathogenic *Escherichia coli*	
1978	First documented outbreak attributed to *Campylobacter*	
1984	First waterborne outbreak attributed to *Cryptosporidium*	

Source: Adapted from Bachmann and Edyvean (2006). Biofilms 2: 197–227; Baker and Taras (1981). The Quest for Pure Water: The History of the Twentieth Century, vols. 1 & 2. American Water Works Association, Denver, CO.; Anonymous (1903). J. Amer. Med. Assoc. 41: 850–853; http://humboldt.edu/arcatamarsh/chlorination.html; http://www.tdsmeter.com/education?id=0002; Lorch, W. (1987). Handbook of Water Purification, 2nd Edition. Ellis Horwood Limited, Chichester, UK; Rochelle, P.A., and J. Clancey (2006). Amer. Water Works Assoc. J. 98 (3): 163–191.

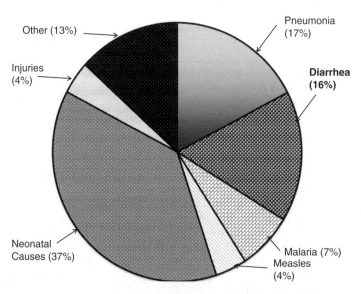

Figure 2.2 Causes of death among children under the age of five. Adapted from WHO/UNICEF (2009). *Diarrhoea: Why children are still dying and what can be done.* World Health Organization, Geneva, Switzerland.

live with inadequate sanitation. In this region, the death rate of children younger than 1 year of age from diarrheal diseases approaches 23 of 1000 children (Montgomery and Elimelech, 2008; WHO/UNICEF, 2004). In Bolama Island (Guinea-Bissau, West Africa), nearly 80% of drinking water wells were fecally contaminated, leading to diarrheal diseases in 11.5% of all medical cases, with children representing 92.5% of diarrheal cases (Bordalo and Savva-Bordalo, 2007). Diarrheal diseases are the primary cause of morbidity and mortality (7900 deaths) among children younger than 5 years (Clasen et al., 2006). Some simple solutions to water and sanitation problems in developing countries are better protection of drinking water wells, household devices, and improved latrine construction and placement at a safe distance from a drinking water well (see Chapter 8 for more information).

In the United States, as of 1997, there were approximately 54,000 public drinking water treatment plants that serve more than 250 million consumers. Eighty percent of the plants use groundwater as source water and serve one-third of the consumers while 20% use surface waters and serve two-thirds of the consumers. A 1997 survey revealed that the estimated per capita water use for industrialized nations varies between 129 L/day for Germany to 382 L/day for United States (Levin et al., 2002). National concern over safety of drinking water began in 1912 when a national regulation was promulgated. After several amendments, the first Safe Drinking Water Act (PL 93-523) was signed into law in 1974. Subsequent amendments added new regulated contaminants to the list (Roberson, 2006).

Advances in drinking water research followed by the establishment of multiple barriers against microbial pathogens and parasites have significantly increased the safety of the water we drink daily, particularly in industrialized nations. This multiple-barrier system includes source water protection, reliable water treatment (pretreatment, coagulation, flocculation, sedimentation, filtration, disinfection), and protection of the water distribution network (Berger and Oshiro, 2002; LeChevallier, 2002).

In this chapter, we will cover the microbiological quality of source waters as well as the microbiological aspects of water treatment processes. Distribution of potable water, with particular emphasis on biofilm formation in distribution lines, will be covered in Chapters 4 and 5.

2.3 MICROBIOLOGICAL QUALITY OF SOURCE WATER

The desirable approach is to withdraw water from the best available and safest source. Sometimes, due to the scarcity of quality source water, communities must use reclaimed water to augment their water supplies. Communities obtain their potable water mostly from surface or underground sources. Both types of water can become contaminated by biological and chemical pollutants originating from point and non-point sources. Point sources, for which the source can be identified, include discharges from wastewater treatment plants, and land disposal of wastewater effluents and biosolids. The wastes are from both human and animal sources (e.g., animal feeding operations). Non-point sources are diffuse sources of pollution and include, for example, urban and agricultural runoffs, water recreation facilities (e.g., swimming, boating), and wildlife (many human pathogens and parasites use animals as reservoirs).

2.3.1 Surface Waters

Although more than 90% of public water systems use groundwater as their source water, it is estimated that two-thirds of the US population obtain their drinking water from surface water sources (lakes, rivers, streams) (U.S. EPA, 2005, 2007b). Surface waters can be a source of pathogens and parasites. For example, a 4-year monitoring of viruses in source waters in the Netherlands showed the presence of enteroviruses (75% of the samples), reoviruses (83% of the samples), noroviruses (45% of the samples), and rotaviruses (48% of the samples) (Lodder et al., 2010). Source waters can also become contaminated by opportunistic pathogens (e.g., *Pseudomonas aeruginosa, Legionella pneumophila, Aeromonas, Mycobacterium, Flavobacterium, Naegleria fowleri, Acanthamoeba* spp.) that are part of the indigenous microorganisms in soil and aquatic environments (Berger and Oshiro, 2002). Surface water quality rapidly changes as a response to changes in the surrounding watershed. Natural disturbances from wildfires and post-fire intervention can impact downstream drinking water treatment. For example, a study of the Lost Creek wildfire in Canada showed increases in turbidity, dissolved organic carbon (DOC), and nutrients downstream (Emelko et al., 2011). Moreover, relatively high concentrations of nutrients (N, P) result in eutrophication of surface waters with excessive algal growth, leading to excessive levels of microorganisms and turbidity in the source water. Surface waters become often contaminated by domestic wastewater, stormwater runoff, cattle feedlot runoff, discharges from food processing plants, and by resuspension of microbial pathogens and parasitic cysts and ova that have accumulated in bottom sediments. Lake destratification leads to a resuspension of microorganisms, humic acids, and turbidity in the water column (Geldreich, 1990). This leads to increased levels of suspended solids, nutrients, biochemical oxygen demand (BOD), and microbial pathogens and parasites in surface waters. In some areas, these surface waters are practically "diluted wastewaters." Problems arise when upstream communities discharge pathogen–laden wastewater effluents into surface waters that become drinking water supplies to downstream communities. The pathogen load downstream from a pollution source depends on the extent of the natural self-purification processes, which are controlled by temperature, solar radiation, dilution, available nutrients, and biological factors which include competition with indigenous microorganisms, protozoan grazing of pathogenic bacteria, or ability to enter the viable but nonculturable (VBNC) state as a survival strategy.

2.3.2 Groundwater Sources

Contamination of groundwater is well documented (Bitton and Gerba, 1984). Concern over these subsurface waters stems from the fact that they supply the drinking water needs of more than 100 million people in the United States. Sixty-five percent of the communities that use groundwater as a source do not disinfect the water. As regard private wells, groundwater is often consumed without any treatment, and health concerns were raised over this source of drinking water (Charrois, 2010; U.S. EPA, 2006). For example, between 1992 and 2003, in England and Wales, the incidence of waterborne infectious diseases was 35 times higher among consumers of groundwater from private wells than among those drinking public water (Smith et al., 2006). In Canada, two-thirds of outbreaks of infectious diseases that occurred between 1974 and 2001 were associated with semi-private or private wells (Schuster et al., 2005). In the United States, groundwater systems accounted for 95.2% of the disease outbreaks in 2007–2008 period (Brunkard et al., 2011).

2.3.3 Roof-Harvested Rainwater

Due to climate change and increasing need for water, roof-harvested rainwater is increasingly being considered worldwide as an alternative source of water for potable and nonpotable uses (e.g., lawn irrigation, toilet flushing, clothes washing) (Despins et al., 2009). However, there is a potential public health risk associated with the presence of microbial pathogens and parasites and toxic chemicals in rainwater. A microbiological study of roof-harvested rainwater in Australia revealed that 58%, 83%, and 46% of the samples were positive, respectively, for *Escherichia coli*, enterococci, and *Clostridium perfringens* spores. Moreover, PCR analysis showed the presence of *Aeromonas hydrophila*, *Campylobacter jejuni*, *L. pneumophila*, *Salmonella* and *Giardia lamblia* (Ahmed et al., 2010a, 2010b). However, an epidemiological study in Adelaide, Australia, showed that drinking untreated rainwater did not contribute significantly to gastroenteritis in healthy adults (Rodrigo et al., 2011).

Source water protection can be achieved by identifying the contamination sources and by taking protective measures. These measures may include the use of physical barriers to exclude humans and animals, wellhead protection, the reduction or elimination of certain activities such as cattle grazing and sewage discharges, and the establishment of land-use restrictions (Robertson and Edberg, 1997).

2.4 OVERVIEW OF PROCESSES INVOLVED IN DRINKING WATER TREATMENT PLANTS

Water contains several chemical and biological contaminants that must be removed efficiently in order to produce drinking water that is safe and aesthetically pleasing to the consumer. The chemical contaminants include nitrate, heavy metals, radionuclides, pesticides, pharmaceuticals, hormonally active chemicals, and other xenobiotics. The finished product must also be free of microbial pathogens and parasites, turbidity, color, taste, and odor (the removal of taste and odor compounds will be discussed in Chapter 5). To achieve this goal, raw water (surface water or groundwater) is subjected to a series of physicochemical and biological treatment processes that will be described in detail. Disinfection alone is sometimes sufficient if the raw water originates from a protected source. More commonly, a multi-barrier approach is taken to produce drinking water. Disinfection is combined with coagulation, flocculation, and filtration. Additional treatments to remove specific compounds may include pre-aeration and activated carbon treatment. The treatment train depends on the quality of the source water under consideration.

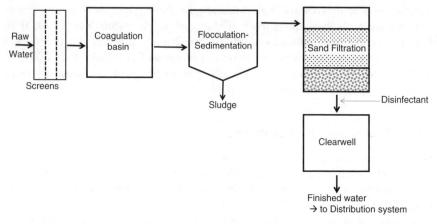

Figure 2.3   Flow diagram of a conventional drinking water filtration plant.

There are two main categories of water treatment plants:

- *Conventional filter plants:* The leading processes in this type of plant are coagulation/flocculation and filtration (Figure 2.3). The raw water is rapidly mixed with aluminum-based or iron-based coagulants. The most used coagulants are aluminum sulfate called alum $Al_2(SO_4)_3.14.3\ H_2O$, ferric sulfate $Fe_2(SO_4)_3$ or ferric chloride $FeCl_3.6\ H_2O$. Sometimes, coagulant aids are required to improve floc settling and strength. They are also synthetic high molecular weight polymers which are commercially available as anionic, cationic, or nonionic polymers. Sometimes, these polymers can be used as primary coagulants. Polyelectrolyte addition leads to less sludge which can be easily dewatered (Hammer and Hammer, 2008). After coagulation, the produced flocs are allowed to settle in a clarifier. Clarified effluents are then passed through sand or diatomaceous earth filters. Water is finally disinfected before distribution.
- *Softening plants:* The leading process in these plants is water softening, which helps remove hardness due to the presence of Ca and Mg in water, and results in the formation of Ca and Mg precipitates. After settling of the precipitates, the water is filtered and disinfected.

2.5 PROCESS MICROBIOLOGY AND FATE OF PATHOGENS AND PARASITES IN WATER TREATMENT PLANTS

2.5.1 Introduction

In water treatment plants, microbial pathogens and parasites can be physically removed by processes such as coagulation, precipitation, filtration, and adsorption, or they can be inactivated by disinfection or by the high pH resulting from water softening.

There are several types of pathogens and parasites of most concern in drinking water (AWWA, 1987). (Chapter 1 should be consulted for more details on these pathogens and parasites):

Viruses: They are occasionally detected in drinking water from conventional water treatment plants that meet the microbiological standard currently used to judge drinking water safety and treatment efficiency (Bitton et al., 1986). For example, enteroviruses were detected at levels ranging from 3 to 20 viruses per 1000 L, in finished drinking water from water treatment plants that included prechlorination, flocculation, sedimentation, sand filtration, ozonation, and final chlorination (Payment et al., 1989; Payment, 1989). In South Korea, tap water processed via flocculation/sedimentation, filtration, and chlorination contained infectious enteroviruses (poliovirus type 1, echovirus type 6, coxsackie B viruses) and adenoviruses (e.g., adenoviruses type 40 and 41) in 39–48% of the samples examined by cell culture assays followed by polymerase chain reaction (PCR) amplification (Lee and Kim, 2002). In other surveys in industrialized countries, no virus was detected in finished drinking water. An example is the water filtration plant in Laval, Canada, where no virus was detected in 162 finished water samples (1000–2000 L per sample (Payment, 1991). A similar finding was reported at a full-scale water treatment plant in Mery sur Oise, France (Joret et al., 1986) and at three water treatment plants in Spain (Jofre et al., 1995).

Cryptosporidium and *Giardia lamblia*: The methodology for concentrating and detecting these parasites has been developed but the skill to routinely monitor their presence in drinking water is not yet available in most water treatment plants. These parasites have been detected in drinking water, and outbreaks have been discussed in Chapter 1.

Opportunistic pathogens: These are waterborne pathogens (e.g., *Pseudomonas*, *Alcaligenes*, *Acinetobacter*, and *Flavobacterium* species), which cause secondary infections in hospitals, particularly among immunocompromised consumers.

Legionella: This bacterial pathogen is an example of a nonenteric microorganism that can be transmitted by inhalation of drinking water aerosols from shower heads or humidifiers. Nosocomial (i.e., hospital-acquired) Legionnaires' disease may be contracted by exposure to *Legionella* from the water distribution system in hospitals (Best et al., 1984).

The U.S. EPA is now examining a list of microbial pathogens in drinking water for potential regulation. These pathogens are transmitted via the waterborne route but other routes may be involved. This list includes *A. hydrophila, Helicobacter pylori, Mycobacterium avium intracellulare* complex (MAIC), adenoviruses, caliciviruses, coxsackieviruses, echoviruses, and *Microsporidia* protozoan parasites) (Hilborn et al., 2002; U.S. EPA, 1998).

Several unit processes and operations are used in water treatment plants to produce microbiologically and chemically safe drinking water. The extent of treatment depends on the source of raw water, with surface waters generally requiring more treatment than is needed for groundwaters. The unit processes designed for water treatment, with the exception of the disinfection step, do not address specifically the destruction or removal of parasites or bacterial and viral pathogens.

2.5.2 Pretreatment of Source Water

Pretreatment is a range of steps that are designed to improve the quality of the source water prior to entry in the water treatment plant:

2.5.2.1 Storage of Raw Water (Off-Stream Reservoirs). Raw water can be stored in reservoirs to minimize fluctuations in water quality. Storage can affect the microbiological water quality which is affected by physical (e.g., settling of solids, evaporation, gas exchange with atmosphere), chemical (e.g., oxidation-reduction, hydrolysis, photolysis), and biological processes (e.g., nutrient cycling, biodegradation, pathogen decay) (Oskam, 1995). The reduction of pathogens, parasites, and indicator microorganisms during storage is variable and is influenced by a number of factors, such as temperature, sunlight, sedimentation, and biological adverse phenomena such as predation, antagonism, and lytic action of bacterial phages. Temperature is a significant factor controlling pathogen survival in reservoirs.

It appears that, under optimal conditions, water storage in reservoirs can lead to approximately 1- to 2-log reduction of bacterial and viral pathogens although higher reductions have been observed. Protozoan cysts are removed by entrapment into suspended solids followed by settling into the sediments.

2.5.2.2 Roughing Filters. Roughing filters contain coarse media (gravel, rocks) which help reduce water turbidity and bacterial concentrations (approximately 1-log reduction).

2.5.2.3 Microstrainers. Microstrainers are made of woven stainless steel or polyester wires with a pore size ranging from 15 to 45 μm. They retain mostly algae and relatively large protozoa. Filamentous or colonial algae are generally more efficiently removed than unicellular algae.

2.5.2.4 River Bank Filtration. This pretreatment has been practiced in Europe since the 1870s and over fifty years in the United States. In Germany, approximately 16% of the drinking water is produced from water processed by river bank filtration (RBF) or infiltration (Kuehn and Mueller, 2000).

RBF is the seepage of water through the bank of a river or lake to the production well of the water treatment plant. RBF involves physical, chemical, and biological processes. This practice provides certain advantages such as removal of pathogens and parasites, removal of algal cells, reduction in turbidity and natural organic matter (NOM), and dilution with groundwater. Particle removal is due to the combined effect of adsorption, straining, and biodegradation. RBF decreases the concentration of assimilable organic carbon (AOC) (see Chapter 6 for more details on AOC), leading

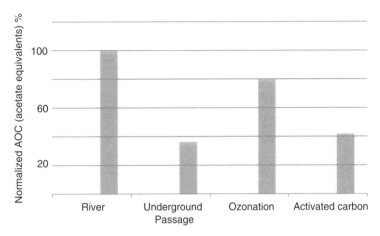

Figure 2.4   Effect of bank filtration on assimilable organic carbon (AOC). *Source*: Kuehn and Mueller (2000). Amer. Water Works Assoc. J. 92 (12): 60-69.

to a reduction of the biological growth potential in the water (Figure 2.4; Kuehn and Mueller, 2000). RBF provides quality source water for water treatment plants as less chemicals are required, and produces quality water for aquaculture, agriculture, and recreation (Ray et al., 2011). A study of five RBFs in the United States showed an efficient removal of *Cryptosporidium* and *Giardia*. In the Netherlands, RBF treatment provided a 4-log removal of viruses and 5- to 6-log removal of F-specific coliphage (Havelaar et al., 1995; Ray et al., 2002).

2.5.2.5 Prechlorination.   A prechlorination step is sometimes included to improve unit processes performance (e.g., filtration, coagulation–flocculation), oxidize color-producing substances such as humic acids and help in the precipitation of iron and manganese. Although prechlorination reduces somewhat the levels of pathogenic microorganisms, its use may lead to increased chances of forming disinfection by-products.

2.5.3 Coagulation-Flocculation-Sedimentation

Coagulation involves the destabilization and inter-particle collisions of colloidal particles (e.g., mineral colloids, microbial cells, virus particles) by coagulants (Al and Fe salts) and sometimes by coagulant aids (e.g., activated silica, bentonite, polyelectrolytes, starch). The most common coagulants are alum (Al sulfate), ferric chloride, and ferrous and ferric sulfate. For example, $Al_2(SO4)_3$ (alum) precipitates as insoluble $Al(OH)_3$ (Al hydroxide) and forms flocs that remove turbidity and microbial contaminants (Percival et al., 2000).

A survey carried out by the American Water Works Association (AWWA) in 2003 showed that 86% of the water treatment plants in the United States and Canada were switching from alum to ferric chloride, ferric sulfate or polyaluminum chloride. Some

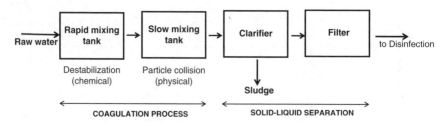

Figure 2.5 Coagulation process. Adapted from Williams, R.B. and G.L. Culp (1986). *Handbook of Public Water Systems*, van Nostrand Reinhold, N.Y.

have suggested the potential use of bioflocculants such as the extracellular polysaccharide from *Klebsiella terrigena* for water treatment (Ghosh et al., 2009) but their use in water treatment plants needs further research. The process of inter-particle contacts and formation of larger particles is called flocculation. After mixing, the colloidal particles form flocs which are large enough to allow rapid settling (Letterman et al., 1999; Williams and Culp, 1986) (Figure 2.5). The pH of the water is probably the most significant factor affecting coagulation/flocculation. Other factors include turbidity, alkalinity, temperature, and mixing regimen. Coagulation is the most important process used in water treatment plants for the clarification of colored and turbid waters. *Enhanced coagulation* consists of increasing the coagulant concentration and/or adjusting the pH of the water. This process targets the removal of both turbidity and NOM which is a precursor of disinfection by-products (see Chapter 3). However, surface-retained hydrophilic organic compounds (e.g., lipopolysaccharides) from cyanobacteria (e.g., *Microcystis aeruginosa*) were found to inhibit the coagulation process during water treatment (Takaara et al., 2010).

Jar tests have shown that the removal of bacteria and protozoan (*Cryptosporidium, Giardia*) cysts may range from 1- to 2-logs. More than 90% of *Cryptosporidium* removal was obtained when using 40–100 mg/L of alum. Alum does not significantly reduce the infectivity of *Cryptosporidium* oocysts even at concentrations as high as 200 mg/L (Keegan et al., 2008). Thus, oocysts will remain infective in sludges from water treatment plants. The removal of viruses is variable and ranges from 1 to more than 3 logs (Bell et al., 2000). Under laboratory conditions, coagulation–flocculation was effective in removing 90–99% of viruses from water. In three Spanish water treatment plants, the prechlorination–flocculation–sedimentation step was most efficient in phage removal (somatic, male-specific and *Bacteroides fragilis*-specific phages) (Jofre et al., 1995). Using enhanced coagulation, bench-scale testing showed that the optimal conditions for the removal of both DOC and viruses were 40 mg/L $FeCl_3$ and pH between 5 and 6. Under these conditions, enhanced coagulation resulted in maximum removals of more than 2 logs of adenovirus type 4 (generally used as a surrogate for adenoviruses 40 and 41), feline calicivirus (used as a surrogate for human noroviruses), and MS2 phage. Lower removals were observed for phages PRd1 and phi-X174 (Abbaszadegan et al., 2007, 2008). Coagulation–flocculation–sedimentation did not remove the avian virus H5N1 significantly as the removal varied between 0 and 1.5 log (Lénès et al., 2010). Although algal cells are well removed by coagulation–flocculation, no efficient removal has been reported for algal toxins (see Chapter 5).

However, pathogen and parasite removal under field conditions may be lower than under controlled laboratory conditions. In a pilot water treatment plant treating water from the Seine River in France, indigenous virus removal by coagulation-flocculation varied by 31–90% with an average removal of 61%. This shows that virus removal obtained with laboratory strains is much higher than with indigenous viruses (Joret et al., 1986).

Endotoxins or lipopolysaccharides (LPS) are components of the outer membrane of most gram-negative bacteria and some cyanobacteria. They have been associated with acute respiratory illnesses, water fever, gastrointestinal disorders, and allergic reactions. The different processes involved in water treatment were shown to remove 59–97% of the endotoxin activity. The highest removal was observed following coagulation, clarification, and rapid sand filtration (Rapala et al., 2002).

Coagulation is sometimes improved by using *coagulant aids*, such as polyelectrolytes, bentonite (a three-layer clay mineral), or activated silica. Polyelectrolytes help form large flocs that settle out rapidly. Their concentration is an important parameter, since excessive dosage can inhibit flocculation.

It is important to add that coagulation merely transfers pathogenic microorganisms from water to the flocculated material, which is incorporated in a sludge that must be disposed of properly.

2.5.4 Water Softening

Water softening was suggested as a water treatment process in the mid-1700s by Francis Home of Scotland, and the first successful water softening plants was built in 1897 in Winnipeg, Canada (Symons, 2006). Hardness is caused by the presence of calcium and magnesium in the water. There are two categories of hardness: *carbonate hardness* which is due to bicarbonates of Ca and Mg, and *noncarbonate hardness* which is due to Ca and Mg chlorides. Hardness is responsible for increased soap consumption and scale formation in pipes. *Water softening* is the removal of Ca and Mg hardness by the lime-soda process or by ion-exchange resins. The lime-soda process consists of adding hydrated lime (calcium hydroxide) or the less expensive quick lime (CaO) to the water. The carbonate hardness is removed according to the following:

$$\underset{\text{Ca-bicarbonate}}{Ca(HCO_3)_2} + \underset{\text{hydrated lime}}{Ca(OH)_2} \rightarrow \underset{\text{Ca-carbonate}}{2CaCO_3} + 2H_2O \tag{2.1}$$

$$\underset{\text{Mg-bicarbonate}}{Mg(HCO_3)_2} + \underset{\text{lime}}{2Ca(OH)_2} \rightarrow \underset{\text{Mg-hydroxide}}{Mg(OH)_2} + CaCO_3 + 2H_2O \tag{2.2}$$

Softening helps in the removal of NOM by calcium carbonate and magnesium hydroxide. The decrease in NOM will lead to lowering the formation of halogenated by-products in drinking water (Kalscheur et al., 2006).

The high pH (>11) generated by water softening with lime leads to an effective inactivation of bacterial and viral pathogens. Poliovirus type 1, rotavirus, and hepatitis A virus (HAV) were effectively removed (>95% removal/inactivation) during water softening (pH = 11) (Rao et al., 1988). Bacterial pathogens are also efficiently reduced following liming to reach a pH that exceeds 11. The inactivation rate is temperature-dependent.

Microbial removal during water softening is due to (1) microbial inactivation at detrimental high pH values (pH ≥ 11) by the loss of structural integrity or inactivation of essential enzymes and (2) physical removal of microorganisms by adsorption to positively charged magnesium hydroxide flocs ($CaCO_3$ precipitates are negatively charged and do not adsorb microorganisms).

As regard ion-exchange resins, Ca and Mg are removed from water by exchange with Na present on the exchange sites on the resin. Viruses are removed by anion-exchange resins but not as much by cation-exchange resins. However, ion-exchange resins cannot be relied upon to remove microbial pathogens.

2.5.5 Filtration

Filtration is defined as the passage of fluids through porous media to remove turbidity (suspended solids, such as clays, silt particles, microbial cells) and flocculated particles. Filtration may include slow and rapid sand filtration, diatomaceous earth filtration, direct filtration, membrane filtration, or cartridge filtration. This process depends on the filter medium, concentration, and type of solids to be filtered out, and the operation of the filter.

Filtration is one of the oldest processes used for water treatment. Back in 1685, an Italian physician, Luc Antonio Porzio, conceived a filtration system for protecting soldiers' health in military installations. The first modern water filtration plant was built in 1804 in Scotland (Symons, 2006). An examination of the occurrence of waterborne outbreaks (e.g., outbreaks of cholera and typhoid fever) around the world clearly shows that filtration has been historically instrumental as a barrier against pathogenic microorganisms and has largely contributed to the reduction of waterborne diseases. Figure 2.6 shows the dramatic reduction of typhoid fever death rate in Albany, NY, after the adoption of sand filtration and chlorination about a decade later (Logsdon and Lippy, 1982; Willcomb, 1923).

2.5.5.1 *Slow Sand Filtration.* In 1830, James first installed a slow sand filter, in London, England. This process was adopted in 1872 in the United States where the first slow sand filters were installed in Poughkeepsie, NY, and Lawrence, MA. Although more popular in Europe than in the United States, slow sand filters serve mostly small communities of less than 10,000 people because capital and operating costs are lower than for rapid sand filters (Slezak and Sims, 1984; Symons, 2006). A slow sand filter (Figure 2.7) contains a layer of sand (60–120 cm depth) supported by a graded gravel layer (30–50 cm depth). The sand grain size varies between 0.15 and 0.35 mm, and the hydraulic loading range is between 0.04 and 0.4 m/h (Bellamy et al., 1985a).

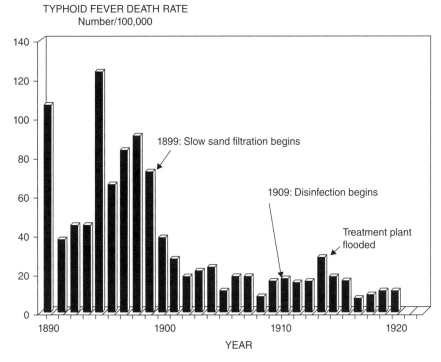

Figure 2.6   Impact of slow sand filtration on the reduction of typhoid fever death rate in Albany N.Y. *Source*: Logsdon and Lippy (1982). Amer. Water Works Assoc. J. 74: 649–655.

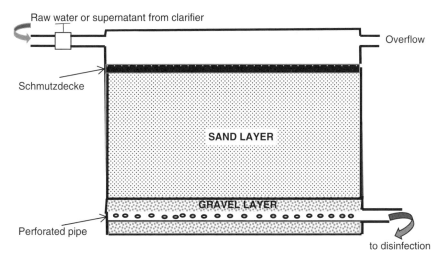

Figure 2.7   Slow sand filter. Adapted from Slezak and Sims (1984). Amer. Water Works Assoc. J. 76: 38–43.

Biological growth inside the filter comprises a wide variety of organisms, including bacteria, algae, protozoa, rotifers, microtubellaria (flatworms), nematodes (round worms), annelids (segmented worms), and arthropods (Duncan, 1988; Hijnen et al., 2007). The buildup of a biologically active layer, called *schmutzdecke*, occurs during the normal operation of a slow sand filter. The top layer is composed of biological growth and filtered particulate matter. However, the buildup of the schmutzdecke is a major contributor to the development of a head loss across the filter (Campos et al., 2006), a problem corrected by removing or scraping the top layer of sand to restore hydraulic conductivity. The length of time between scrapings depends on the turbidity of the raw water and varies from 1 to 2 weeks to several months (Logsdon and Hoff, 1986). Scraping is followed by replenishing of the filter bed with clean sand, an operation called *resanding*. There is a deterioration of water quality for some days after scraping, but it later improves during the ripening period (Cullen and Letterman, 1985). Temperature was found to affect the time period for the recovery of *E. coli* removal efficiency following scraping (Unger and Collins, 2008). Column experiments on the distribution of *E. coli* in sand columns showed that most of the bacteria were retained in the schmutzdecke, down to 6 cm depth (Figure 2.8,

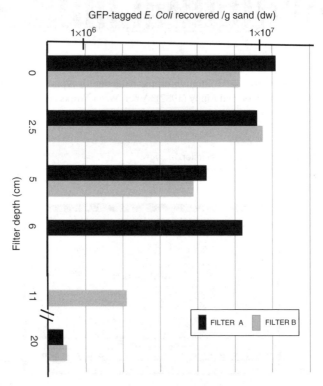

Figure 2.8 Distribution of challenge *E. coli* recovered from different depths of sand filter operated at 0.6 m/hr. *Source*: Unger and Collins (2008). Amer. Water Works Assoc. J. 100 (12): 60–72.

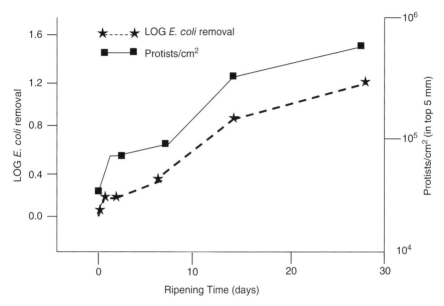

Figure 2.9   Protists abundance and *E. coli* removal in sand columns after various ripening times. *Source*: Unger and Collins (2008). Amer. Water Works Assoc. J. 100 (12): 60–72.

Unger and Collins, 2008). Bacterial activity in the schmutzdecke also helps remove assimilable organic compounds, some of which are precursors of chlorinated organics such as trihalomethanes (THM) (Collins et al., 1992; Fox et al., 1984).

Bacterial removal efficiency is influenced by operational parameters such as temperature, sand grain size, filter depth, and empty bed contact time (EBCT which is equal to L/HLR where L = filter depth, HLR = hydraulic loading rate) (Bellamy et al., 1985a, 1985b; Unger and Collins, 2008). Modeling the removal of viruses and bacteria by slow sand filtration showed that the most important operational parameters were temperature and the schmutzdecke age (Schijven et al., 2013). In the schmutzdecke, protozoan predation on bacteria also affects their removal in slow sand filters. Since protozoan numbers are higher in ripened filters, a relationship was found between the length of the ripening period and *E. coli* removal (Figure 2.9; Unger and Collins, 2008). Removal of *Giardia* and total coliforms exceed 99%, even at the highest loading rate of 0.4 m/h. Pilot plant studies showed that slow sand filtration achieves a 4- to 5-log reductions of coliforms (Fox et al., 1984). A survey of slow sand filters in the United States showed that most of the plants reported coliform levels of 1 per 100 mL or less (Slezak and Sims, 1984). In water treatment plants in Paris, France, slow sand filters were found to be more efficient in mycobacteria removal than rapid sand filters (Le Dantec et al., 2002b).

No *Cryptosporidium* oocysts were detected in treated water from two water treatment plants using slow sand filters, despite the frequent isolation of this parasite in the raw waters entering the plants (Chauret et al., 1995). Laboratory experiments showed that slow sand filters removed 4.7-log oocysts (Hijnen et al., 2007). Zooplankton in the filter bed possibly predates upon the retained oocysts but this awaits further research. The microbial community developing on the sand filter greatly influences the removal of bacteria, protozoan cysts, and turbidity by slow sand filters (Bellamy et al., 1985a; Cleasby et al., 1984; Huisman and Wood, 1974).

With regard to viruses, a removal exceeding 99.999% was observed in a slow sand filter operated at 11°C at a rate of 0.2 m/h (Poynter and Slade, 1977). Slow sand filtration achieved from 2.05 $\log_{10}$ to 3.19 $\log_{10}$ removal of indigenous somatic phage at 90 cm filter depth (Bauer et al., 2011). An established filter appears to perform better than a new filter with regard to virus removal (Wheeler et al., 1988). As mentioned for bacteria, the factors controlling virus removal by slow sand filters include filter depth, flow rate, temperature, and the presence of a well-developed biofilm on the filter.

2.5.5.2 Rapid Sand Filtration.

In the 1890s, rapid sand filtration became more popular than slow sand filtration and is by now adopted by many municipalities. A rapid sand filter consists of a layer of sand supported by a layer of anthracite, gravel, or calcite. The rapid sand effluent is collected by an underdrain system. Rapid sand filters are operated at filtration rates of 5–24 m^3/h m^2 as compared with 0.1-1 m^3/h m^2 for slow sand filters (Huisman and Wood, 1974; Logsdon and Hoff, 1986). These filters are periodically cleaned by backwashing (i.e., by reversing the flow) at a sufficient flow rate to allow a thorough cleaning of the sand.

This process appears to be less effective for the removal of bacteria, viruses, and protozoan cysts unless it is preceded by coagulation–flocculation. The removal of *Salmonella* and *Shigella* by coagulation and filtration is similar to that of coliforms. The breakthrough is more pronounced during winter months at temperatures below 5°C (Hibler and Hancock, 1990). Significant changes in flow rates may result in deterioration of water quality due to the release of retained particles, including protozoan cysts. The combination of filtration with coagulation–flocculation is particularly important with respect to *Giardia* cysts and *Cryptosporidium* oocysts removal from water (Logsdon et al., 1981). For low turbidity water, proper chemical coagulation of the water before rapid sand filtration is necessary to achieve a good removal of turbidity and protozoan cysts and oocysts. In a plant in Japan, coagulation–flocculation–sedimentation followed by rapid sand filtration removed approximately 2.5-$\log_{10}$ of *Cryptosporidium* and *Giardia* (Hashimoto et al., 2002). Some suggested that the removal of turbidity could serve as a surrogate indicator of *Giardia* cysts removal in water with low turbidity (Al-Ani et al., 1986). However, *Giardia* and *Cryptosporidium* monitoring in filtered drinking water samples indicated that turbidity removal was a good predictor of *Cryptosporidium* oocysts removal but not *Giardia* cysts (LeChevallier et al., 1991b). Similarly, oocyst-sized polystyrene microspheres also appear to be reliable surrogates for *Cryptospodium parvum* oocyst removal by sand filtration (Emelko and Huck, 2004). Surface coating of sand particles with a hydrous iron aluminum oxide was found to improve the removal of *Cryptosporidium*. This surface coating increased the zeta potential of the sand from -40 to $+45$ mV at pH = 7.0, thus favoring its interaction with the negatively charged oocysts (Shaw et al., 2000). However, the relatively long-term effect of the coating on oocyst removal is not known.

Because sand particles are essentially poor adsorbents toward viruses, pathogen removal by sand filtration is variable and often low. However, coagulation, settling, and sand filtration remove more than two to three logs of HAV, simian rotavirus (SA-11), and poliovirus (Rao et al., 1988). A pilot plant, using the Seine River water which contained 190–1420 PFU/1000 L removed 1–2 logs of viruses after coagulation–flocculation followed by sand filtration (Joret et al., 1986). A coagulation-rapid sand filtration process, using polyaluminum chloride or ferric chloride at pH = 5.8 as coagulants, achieved a 3-log removal of recombinant norovirus particles (Shirasaki et al., 2010).

Thus, optimum coagulation–flocculation followed by filtration in water treatment plants is essential for parasites and pathogens control. Parameters that adversely affect filter operation are sudden changes in water flow rates, interruptions in chemical feed, inadequate filter backwashing, and the use of clean sand which has not been allowed to undergo a ripening period (Logsdon and Hoff, 1986).

Because of their limited capacity to retain solids, rapid sand filters must be *backwashed* to remove the trapped solids from the filter matrix. Approximately 3–5% of the water treatment plant finished water is generally used as backwash water (Cornwell and MacPhee, 2001). However, the spent filter backwash water (SFBW) must be adequately treated prior to reuse within the plant or disposal into a receiving stream (Arora and LeChevallier, 2002). The recycling of SFBW, mostly by treatment plants that use surface water as the source water, is regulated by the U.S. EPA, due to the increased levels of biological (protozoan cysts, bacterial, and viral pathogens) and chemical (disinfection by-products, metals) contaminants in SFBW. SFBW must be treated to reduce the risk of breakthrough of pathogens and parasites into finished water. The treatments include sedimentation (preferably in the presence of coagulants), dissolved air flotation, filtration through granular media or membrane filters, and disinfection, preferably after clarification.

Dual-stage filtration (DSF) is an alternative to rapid sand filtration for small water treatment plants. This process consists of chemical coagulation followed by a filter assembly consisting of two tanks: a depth clarifier and a depth filter. This treatment removed more than 99% of *Giardia* cysts from water with an effluent turbidity less than 1 nephelometric turbidity unit (NTU) (Horn et al., 1988).

2.5.5.3 *Diatomaceous Earth Filtration.*   Diatoms are eukaryotic algae that are ubiquitous in marine and freshwater environments. Their cell wall typically contains silica and, upon their death, they form geological deposits of diatomite which is made of remains of silica shells and is sold in a variety of grades. The mean particle size is 23 μm and the average pore size is 7 μm (Fulton, 2000). During World War II, the US Army developed a diatomaceous earth (DE) filter for the rapid removal of amoebic cysts (*Entamoeba histolytica*) from canteen water. Later on, DE filtration, also known as precoat filtration, was applied to swimming pools and later on to drinking water treatment. The earliest DE filter was installed in Illinois more than 60 years ago.

The DE filter consists of a porous septum which serves as a support for a 0.3 cm layer of diatomite called precoat. Following formation of the precoat layer, raw water containing a small concentration of diatomite is continuously added to reduce plugging of the precoat and to extend the filter run. This additional diatomite is called body feed. The filtration rates range between 0.5 and 2 gpm/ft^2.

There are two configurations of DE filters: pressure filters which are housed in a pressure vessel and where the water is introduced at the inlet and the vacuum filters with a suction pump at the effluent side and are open to the atmosphere. DE filtration is suitable for low turbidity waters and meets the requirements of the EPA Surface Water Treatment Rule (SWTR).

The main factors controlling the performance of DE filtration are the diatomite grade, bodyfeed rate, and the filtration rate. DE with size ranging from 15 to 26 μm and at filtration rate of 4.88 m/h removed more than 3 logs of *Giardia* cysts and *Cryptosporidium* oocysts. DE performance regarding *Cryptosporidium* and coliform removal is improved following addition of a chemical coagulants (Lange et al., 1986; Schuler and Ghosh, 1990). Approximately 6-log reduction of *Cryptosporidium* oocysts was obtained with diatomaceous grades ranging from 16 to 42 μm medium size and at filtration rates between 1 and 2 gpm/ft^2 (Ongerth and Hutton, 1997). Virus removal by DE is generally low but can be enhanced by coating the diatomite medium with Fe and Al salts or with cationic polyelectrolytes (Brown et al., 1974).

2.5.6 Activated Carbon

Activated carbon is an adsorbent derived from wood, bituminous coal, lignite, or other carbonaceous materials and is the most widely utilized adsorbent for the treatment of water and wastewater. It is activated by a combustion process to increase its internal surface area. It offers a total surface area of 500 to more than 1000 m^2/g (see Huang et al., 2007) for the adsorption of taste, odor, and color compounds, excess chlorine, toxic and mutagenic substances (e.g., chlorinated organic compounds, including THM), trihalomethane precursors, pesticides, phenolic compounds, dyes, and toxic metals (Allen, 1996; Najm et al., 1991).

Activated carbon may be used in the form of granular activated carbon (GAC) applied after sand filtration and before chlorination or in the form of powdered activated carbon (PAC), which has a smaller particle size than GAC and can be applied at various points in water and wastewater treatment plants, mainly prior to filtration. PAC treatment is less costly than GAC treatment because the powdered carbon is applied only when needed, leading to lower amounts of carbon used.

Activated carbon has been known for its purification properties since the Antiquity (i.e., early Egyptians) but its microbiological aspects have only recently been investigated (Gibert et al., 2013; LeChevallier and McFeters, 1990; Weber et al., 1978). The functional groups on the carbon surface help in the adsorption of microorganisms to form a biofilm which harbors bacteria (rods, cocci, and filamentous bacteria), fungi, algae, and protozoa. High resolution field-emission scanning electron microscopy (FESEM) shows the colonization of carbon surface by microorganisms

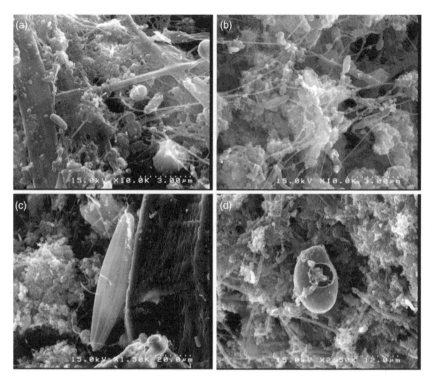

Figure 2.10 FESEM micrographs showing (a and b) organic filaments closely resembling fungal hyphae with visible isolated bacterial cells, and (c and d) siliceous diatom skeletons observed on a GAC sample collected from the top of the column at the end of the experiment. *Source*: Gibert et al. (2013). Water Res. 47: 1101–1110.

(polysaccharide-producing bacteria, fungi, algae, stalked protozoa) (Figure 2.10, Gibert et al., 2013). The dominant bacterial genera identified on GAC particles or in interstitial water were *Pseudomonas, Alcaligenes, Aeromonas, Acinetobacter, Arthrobacter, Enterobacter, Flavobacterium, Chromobacterium, Bacillus, Corynebacterium, Micrococcus, Paracoccus*, and *Moraxella* (Camper et al., 1986; Rollinger and Dott, 1987; Wilcox et al., 1983). More recently, bacterial isolates were identified, following enrichment in dilute media, in 21 GAC filters from nine water treatment plants in the Netherlands. Genomic fingerprinting and 16S rDNA sequence analysis showed that 68% of the isolates belonged to the betaproteobacteria (the predominant genera were *Polaromonas* and *Hydrogenophaga*) while 25% were alphaproteobacteria (the predominant genera were *Sphingomonas* and *Afipia*) (Magic-Knezev et al., 2009).

A vertical stratification of microorganisms with higher biomass at the top of the activated carbon column was observed, using confocal laser scanning microscopy (CLSM) counts and DNA analysis. This stratification disappeared with time (Gibert et al., 2013).

Activated carbon has the ability to remove NOM via adsorption and biodegradation. The organic materials serve as carbon sources that support bacterial growth in the filter matrix (Servais et al., 1991). The magnitude of biological assimilation depends on various factors, including temperature, running time and intensity, and frequency of filter backwashing (van der Kooij, 1983). Removal of organic compounds results from microbial proliferation in the activated carbon column, leading to a bacterial biomass estimated at 60–1250 g/m^3 of GAC (Magic-Knezev and van der Kooij, 2004). Some of these bacteria may, however, produce endotoxins, which may enter into the treated water. Treatment to enhance bacterial growth on GAC produces biological activated carbon (BAC), which has been recommended for increasing the removal of organics, as well as extending the lifetime of GAC columns (Simpson, 2008). Bacterial growth in activated carbon columns is enhanced by ozone which degrades NOM to easily assimilable low molecular weight compounds such as organic acids, aldehydes, and ketones (Hammes et al., 2006).

Pathogenic bacteria may be successful in colonizing mature GAC filters (Grabow and Kfir, 1990; LeChevallier and McFeters, 1985b). However, pathogenic and indicator bacteria can be inhibited by biofilm microorganisms on activated carbon. This phenomenon is attributable to nutritional competitive inhibition or to the production of bacteriocin-like substances by the filter microbial community (Camper et al., 1985; LeChevallier and McFeters, 1985b; Rollinger and Dott, 1987).

Problems arise when bacterial cells or bacterial microcolonies attached to carbon filters are sloughed off the filter or when bacteria-coated carbon particles penetrate the distribution system (Camper et al., 1986). Some of these carbon fines can become associated with biofilms and may serve as inoculants that induce the regrowth of potential pathogens in water distribution systems (Morin et al., 1996). An examination of 201 samples showed that heterotrophic bacteria and coliform bacteria are associated with the carbon particles. The attached bacteria display increased resistance to chlorination (Camper et al., 1986; LeChevallier et al., 1984; Stewart et al., 1990). Operational variables contributing to the release of particles from activated carbon beds include filter backwashing, increase in bed depth, and filtration rate (Camper et al., 1987). In general, particle-associated bacteria (PAB) in drinking water are of concern due to their ability to serve as seeds for bacterial regrowth in distribution pipes, their higher resistance to disinfectants, and the potential underestimation of bacterial numbers in traditional culture techniques. An average 20–50 PAB were found to be attached to a single particle in drinking water from three water treatment plants in the Netherlands. Among PAB, we note the presence of pathogens such as *Legionella* as well as Fe- and Mn-oxidizing bacteria such as *Gallionella* and *Crenothrix* (Liu et al., 2013).

Viruses are adsorbed to activated carbon by electrostatic forces and the interaction is controlled by pH, ionic strength, and organic matter content of the water. The competition of organics with viruses for the attachment sites on the carbon surface makes this material an unreliable sorbent for removing viruses from water (Bitton, 1980). However, bacterial and potentially viral removal by activated carbon can be enhanced by modifying the surface with Al and Fe hydroxides or aluminum hydroxychloride. Moreover, antibacterial compounds, such as nanosized AgBr crystals, can be immobilized on the carbon surface to prevent bacterial growth in the filter (Pal et al., 2006).

The following are some disadvantages of carbon filters:

1. Clogging, leading to head loss and increase in effluent turbidity. It is relieved by periodic (weekly in general) backwashing which removes organic matter and loosely bound microorganisms from the biofilm surface. In a BAC filter, backwashing led to approximately one-third reduction of the bacterial abundance at the top of the filter and caused changes in the microbial composition of the biofilm (Kasuga et al., 2007).

   Other measures used to control biofilm thickness are the control the nutrient loading to the biofilm, manipulation of dissolved oxygen and pH levels of the filter influent, and disinfection with chemical oxidants such as chlorine, chlorine dioxide, or ozone (Simpson, 2008).

2. Production and release of endotoxins

3. Creation of anaerobic conditions inside filters with subsequent production of odorous compounds (e.g., H_2S).

4. Production of effluents with high colony counts

5. Occasional growth of zooplankton in carbon filters and their release in the filter effluents.

2.5.7 Membrane Filtration

In the early 1980s, Francois Fiessinger (France), Dennis Hanley (Australia), and Andrew Benedek (Canada) predicted that membrane filtration would gradually replace granular media filtration in the production of safe drinking water (Trussell, 2006). Membranes can be used to treat water to remove a wide range of contaminants. Membrane processes include reverse osmosis (removal of cations, anions, metals, organics, and microorganisms), nanofiltration (NF) (mostly removing ions like calcium and magnesium, small molecules and microorganisms), ultrafiltration (UF) (removes colloidal particles, macromolecules and microorganisms), and microfiltration (MF) for removing micron or submicron-size particles (Pizzi, 2002). The various pressure-driven membrane processes are illustrated in Table 2.2 (AWWA Committee Report, 2008; van Reis and Zydney, 2007). For pressure-driven membrane systems such as NF and reverse osmosis, organic solute rejection is controlled by the following factors (Bellona et al., 2004):

- Solute characteristics (molecular weight and size, pKa, log K_{ow}, and diffusion coefficient)
- Membrane characteristics (pore size, molecular weight cutoff, surface charge and hydrophobicity)
- Suspending medium chemistry (pH, ionic strength, presence of organic matter)

TABLE 2.2   Comparison of removal characteristics of different pressure-driven membrane processes[a]

	Microfiltration	Ultrafiltration	Nanofiltration	Reverse Osmosis
Components retained by membrane	Intact cells Bacteria Algae Protozoa	Viruses Proteins	Divalent ions Amino acids Antibiotics	Amino acids Sugars salts
		Membrane Matrix		
Components passing through membrane	Viruses Colloids Proteins Salts	Amino acids Antifoam Salts	Salts Water	Water

[a] Adapted from Van Reis and Zydney. 2007. *J. Membr. Sci.* 297:16–50.

Microfiltration and UF are used in water and wastewater treatment or as a post-treatment to remove suspended solids, algae, bacterial pathogens as well as cysts and oocysts of protozoan parasites. Marketed membranes are made of cellulose acetate, polypropylene, polyvinylidene fluoride, polyethersulfone, polysulfone, and other proprietary materials, and have various nominal pore sizes ranging from 0.01 to 0.2 µm (AWWA Membrane Process Committee, 2008). The removal of dissolved organic and inorganic constituents necessitates a pretreatment process such as activated carbon, coagulation, or oxidation.

Membrane challenge pilot studies showed that membranes remove from 4.9 to 5.8 $\log_{10}$ units for *Giardia* cysts, and from 5.8 to 6.8 $\log_{10}$ units for *Cryptosporidium* oocysts and the removal mechanism is probably physical straining (Jacongelo et al., 1995; States et al., 2000). Microfiltration was able to achieve a 4-log removal of *Bacillus subtilis* spores from water (Huertas et al., 2003). A hydrophobic microfiltration membrane with a nominal pore size of 0.22 µm achieved 91% to near 100% rejection of poliovirus 1 (Madaeni et al., 1995). UF (5–20 nm pore size) showed more than 4-log reduction of indigenous noroviruses from wastewater effluents (Sano et al., 2006) and the avian virus H5N1 from water (Lénès et al., 2010). Polysulfone ultrafilters with a molecular weight cut-off of 30 kD completely removed poliovirus 1 (Madaeni et al., 1995). For phage MS2, an average log retention value of more than 6.7 was achieved when membrane filtration (UF or MF) was preceded by coagulation/flocculation using alum or polyaluminum chloride (Fiksdal and Leiknes, 2006). Submerged UF membrane bioreactors can be combined with coagulation with polyaluminum chloride for drinking water treatment. When compared to traditional membrane bioreactors, they achieved a higher removal of organic matter, THM and haloacetic acids (HAA) precursors, and phosphate. However, the removal of biodegradable dissolved organic carbon (BDOC) and AOC was only slightly higher. In these membrane bioreactors, organic matter removal is due to the combined effects of rejection, biodegradation, and coagulation (Tian et al., 2008).

UF membranes may sometimes underperform, allowing the passage of virus particles. The breakthrough may be due to imperfections during the membrane polymerization process or due to high transmembrane pressures (TMPs) which may cause pore enlargement (Arkhangelsky and Gitis, 2008).

NF retains colloidal particles and macromolecules by physical sieving while charge effects are involved in ion separation. NF advantage over reverse osmosis is operation at lower pressure, thus lowering operational cost (Shon et al., 2013). Four NF polyamide membranes were tested for their performance in improving the microbiological quality of groundwater impacted by wastewater and serving as drinking water supply in Mexico. The membranes mostly eliminated total and fecal coliforms and removed at least 99.96% of somatic phages (Aguilar et al., 2008).

However, membranes are subject to fouling by particulate and colloidal particles, NOM, and biofouling (i.e., biofilm formation), leading to pore blocking and subsequent cake formation on the membrane surface. Fouling can be reversible or irreversible. The NOM fraction contributing to fouling comprises small, neutral, hydrophilic compounds (Carroll et al., 2000).

A two-step process was proposed to describe the irreversible membrane fouling (Yamamura et al., 2007): (1) small hydrophobic humic-like compounds first adsorb to the membrane surface and (2) they are followed by the adsorption of larger hydrophilic, carbohydrate-like compounds.

The pretreatment processes for reducing biofouling include coagulation, sedimentation, dissolved air flotation, lime softening, or 500-μm screens. UF membranes impregnated with silver nanoparticles were also shown to resist biofouling in addition to their antibacterial and antiviral properties. However, better fixation techniques for the silver nanoparticles are needed for the long-term performance of these membranes (Zodrow et al., 2009).

The integrity of low pressure membrane filters (microfilters and ultrafilters) may be compromised during operation, leading to the breakthrough of pathogenic microorganisms. Several membrane integrity tests are available and include direct (e.g., pressure driven tests, diffusive air flow) and indirect tests (e.g., particle and turbidity monitoring, surrogate challenge tests) (Guo et al., 2010).

Several water treatment plants are now considering membranes to add to or to replace traditional granular filtration media. Water treatment/reclamation plants using reverse osmosis membranes have been built and treat from 4000 m^3/day in a small town in N.E. United States to 375,000 m^3/day in a water reclamation plant in Kuwait (Wright et al., 2008; http://www.water-technology.net/projects/sulaibiya/).

Other advantages of membrane processes are treatment without addition of chemicals and their relatively small footprint.

2.5.8 Nanotechnology in Water Treatment

Nanofibers with a high surface-to-volume ratio are produced via electrospinning of polymers such as polyurethane, polycarbonate, polystyrene, polyamide, cellulose acetate, and others. Their surface may be modified by attaching biocides for the inactivation of pathogens (Yao et al., 2008). Silver nanoparticles can be incorporated into

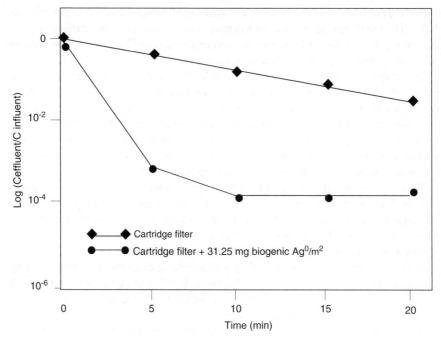

Figure 2.11   Inactivation of bacteriophage UZ1 during continuous filtration through Nano-Ceram cartridge filters homogenously coated with 31.25 mg biogenic silver. Adapted from De Gusseme et al. (2010). Appl. Environ. Microbiol. 76: 1082–1087.

polymers to produce nanofilters with antimicrobial properties and increased resistance to fouling (Jain and Pradeep, 2005; Zodrow et al., 2009). Silver nanoparticles can also be anchored onto methacrylic acid copolymer beads which can be used for disinfection (Gangadharan et al., 2010). One example is the incorporation of biogenic Ag^0 (i.e., Ag^0 associated with dead *Lactobacillus fermentum* cells which serve as a carrier) onto a NanoCeram cartridge filter (Sanford, FL; NanoCeram is an electropositive filter containing alumina nanofibers in a microglass fiber matrix) for continuous inactivation of phage UZ1. Figure 2.11 (De Gusseme et al., 2010) shows that, after 10 HRT (Hydraulic retention time = 2 minutes), the cartridge with biogenic Ag^0 removed 3.8 $\log_{10}$ of phage as compared to 1.5 $\log_{10}$ decrease in the cartridge without silver. Other nanoparticles considered for incorporation into filter media owing to their antimicrobial properties are zinc oxide, copper and copper oxide, titanium dioxide, nano alumina, aluminum hydroxide fibers, fullerenes (C_{60}) and other carbonaceous nanoparticles, and natural antimicrobial substances such as chitosan or chitosan–metal composites. Potential problems associated with the incorporation of nanobiocides into filters are cost and leaching of the biocides from the filter material with potential consequences for human and animal health and the environment. Nanomaterials are also promising for the removal of chemical pollutants from drinking water (Botes and Cloete, 2010; Theron et al., 2008). Another approach for removing microbial pathogens from water is the use of amine-functionalized magnetic nanoparticles. The capture efficiency of various gram-positive and gram-negative bacteria varied between 55% and 97.1% (Huang et al., 2010). Magnetic iron oxide nanoparticles removed more than 90% of turbidity. The removal took 12 minutes as compared to 180 minutes with alum (Okoli et al., 2012). Most of the nanotechnology-based treatments have been tested under laboratory conditions and more research is needed to evaluate their performance under field conditions. Of particular concern are their cost effectiveness and the impact of nanomaterials on human and environmental health (Boxall et al., 2007; Klaine et al., 2008; Mielke et al., 2013; Qu et al., 2013).

A summary of bacterial, viral, and protozoan reductions by the various water treatment processes discussed above is shown in Table 2.3 (WHO, 2011c).

2.5.9 Disinfection

Disinfection is the last barrier against the entry of pathogens and parasites into our drinking water. Disinfection addresses specifically the *inactivation* of disease-causing organisms. For a detailed discussion of this topic, the reader is referred to Chapter 3.

2.6 WASTE RESIDUALS FROM WATER TREATMENT PLANTS

Water treatment plants generate liquid (e.g., filter backwash water, sedimentation tanks wash water, brine) and solid (e.g., sludges, sloughed off biofilms). These residuals harbor pathogens and parasites and may contribute to disease outbreaks. Hence, they must be disposed of properly. The liquid residuals are disposed of via direct discharge to surface waters, indirect discharge into sanitary sewers, underground injection, or land disposal. The solid residuals are disposed of via solid waste and hazardous wastes landfills, land application, or incineration (U.S. EPA, 2007).

2.7 DRINKING WATER QUALITY AT THE CONSUMER'S TAP

Drinking water quality at the consumer's tap is affected by the service lines, indoor plumbing, and the home devices occasionally installed by the consumer to improve taste and odor problems. Water quality indoors is not covered by U.S. EPA regulations except for the Lead and Copper Rule (LCR) dealing with Pb and Cu monitoring at the tap.

2.7.1 Effect of Service Lines and Indoor Plumbing on Drinking Water Quality

In addition to water distribution pipes, service lines and house plumbing system can also influence the microbiological and chemical quality of drinking water (Eboigbodin

TABLE 2.3   **Removal/inactivation range of bacteria, viruses, and protozoa at various water treatment plants for large communities**

Treatment Process	Enteric Pathogen Group	Log_{10} Reduction Range	Remarks
Pretreatment			
Roughing filters	Bacteria	0.2–2.3	Depends on filter medium and coagulant
Storage reservoirs	Bacteria	0.7–2.2	Residence time >40 days
	Protozoa	1.4–2.3	Residence time >160 days
Bank filtration	Bacteria	2 to >6	Depends on travel distance, soil
	Viruses	2.1–8.3	type, pumping rate, pH, and ionic
	Protozoa	1–>2	strength
Coagulation–Flocculation–Sedimentation			
Conventional clarification	Bacteria	0.2–2	Depends on coagulation conditions
	Viruses	0.1–3.4	
	Protozoa	1–2	
High rate clarification	Protozoa	>2 to 2.8	Depends on blanket polymer
Dissolved air flotation	Protozoa	0.6–2.6	Depends on coagulant dose
Lime softening	Bacteria	1–4	Depends on pH and settling time
	Viruses	2–4	
	Protozoa	0–2	
Filtration			
Granular high rate filtration	Bacteria	0.2–4.4	Depends on filter media and
	Viruses	0–3.5	coagulation pretreatment
	Protozoa	0.4–3.3	
Slow sand filtration	Bacteria	2–6	Depends on grain size, flow rate,
	Viruses	0.25–4	pH, temperature and presence of
	Protozoa	0.3 to >5	Schmutzdecke
Precoat filtration	Bacteria	0.2–2.3	Depends on Chem. Pretreatment
	Viruses	1–1.7	If filter cake is present
	Protozoa	3–6.7	Depends on media grade and filtration ratet
Membrane filtration (microfiltration, ultrafiltration, nanofiltration, reverse osmosis)	Bacteria	1 to >7	Varies with pore size, integrity of
	Viruses	<1 to >6.5	filter medium, and resistance to
	Protozoa	2.3 to >7	chemical and biological degradation

Source: Adapted from WHO (2011a). Guidelines for Drinking-Water Quality. 4th Ed. World Health Organization, Geneva, Switzerland. http://www.who.int/water_sanitation_health/publications/2011/dwq_guidelines/en/.

et al., 2008). Conditions prevailing in service lines and indoor plumbing that are conducive to bacterial proliferation include long stagnation time, backflow problems, absence of chlorine residual, higher temperature, and high surface-to-volume ratio (ratio of pipe surface to volume of water). In unchlorinated drinking water in the Netherlands, opportunistic pathogens were more often detected in indoor plumbing water than in the distribution system water, confirming their growth in indoor plumbing (van der Wielen and van der Kooij, 2013). Total bacterial numbers, as measured by epifluorescence microscopy, are generally much higher in the first flush than in the distribution system (Prévost et al., 1997). Similarly, a 3-minute flushing decreased the gene copy numbers of opportunistic pathogens such as *Legionella* spp., *Mycobacterium* spp., *Hartmanella vermiformis* to levels found in the distribution system (Figure 2.12; Wang et al., 2012b). These results demonstrate the growth of opportunistic pathogens in the premise plumbing. Bacterial numbers, as measured by heterotrophic plate counts, flow cytometry or ATP, decreased from their highest values at 8 A.M. to their lowest values around 10 A.M. and stabilize for the rest of the day (Siebel et al., 2008). In addition to an increase in bacterial cell numbers, a change in community composition was also observed following stagnation of drinking water in indoor plumbing (Lautenschlager et al., 2010).

Depending on the distance of the home to the water treatment plant, a flushing time of at least 2–10 minutes is necessary to lower bacterial numbers to background levels. Similarly, lead and copper concentrations were high in the first flush and decreased to background levels within 2 minutes of flushing (Prévost et al., 1997).

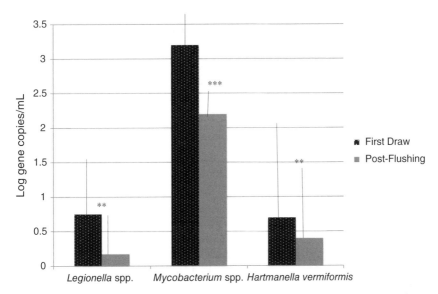

Figure 2.12   Average copy numbers of *Legionella* spp., *Mycobacterium* spp., *H. vermiformis*, and the 16S rRNA gene in first-draw and postflushing samples. ** and ***, significant differences according to paired Wilcoxon rank sum testing. Adapted from Wang et al. (2012b). Appl. Environ. Microbiol. 78: 6285–629.

2.7.2　Point-of-Use Devices for Indoor Water Treatment

Household water treatment comprises technologies, devices, or methods to treat water at the household or at point-of-use (POU) in schools, hospitals, and other facilities (WHO, 2011b). The public at large is interested in POU home devices to remove microbial pathogens and parasites as well as toxic chemicals, and improve the aesthetic quality of drinking water (removal of taste and odor, turbidity, color). POU devices are particularly useful in rural areas not served by centralized systems and their use is increasingly being considered by consumers. POU devices and methods used in developing countries will be discussed in Chapter 8.

The contaminants of concern are pathogenic bacteria, viruses, protozoan cysts (e.g., *Giardia*), and toxic metals (e.g., cadmium, mercury, lead, arsenic, iron, manganese, organic substances of potential health significance, particulates, color, odor, and chlorine taste (Geldreich and Reasoner, 1990). As to microbial contaminants, the U.S. EPA requires a minimum removal capacity of 99.9%, 99.99%, and 99.9999% for *Giardia* cysts, viruses, and bacteria, respectively.

Treatment is accomplished by filtration, adsorption, ion exchange, reverse osmosis, distillation, or UV irradiation (Geldreich and Reasoner, 1990; Reasoner et al., 1987; Stauber et al., 2006). Table 2.4 summarizes the basic processes involved in water treatment by home units (Reasoner, 2002; U.S. EPA, 2007b). The most frequently used process is filtration through activated carbon. POU devices may be installed in a home as faucet add-on units consisting of small activated carbon cartridges, in-line devices (filters, reverse osmosis units) that are installed under the kitchen sink or pour-through pitchers with an activated carbon filter (Figure 2.13; Reasoner, 2002). There are also point-of-entry (POE) devices for treating the entire home water supply. However, there are some problems associated with the use of activated carbon-based filters. Heterotrophic bacteria and, possibly, pathogenic microorganisms, colonize the activated carbon surface, leading to the occurrence of high bacterial levels in the product water (Reasoner et al., 1987; Geldreich et al., 1985; Taylor et al., 1979; Wallis et al., 1974). Static conditions overnight or following vacation periods as

TABLE 2.4　Processes involved in point of use (POU) and in point of entry (POE) devices

Process	Contaminant(s) Removed
Adsorption (activated carbon)	Chlorine
Mechanical filtration	Particulates, color, turbidity, asbestos fibers, cysts, and oocysts
Reverse osmosis	Total dissolved solids, metals, nitrate, bacteria, viruses, cysts, and oocysts
Water softening (cationic)	Ca, Mg, Fe, Mn, Ba, Ra
Water softening (anionic)	Sulfate, nitrate, bicarbonate, chloride, arsenic
Distillation	Inorganics, dissolved solids, organics
Disinfection (chemical or UV)	Bacteria, viruses, cysts, and oocysts

Source: Adapted from D.J. Reasoner (2002). In: Encyclopedia of Environmental Microbiology, Gabriel Bitton, editor-in-chief, Wiley-Interscience, N.Y., pp. 1563–1575.

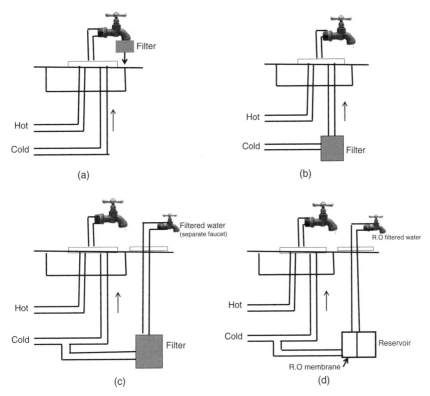

Figure 2.13 Point-of-use devices installation: (a) Faucet add-on filter; (b) Under-sink-unit: cold water line; (c) Under-sink unit: cold water bypass to separate faucet; (d) Reverse osmosis: cold water bypass with reservoir to separate faucet. Adapted from Reasoner (2002). In: *Encyclopedia of Environmental Microbiology*, Gabriel Bitton, editor-in-chief, Wiley-Interscience, N.Y.

well as favorable temperatures and nutritional conditions provide opportunity for bacterial growth in the treatment device. An increase in heterotrophic plate count is generally observed in the first-draw filtered water (Snyder et al., 1995). Pathogenic bacteria (e.g., *P. aeruginosa* or *Klebsiella pneumoniae*) or opportunistic pathogens (e.g., *Mycobacterium avium*) are also able to colonize the filter surface (Geldreich et al., 1985; Reasoner et al., 1987; Rodgers et al., 1999; Tobin et al., 1981) and may pass in the product water. However, the epidemiological significance of these findings is unknown. It is thus advisable to flush the unit for 1–3 minutes prior to use in the morning or after returning from vacation.

2.7.3 Modified Carbon Filters and Other Devices

2.7.3.1 *Ag-Impregnated Carbon Filters.* Silver-impregnated filters are sold by some manufacturers to control bacterial growth inside the filter. The antibacterial effect of silver is rather slow but significant bacterial reductions are obtained only

after hours of contact (Bell, 1991; Muhammad et al., 2008). Furthermore, at the concentrations used in some POU devices, silver does not exert any significant detrimental effect on heterotrophic bacterial growth (possibly due to the selection of silver-resistant bacteria such as nontuberculous mycobacteria) and has no significant antiviral effect (Gerba and Thurman, 1986; Rodgers et al., 1999; Tobin et al., 1981). Despite the presence of silver, GAC filters are prone to colonization by bacteria. Furthermore, because of concern over consumers' health, the silver released in the treated water cannot exceed 100 µg/L (Geldreich et al., 1985).

More than 6-log *E. coli* removal was obtained when using 30 g of activated carbon modified by aluminum hydroxychloride and impregnated with nanosized AgBr crystals and challenged with 1000 L of water with an *E. coli* load of 10^7 CFU/mL (Pal et al., 2006). Furthermore, columns of silver nanoparticles–alginate composite beads removed 5 logs of *E. coli* with 1-minute hydraulic retention time and could potentially serve as a POU device (Lin et al., 2013).

2.7.3.2 *Ag-Impregnated Blotting Paper.*   Blotting paper impregnated with silver nanoparticles was considered for use as an inexpensive POU method for drinking water treatment in emergency situations or in homes with no connection to a network. At a silver concentration of 2–3 mg/g paper, this filter helped achieve a bacterial reduction of more than 7 logs for *E. coli* and more than 3 logs for *E. faecalis* (Figure 2.14; Dankovich and Gray, 2011). The Ag concentration in treated water was below 100 ppb, the WHO and U.S. EPA limit for drinking water.

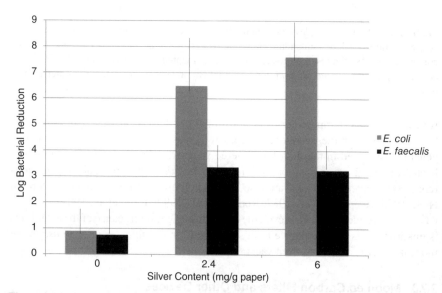

Figure 2.14   Log reduction of *E. coli* and *E. faecalis* bacterial count after permeation through the silver nanoparticle paper, at different silver contents in paper. Adapted from Dankovich and Gray (2011). Environ. Sci. Technol. 45, 1992–1998.

Due to their large surface area, carbon nanotubes can also be incorporated in POU devices to remove/inactivate bacterial and viral pathogens (Brady-Estevez et al., 2010; Upadhyayula et al., 2009). Application of a small voltage (2–3 volts) to a carbon nanotube filter drastically improved the filter performance. More than 7-log removal/inactivation of phage MS2 was achieved in the presence of NOM over a range of pH and ionic strength values (Rahaman et al., 2012).

2.7.3.3 *Reverse Osmosis.*

Reverse osmosis (RO) is a process where a high pressure (20–100 bar) is applied to move water through a semipermeable membrane from a region of high solute concentration to a region of low solute concentration. The RO membrane is thus capable of retaining bacterial cells, viruses, and ions. This process is used for seawater desalination and drinking water purification. As a household device, RO can be incorporated in a system that may include a sediment filter, an activated carbon filter, and a UV unit. A prospective epidemiological study undertaken in a Montreal, Canada, suburban area, showed a correlation between drinking water bacterial counts grown at 35°C and the incidence of gastrointestinal symptoms in 600 households consuming water treated by RO. The level of gastrointestinal illnesses in the group consuming regular tap water was 30% higher that in the group consuming water treated by RO (Payment et al., 1991a, 1993). A follow-up study showed that 14–40% of gastrointestinal illnesses were due to consumption of tap water meeting microbiological standards. Children (up to 5-year old) were the most affected by gastrointestinal illnesses. Figure 2.15 (Nwachcuku and Gerba, 2004; Payment et al., 1997) illustrates the higher susceptibility of young children (<5 years) and the lower incidence of gastrointestinal symptoms for consumers of RO-filtered drinking water.

2.7.3.4 *Other POU Devices.*

A novel POU device, based on inactivation of pathogens by pulsed broad spectrum white light, achieved inactivation exceeding 4 log_{10} for viruses (poliovirus and rotavirus) and *C. parvum*, and 7 log_{10} for *Klebsiella terrigena*. The device was operated at a flow rate of 4 gallons/min (Huffman et al., 2000). Two patented POU devices were investigated with regard to the removal of *E. coli*, phage MS2, *B. subtilis* endospores, and *Cryptosporidium* oocysts. Total removal was observed for all microorganisms. It was also found that bacterial spores are suitable surrogates for *Cryptosporidium* oocysts removal (Brown and Cornwell, 2007; Muhammad et al., 2008). Another POU novel technology for water disinfection in emergency situations is the use of polysodium acrylate cryogels doped with silver nanoparticles. These cryogels helped achieve a 3-log reduction of *E. coli* and *B. subtilis* after 15 seconds contact time. Ag release from the cryogels was less than 100 µg/L, the limit recommended by WHO for drinking water (Loo et al., 2013).

The use of POU devices by consumers was considered by some experts as a potential health hazard, and some strongly believe that water treatment should be carried out by trained professionals and not by poorly trained consumers (Geldreich et al., 1985).

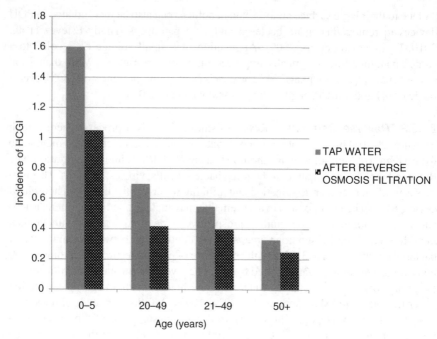

Figure 2.15 Relative incidence of gastroenteritis by age in individuals drinking tap water and those drinking tap water after filtration by a reverse-osmosis filter. HCGI = highly credible gastrointestinal symptoms. Adapted from Nwachuku and Gerba (2004). Current Opinion Microbiol. 7: 206–209.

WEB RESOURCES

http://water.epa.gov/drink/tour/
(virtual tour of a drinking water plant: from U.S. EPA)

http://www.tampabaywater.org/tampa-bay-regional-surface-water-treatment-plant.aspx
(Tampa Bay water treatment plant)

http://www.thewaterq.com
(residential drinking water. Several videos on drinking water treatment)

http://www.a2gov.org/government/publicservices/water_treatment/Pages/treatment.aspx
(water treatment plant, Ann Arbor, MI)

http://www.cdc.gov/safewater/sand-filtration.html
(slow sand filtration from CDC)

http://www.who.int/water_sanitation_health/publications/ssf/en/
(slow sand filtration from WHO)

http://www.youtube.com/channel/HCjAtHXZNsWxA
(Youtube videos on slow sand filtration)

http://www.bluefuturefilters.com/home.html
(home slow sand filters)

http://iaspub.epa.gov/tdb/pages/treatment/treatmentOverview.do?treatmentProcess
Id=1942020127
(diatomaceous earth filtration: U.S. EPA)

http://www.msue.msu.edu/objects/content_revision/download.cfm/revision_id.
499668/workspace_id.-4/01500610.html/
(activated carbon: Michigan State University)

http://www.waterprofessionals.com/process-water/activated_carbon_filters.html
(activated carbon filters)

http://www.awwa.org
(American Water Works Association)

http://www.epa.gov/ogwdw000/sdwa/sdwa.html
(Safe Drinking Water Act)

http://www.epa.gov/ogwdw000/protect.html
(EPA source water protection)

http://www.epa.gov/fedrgstr/
(Federal Register Environmental documents)

http://www.water-ed.org/watersources/subpage.asp?rid=&page=382
(primer on drinking water treatment)

FURTHER READING

Botes, M. and T.E. Cloete. 2010. The potential of nanofibers and nanobiocides in water purification. *Crit. Rev. Microbiol.* 36(1):68–81.
Edzwald, J.K. (Ed.). 2011. *Water Quality and Treatment: A Handbook on Drinking Water*, 6th edition. AWWA, Denver, CO.
Geldreich, E.E. 1990. Microbiological quality of source waters for water supply. In: *Drinking Water Microbiology*, G.A. McFeters (Ed.). Springer, New York, pp. 3–31.
Gray, N.F. 1994. *Drinking Water Quality: Problems and Solutions*. John Wiley & Sons, Chichester, UK, 315 pp.

LeChevallier, M.W. 2002. Microbial removal by pretreatment, coagulation and ion exchange. In *Encyclopedia of Environmental Microbiology*, G. Bitton (editor-in-chief). Wiley-Interscience, New York, pp. 2012-2019.

LeChevallier, M.W., and K.K. Au. 2004. *Water Treatment and Pathogen Control: Process Efficiency in Achieving Safe Drinking Water*. IWA Publishing, London, UK. ISBN: 1 84339 069 8.

Levin, R.B., P.R. Epstein, T.E. Ford, W. Harrington, E. Olson, and E.G. Reichard. 2002. U.S. drinking water challenges in the twenty-first century. *Environ. Health Perspect.* 110(suppl. 1):43–52.

Logsdon, G.S., M.B. Hosley, S.D.N. Freeman, J.J. Neemann, and G.C. Budd. 2006. Filtration processes – A distinguished history and a promising future. *Am. Water Works Assoc. J.* 98(3):150–162.

McFeters, G.A. (Ed.). 1990. *Drinking Water Microbiology*. Springer, New York, 502 p.

Pizzi, N.G. 2002. *Water Treatment Operator Handbook*. American Water Association, Washington, DC.

Qu, X., P.J.J. Alvarez, and Q. Li. 2013. Applications of nanotechnology in water and wastewater treatment. *Water Res.* 47:3931–3946.

Van Reis, R., and A. Zydney. 2007. Bioprocess membrane technology. *J. Membr. Sci.* 297:16–50.

Reasoner, D.J. 2002. Home treatment devices. Microbiology of point of use and point of entries devices. In: *Encyclopedia of Environmental Microbiology*, G. Bitton (editor-in-chief). Wiley-Interscience, New York.

Rochelle, P., and J. Clancey. 2006. The evolution of microbiology in the drinking water industry. *Am. Water Works Assoc. J.* 98(3):163–191.

Shannon, M.A., P.W. Bohn, M. Elimelech, J.G. Georgiadis, B.J. Mariñas, and A.M. Mayes. 2007. Science and technology for water purification in the coming decades. *Nature* 452:301–310.

Simpson, D.R. 2008. Biofilm processes in biologically active carbon water purification. *Water Res.* 42:2839–2848.

Symons, G.E. 2006. Water treatment through the ages. *Am. Water Works Assoc. J.* 98(3):87–98.

Theron, J., J.A. Walker and T.E. Cloete. 2008. Nanotechnology and water treatment: applications and emerging opportunities. *Crit. Rev. Microbiol.* 34:43–69.

U.S. EPA. 2003. *Small Drinking Water Systems Handbook: A Guide to Packaged Filtration and Disinfection Technologies with Remote Monitoring and Control Tools*. EPA-600-R-03-041.

U.S. EPA. 2007. *Small Drinking Water Systems: State of the Industry and Treatment Technologies to Meet the Safe Drinking Water Act Requirements*. EPA/600/R-07/110.

World Health Organization. 2003. *The Right to Water*. WHO, Geneva, Switzerland.

3

DRINKING WATER DISINFECTION

3.1 INTRODUCTION

In the preceding chapter, we have discussed the various processes to reduce/remove microbial pathogens and parasites in drinking water treatment plants.

This chapter deals with disinfection with the goal of destroying or inhibiting microbial pathogens and parasites through the use of chemical or physical biocides which interact with one or several targets in the microbial cell. The main targets are shown in Figure 3.1 (Russel et al., 1997). The disinfection step is an essential and final barrier against human exposure to disease-causing pathogenic microorganisms, including viruses, bacteria, and protozoan parasites. Emerging pathogens (e.g., *Mycobacterium avium, Cryptosporidium parvum*) present new challenges to disinfection experts. Disinfection of drinking water is probably the most significant preventive measure in human history. Following the discovery of the "germ theory" by Louis Pasteur and Robert Koch in the 1880s, chlorination was initiated at the beginning of the twentieth century to provide an additional safeguard against pathogenic microorganisms. The United States proceeded in 1908 with the large-scale use of chlorine for water disinfection in Jersey City, NJ (Symons, 2006).

Disinfection was crucial in reducing waterborne and foodborne diseases. According to the U.S. EPA, disinfection of drinking water should result in the 99.99% destruction of bacterial and viral pathogens and 99.9% of protozoan parasites. In addition to their use for pathogen and parasite destruction, some of the disinfectants (e.g., ozone, chlorine dioxide) are also employed for oxidation of organic matter, iron and manganese, improving coagulation and filtration efficiency, controlling biofilm growth in water distribution systems, and for controlling taste and odor problems and algal growth (see Chapter 5). The major factors controlling the disinfection efficiency of pathogens and parasites include the type of disinfectant, type of pathogen or protozoan parasite, disinfectant concentration and contact time, pH, temperature,

Microbiology of Drinking Water Production and Distribution, First Edition. Gabriel Bitton.
© 2014 John Wiley & Sons, Inc. Published 2014 by John Wiley & Sons, Inc.

**TARGET SITES OF BIOCIDES IN
MICROBIAL CELLS**

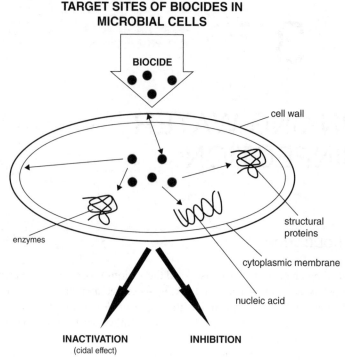

INACTIVATION   INHIBITION
(cidal effect)

Figure 3.1   Target sites of biocides in microbial cells. Adapted from Russel et al. (1997). Amer. Soc. Microbiol. News 63: 481–487.

interfering agents such as turbidity, protective effect of microinvertebrates, and other factors such as cell clumping and aggregation (Bichai et al., 2011; Bohrerova and Linden, 2006; Templeton et al., 2008).

This chapter covers the disinfectants traditionally used by the water treatment industry (e.g., chlorine, chlorine dioxide, ozone, UV radiation) as well as other emerging technologies to inactivate or physically remove microbial pathogens and parasites.

3.2 CHLORINE

Chlorine is the most widely used disinfectant in water treatment plants. It is mostly used as gas or hypochlorite solution. A survey showed that about 64% of water treatment plants use chlorine gas (Committee Report, 2008).

3.2.1 Chlorine Chemistry

Chlorine gas (Cl_2), when introduced in water, hydrolyzes according to the following equation:

$$\underset{\text{Chlorine Gas}}{Cl_2} + H_2O \leftrightarrow \underset{\text{Hypochlorous Acid}}{HOCl} + H^+ + Cl^- \qquad (3.1)$$

Hypochlorous acid dissociates in water according to the following:

$$\underset{\text{Hypochlorous Acid}}{\text{HOCl}} \leftrightarrow \text{H}^+ + \underset{\text{Hypochlorite Ion}}{\text{OCl}^-} \tag{3.2}$$

Both HOCl and OCl^- are considered as free available chlorine. The combination of HOCl with ammonia or organic nitrogen compounds leads to the formation of chloramines that are combined with available chlorine.

HOCl reacts with ammonia and forms inorganic chloramines according to the following equations (Snoeyink and Jenkins, 1980):

$$\text{NH}_3 + \text{HOCl} \rightarrow \underset{\text{monochloramine}}{\text{NH}_2\text{Cl}} + \text{H}_2\text{O} \tag{3.3}$$

$$\text{NH}_2\text{Cl} + \text{HOCl} \rightarrow \underset{\text{dichloramine}}{\text{NHCl}_2} + \text{H}_2\text{O} \tag{3.4}$$

$$\text{NHCl}_2 + \text{HOCl} \rightarrow \underset{\text{trichloramine}}{\text{NCl}_3} + \text{H}_2\text{O} \tag{3.5}$$

The type of chloramine is dependent on the pH of the water. Monochloramine is the predominant chloramine formed at the pH range usually encountered in water treatment plants (pH = 6–9). The presence of monochloramine is desirable because dichloramines and trichloramines impart unpleasant taste to the water. Chloramines exert a chlorine demand. Breakpoint chlorination is reached when the chlorine to ammonia-N ratio is between 7.5:1 and 11:1, leading to the oxidation of ammonia to nitrogen gas:

$$2\text{NH}_3 + 3\text{HOCl} \rightarrow \text{N}_2 + 3\text{H}_2\text{O} + 3\text{HCl} \tag{3.6}$$

Addition of chlorine beyond *breakpoint chlorination* ensures the existence of a free available chlorine residual.

3.2.2 Inactivation of Microorganisms by Chlorine and Chloramines

Disinfectant effectiveness is generally expressed as Ct, C being the disinfectant concentration, and t the time required to inactivate a certain percentage of the population under specific conditions (pH and temperature). The relationship between disinfectant concentration and contact time is given by the Watson's law (Clark et al., 1989):

$$K = C^n t \tag{3.7}$$

where

K = constant for a given microorganism exposed to a disinfectant under specific conditions,
C = disinfectant concentration (mg/L),
t = time required to kill a certain percentage of the population (min), and
n = constant also called the "coefficient of dilution."

TABLE 3.1   Microbial inactivation by chlorine: some Ct values reported in the literature

Microorganism	Chlorine Concentration, mg/L	Inactivation Time, min	Ct
Escherichia coli[a]	0.1	0.4	0.04
Adenovirus type 2[b]			0.023–0.027
Adenovirus type 3[b]			0.027–0.067
Poliovirus 1[a]	1.0	1.7	1.7
Human rotaviruses[e]			5.55–5.59
Entamoeba histolytica cysts[a]	5.0	18	90
Giardia lamblia cysts[a]	1.0	50	50
	2.0	40	80
	2.5	100	250
G. muris cysts[a]	2.5	100	250
Cryptosporidium parvum[c]			3700
Cladosporium tenuissimum[d]			71
Aspergillus terreus[d]			1404

[a]Conditions: 5°C; pH = 6.0 (Hoff and Akin (1986); *Environ. Health Perspect.* 69:7–13).
[b]Conditions: 4°C; pH = 7 (Page et al. (2009). *Water Res.* 43:2916–2926).
[c]Conditions: 20°C; pH = 6 (Driedger et al. (2000). *Water Res.* 34:3591–3597).
[d]Conditions: 25°C; pH = 7 (Pereira et al. (2013). *Water Res.* 47:517–523).
[e]Conditions: 20°C; pH = 7.2 (Xue et al. (2013a). *Water Res.* 47:3329–3338).

Of all the chlorine species discussed above, hypochlorous acid (HOCl) is the most efficient in the inactivation of pathogens. In water treatment plants, less than 1ppm of residual chlorine for 30min is enough to significantly reduce bacterial pathogens. In general, resistance to disinfection increases along the following order:

Nonspore-forming bacteria < enteric viruses < spore-forming bacteria
< protozoan cysts

For example, in the presence of HOCl at pH = 6, the Ct for *Escherichia coli* is 0.04 as compared to a Ct value of 1.05 for poliovirus type 1 and a Ct value of 80 for *Giardia lamblia* cysts (Logsdon and Hoff, 1986). The Ct values for fungi (25°C; pH = 7; $C = 1$ mg/L) vary between 71 for *Cladosporium tenuissimum* and 1404 for *Aspergillus terreus* (Pereira et al., 2013). Other Ct values reported in the literature are shown in Table 3.1. *Cryptosporidium* oocysts are extremely resistant to chlorine disinfection as the oocysts can remain viable for 3–4 months in 2.5% potassium dichromate solution. The reported Ct value for 99% inactivation of oocysts is higher than 7200 (Maier et al., 2009). *Cryptosporidium* would thus be extremely resistant to disinfection as carried out in water and wastewater treatment plants (Campbell et al., 1982; Current, 1988; Korich et al., 1989).

3.2.3 Adverse Effects of Chlorine on Pathogens

Chlorine damages bacteria by:

- *Disrupting cell permeability*: Changes in cell permeability eventually lead to cell death following leakage of proteins, RNA, DNA, decrease in K uptake, and in protein and DNA synthesis (Haas and Engelbrecht, 1980; Venkobachar et al., 1977).
- *Damage to nucleic acids and enzymes*
- *Repression of gene transcription*: Whole genome analysis of *Staphylococcus aureus* exposed to hypochlorous acid (HOCl) has shown that the disinfectant led to repression of the transcription of genes controlling cell wall synthesis, protein synthesis, membrane transport, and primary metabolism. However, HOCl induced genes encoding for virulence factors in *S. aureus* (Chang et al., 2007). Exposure of *Legionella pneumophila* to sublethal concentrations of chlorine induces the expression of genes involved in virulence, stress, metabolism, information pathways, and transport. The response to stress involves the induction of antioxidant and stress proteins, and of glutathione S transferase activity (Bodet et al., 2012).
- *Other effects*: Hypochlorous acid also oxidizes sulfhydryl groups, damage iron–sulfur centers, disrupts nutrient transport, inhibits cell respiration, and impairs ATP production (Barrette et al., 1988).

Thus, chlorine and other chemical and physical agents cause cell injury in bacteria in drinking water (LeChevallier and McFeters, 1985a; Singh and McFeters, 1990). However, the injured cells may be able to recover in the gastrointestinal (GI) tract and retain their pathogenicity (Singh and McFeters, 1986).

As regard viruses the mode of action of chlorine depends on the type of virus. The site of action may be the protein coat (e.g., VP4 in human rotaviruses) or the nucleic acid (Nuanualsuwan and Cliver, 2003; Vaughn and Novotny, 1991; Xue et al., 2013a).

3.2.4 Disinfection By-Products

The risks from chemicals in drinking water were not as well defined as those from pathogenic microorganisms and parasites. In 1974, Bellar and Lichtenberg in the United States and Rook in the Netherlands first detected four trihalomethanes (THMs) (chloroform, dichlorobromomethane, monochlorodibromomethane, bromoform) in water following chlorination (Trussell, 2006).

Disinfection by-products (DBPs) are formed following the reaction of chlorine with precursors such as natural organic matter (mainly humic and fulvic acids) and extracellular products from microorganisms such as algal cells (especially blue-green algae and diatoms) (Chow, 2006; LeChevallier et al., 1992; Plummer and Edzwald, 2001). DBPs include THMs (e.g., chloroform ($CHCl_3$), bromodichloromethane, dibromochloromethane, bromoform), haloacetic acids (HAAs) (e.g., monochloroacetic acid, monobromoacetic acid, dichloroacetic acid, dibromoacetic acid, trichloroacetic acid), and haloacetonitriles.

Some DBPs are suspected mutagens/carcinogens or teratogens. There is a potential association between chlorination of drinking water and increased risk of bladder, kidney, and colorectal cancers. As regard bladder cancer, human epidemiological studies have shown a measurable effect of DBPs with an odds ratio of 1.2 to 2 (in a retrospective epidemiological study, the odds ratio is the ratio of odds of cases in the exposed population to the odds of cases in the control population) (Hrudey, 2009). Other factors increasing the risk of bladder cancer are smoking and the presence of arsenic in drinking water. Increased intake of liquids appears to help in reducing the risk of bladder cancer (Moyad, 2003). Exposure to high HAA levels in drinking water is associated with increased cancer risks in animals and humans. Chlorohydroxyfuranone, particularly 3-chloro-4-(dichloromethyl)-5-hydroxy-2(5H)-furanone (designated as MX) has also been identified as a chlorination by-product. MX is a potent mutagen and a suspected carcinogen but its impact on human health remains to be elucidated (Meier et al., 1987). Chlorinated and brominated analogues of MX were detected in finished waters in several plants in the United States, with MX levels as high as 310 ng/L. The levels were as high as 160 ng/L in a Russian drinking water. Other mutagens are formed following the chlorination of azo dyes (Oliveira et al., 2006). There is also the possibility of an association of water chlorination with increased risk of cardiovascular diseases (Craun, 1988). In Perth, Australia, the total THMs, 92% of which were brominated, ranged from 36 to 190 µg/L. Increases in birth defects were significantly associated with living in areas with high brominated THM levels (Chisholm et al., 2008).

In swimming pool water, many compounds can serve as precursors of DBPs. These include sunscreens, swimmers' perspiration, mucus, skin, saliva, urine, leaves, and algal exudates (Zwiener et al., 2006). Some sunscreens and their halogenated by-products are potential hormonally active agents (Schlumpf et al., 2001). Swimmers can become exposed to DBPs (THMs, HAAs, haloketones, etc) via dermal absorption or inhalation. Inhalation of volatile DBPs is a major exposure route for swimmers and nonswimmers in indoor pools. HAAs are mainly absorbed via accidental ingestion of pool water. The volatile trichloramine (NCl_3) can also be present in swimming pool water and can lead to respiratory problems, including asthma (Li and Blatchley, 2007). Some measures to reduce the load of DBPs precursors in swimming pools include adequate filtration, showering before entering in the pool to wash off sunscreens and other compounds, and increased air circulation in indoor pools (Zwiener et al., 2006).

Following are some suggested approaches for reducing or controlling DBP in drinking water (Wolfe et al., 1984):

1. Removal or reduction of DBP precursors (e.g., natural organic matter, extracellular products from microorganisms), prior to disinfection. A good relationship was found between total organic carbon and THM formation potential (LeChevallier et al., 1992). Organic carbon concentrations can be reduced by enhanced coagulation, granular activated carbon (GAC), membrane filtration, iron oxide-coated filtration media, water softening, or nanofiltration (Black

et al., 1996; Chang et al., 1997; Kalscheur et al., 2006; Kim and Kang, 2008; de la Rubia et al., 2008). In water treatment plants, HAA and THM and chlorinated furanones (e.g., MX) concentration can also be reduced via adsorption to activated carbon and biodegradation (Kim and Kang, 2008; McRae et al., 2004; Onstad et al., 2008; Tung et al., 2006) or via biodegradation in sand filters and biofilms in water distribution pipes (Rodriguez et al., 2007b; Zhang et al., 2009a). In homes, THMs can be volatilized upon boiling tap water.

2. Preozonation reduces the formation of THMs, HAAs, and total organic halogen (TOX).

3. Posttreatment aeration (spray or diffused aeration) of drinking water was found to be a cost-effective technique to remove THMs from water. For example, spray aeration removed from 20% to >99.5% of THMs. The removal depends on temperature, air-to-water ratio, and THM species (Brooke and Collins, 2011).

4. Use of combined or alternative disinfectants that generate less THMs (e.g., chloramination, ozone, or ultraviolet irradiation). However, some of the alternative disinfection processes may result in undesirable by-products such as carcinogenic nitrosamines following chloramination or increased levels of MX following disinfection with chlorine dioxide (Hrudey, 2009).

5. Ultimately, a sound management and control strategy should involve the control of NOM and bromide in the source water (Singer, 2006).

The U.S. EPA regulates THMs in water, and in 1979, it established a maximum contaminant level (MCL) of 80 μg/L for total THM and 60 μg/L for HAAs (the sum of five HAAs: mono-, di-, and trichloroacetic acids, and mono- and dibromoacetic acids) in finished drinking water. Health Canada has set a guideline of 100 μg/L for total THMs. European countries such as Italy have adopted a much lower MCL of 30 μg/L. It was suggested that DBPs should be regulated by individual species (Singer, 2006). Indeed, WHO (2005) established different guidelines for each of the four THM species. Because water treatment with chloramines does not produce THMs, consumers drinking chloraminated water appear to experience less bladder cancers than do those consuming chlorinated water (Zierler et al., 1987).

3.2.5 Chloramination of Drinking Water

Approximately 30–40% of US drinking water plants use chloramination in their distribution systems (AWWA, 2000; Betts 2002; Committee Report, 2008; Singer, 2006). An example is the Denver Water Department which has been successfully using chloramination for decades (Dice, 1985). Chloramines offer several advantages such as lower THMs and HAAs, better control of biofilm microorganisms in distribution pipes, improved maintenance of disinfectant residual, and lower chlorine taste and odor (Norton and LeChevallier, 1997). Hospitals using free chlorine as residual disinfectant are 10 times more likely to experience outbreaks of Legionnaires' disease than those using monochloramine (Kool et al., 2000). Thus, it was suggested to use free chlorine as a primary disinfectant in water distribution systems and to convert the residual to monochloramine if biofilm control is the goal (LeChevallier et al., 1990). However, chloramination may promote the growth of nitrifying bacteria which convert ammonia to nitrite and nitrate (Wilczak et al., 1996; see Chapter 5 for further details). Nitrification can be controlled by switching periodically to free chlorine residual. A 2007 survey showed that 23% of US water utilities switch to free chlorine residual to control nitrification (Committee Report, 2008).

Dichloramine and trichloramine have offensive odors with threshold odor concentrations of 0.8 and 0.02 mg/L, respectively (Kreft et al., 1985). Chloramines cause hemolytic anemia in kidney hemodialysis patients (Eaton et al., 1973) and are toxic to aquatic organisms. Of concern is the reaction of monochloramine with dimethylamine to form N-nitrosodimethylamine (NDMA) which has been classified as a human carcinogen by the U.S. EPA (Choi and Valentine, 2002; Najm and Trussell, 2001; Schmidt and Brauch, 2008). The United States established a public health goal of 3 ng/L for NDMA while Ontario, Canada, established a standard of 9 ng/L. NDMA has been detected in drinking water wells at levels ranging from 70 to 3000 ng/L. It was also detected in swimming pools and hot tubs. The median levels of NDMA in indoor pools, outdoor pools, and hot tubs were 32, 5.3, and 313 ng/L, respectively. The relatively high concentrations are explained by the fact that swimming pools and hot tubs contain amine precursors from swimmers' urine and sweat. Other nitrosamine carcinogens detected at lower levels are N-nitrosobutylamine and N-nitrosopiperidine (Walse and Mitch, 2008). Epidemiological studies suggest an increased risk of bladder cancer due to the presence of these nitrosamines in water (Villanueva et al., 2007). In water treatment plants, NDMA biodegradation occurs in biologically active treatment processes such as sand or activated carbon filtration (Schmidt and Brauch, 2008).

3.3 CHLORINE DIOXIDE

Chlorine dioxide (ClO_2) use as a disinfectant in water treatment is becoming widespread because it forms much less THMs and HAAs than free chlorine and does not react with ammonia to form chloramines (Rand et al., 2007). A 2007 survey revealed that 8% of water utilities use chlorine dioxide as a disinfectant (Committee Report, 2008). However, chlorine dioxide must be generated at the site as follows:

$$2\,NaClO_2 + Cl_2 \rightarrow 2\,ClO_2 + 2\,NaCl \tag{3.8}$$

Chlorine dioxide is generally a more powerful disinfectant than chlorine for the inactivation of bacterial and viral pathogens and protozoan parasites in drinking water and wastewater (Aieta and Berg, 1986; Chen et al., 1985; Longley et al., 1980; Narkis and Kott, 1992; Xue et al., 2013a). Sensitivity of health-related microorganisms to ClO_2, expressed as Ct values, is generally much higher than sensitivity to chlorine (Chauret et al., 2001; Radziminski et al., 2002; Sobsey, 1989; Taylor et al., 2000). The Ct for *M. avium* ranged from 2 to 11 while the Ct for *C. parvum* varied from

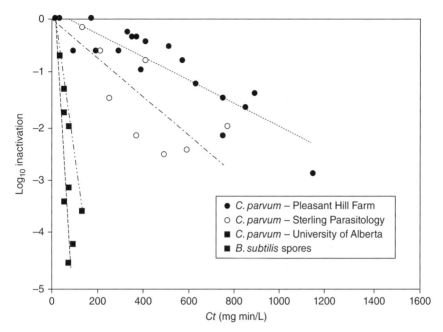

Figure 3.2 Chlorine dioxide inactivation of *Cryptosporidium parvum* oocysts from various sources, and *Bacillus subtilis* spores. Adapted from Chauret et al. (2001). Appl. Environ. Microbiol. 67: 2993–3001.

75 to more than 1000, depending on the source of the oocysts (Figure 3.2; Chauret et al., 2001). *Giardia* is more sensitive to ClO2 and displays a Ct of 15.

Pathogen and parasite resistance to ClO_2 follows the general trend:

Protozoan cysts and oocysts and bacterial spores > viruses > bacteria

Chlorine dioxide is also superior to chlorine for inactivating biofilm bacteria. The inactivation can be enhanced by combining ClO_2 with UV pretreatment (Rand et al., 2007).

ClO_2 inactivates bacterial pathogens by disrupting the outer membrane of gram-negative bacteria or by interfering with protein synthesis (Berg et al., 1986; Russel et al., 1997). As regard viral pathogens, the targets of ClO_2 lethal action may be the protein coat or the viral genome (Alvarez and O'Brien, 1982; Noss et al., 1986; Taylor and Butler, 1982).

Chlorine dioxide is rapidly reduced in water to form two inorganic DBPs, chlorite (ClO_2^-) and chlorate (ClO_3^-):

$$2\,ClO_2 + 2OH^- \rightarrow ClO_2^- + ClO_3^- + H_2O \qquad (3.9)$$

Chlorite, the predominant species formed in water treatment plants, is of greater health concern than chlorate and is regulated by the U.S. EPA at 0.8 mg/L in drinking water, although some European countries mandate lower levels. WHO has established a guideline of 0.7 mg/L for chlorite and chlorates in drinking water. Both chlorite and chlorate may combine with hemoglobin to cause methemoglobinemia (i.e., blue baby syndrome), as shown for nitrite.

As shown for THMs, chlorite can be removed with GAC. Despite the competition between natural organic matter with chlorite for the adsorption sites on GAC, the efficiency of GAC for chlorite removal can be enhanced by regenerating GAC with a base-acid solution (Collivignarelli et al., 2006). Reduced sulfur compounds and ferrous ions can also be used to remove chlorite from water and wastewater (Katz and Narkis, 2001).

3.4 OZONE

3.4.1 Introduction

Ozone O_3 was first introduced as a strong oxidizing agent for the removal of taste, color, odors, and for the oxidation of iron and manganese. Later on, in 1893, it was first used as a disinfectant to inactivate pathogens (see Table 2.1, Chapter 2). The first water treatment plant using ozone started operations in 1906 in Nice, France. Modern ozone plants produce ozone from pure oxygen and require as little as 5 kW.h/lb of generated ozone (Trussell, 2006). Ozone is used as a disinfectant in approximately 10% of water treatment plants surveyed (AWWA Committee Report, 2008). The integration of ozone in biological treatment systems is also quite useful for the removal of persistent organic compounds such as industrial chemicals, endocrine disrupters, and pharmaceuticals, (Ried et al., 2007). Preozonation also lowers THM formation potential and promotes particle coagulation during water treatment (Chang and Singer, 1991). However, ozone breaks down complex compounds into simpler low molecular weight compounds such as organic acids, aldehydes, and ketones (Bazri et al., 2012; Hammes et al., 2006; Vital et al., 2012), which may serve as substrates for microbial growth in water distribution systems (Bancroft et al., 1984). Ozone is a powerful disinfectant that does not interact with ammonia and its effectiveness is not controlled by pH (Driedger et al., 2001). However, ozone is more expensive than chlorine and does not leave any residual in water. Ozone use as a primary disinfectant is sometimes combined with postchlorination to maintain a chlorine residual.

3.4.2 Inactivation of Pathogens and Parasites

Ozone is a much more powerful oxidant than chlorine. The Ct values for 99% inactivation by ozone are quite low and range between 0.001 and 0.2 for *E. coli* and between 0.04 and 0.42 for enteric viruses (Engelbrecht, 1983; Hall and Sobsey, 1993).

Ozone is more effective against rotaviruses than chlorine, monochloramine, or chlorine dioxide (Chen and Vaughn, 1990; Korich et al., 1990). Moreover, certain pathogens such as *Mycobacterium fortuitum* are more resistant to ozone than

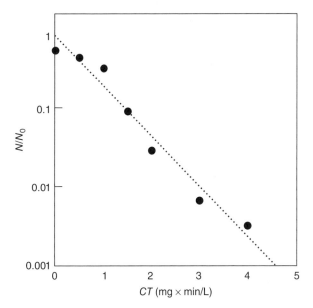

Figure 3.3 Inactivation of *Cryptosporidium parvum* oocysts with ozone *Source*: Driedger et al. (2000). Water Res. 34: 3591–3597.

poliovirus type 1, *Salmonella typhimurium* or *E. coli* (Farooq and Akhlaque, 1983). Ozone is also quite effective against protozoan cysts and oocysts and against *Bacillus subtilis* spores (Driedger et al., 2000; Jung et al., 2008; Peeters et al., 1989). A 3-log inactivation of *C. parvum* oocysts was reached at a *Ct* of 4.5 (Figure 3.3; Driedger et al., 2000).

As shown for chlorine, pathogen inactivation by ozone is reduced in the presence of suspended solids (Kaneko, 1989) and at low temperatures (Driedger et al., 2001; Rennecker et al., 2000; Wickramanayake et al., 1985).

3.4.3 Mechanisms of Inactivation by Ozone

Ozone inactivates microorganisms via production of hydroxyl and superoxide-free radicals. Ozone affects the permeability, enzymatic activity, and DNA of bacterial cells where guanine or thymine residues appear to be the most susceptible targets (Hamelin et al., 1978; Ishizaki et al., 1984, 1987). In viruses, ozone damages the protein coat or the nucleic acid core (Roy et al., 1981; Sproul et al., 1982) or both as in the case of rotaviruses (Chen et al., 1987).

3.4.4 Ozonation By-Products

Ozonation by-products of concern are bromate (BrO_3^-), as well as aldehydes and keto acids. Bromate is a mutagen and potential carcinogen produced by the reaction

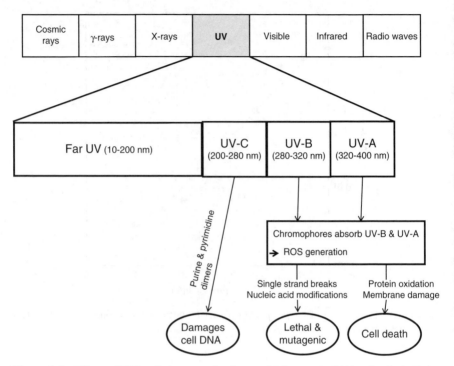

Figure 3.4   Effects of UV radiation on cells. *Source*: Malato et al. (2009). Catalysis Today 147: 1–59.

of bromide ion with molecular ozone and hydroxyl radicals. Cell culture studies showed that an increase in bromate concentration caused an increase in the number of abnormal chromosomes in Chinese Hamster cells (Echigo et al., 2004; Najm and Krasner, 1995).

3.5 ULTRAVIOLET LIGHT

3.5.1 Introduction

Ultraviolet (UV) light is the portion of the electromagnetic spectrum that is between visible light and X-rays. Ultraviolet radiation is broken down into UV-A (black light; 320–400 nm), UV-B (280–320 nm), UV-C (200–280 nm), and vacuum UV (10–200 nm). (Figure 3.4; Malato et al., 2009).

UV was first used in 1916 for disinfecting drinking water in Henderson, Kentucky, but was later abandoned in favor of chlorination. Owing to technological improvements, this disinfectant is now regaining popularity (Wolfe, 1990; Wright et al., 2006). The advantages of UV disinfection are the following: no production of toxic or mutagenic/carcinogenic by-products unless UV irradiation is followed by chlorination (Choi and Choi, 2010; Dotson et al., 2010), no taste and odor problems,

TABLE 3.2   Average inactivation of *Escherichia coli*, and phages T4 and T7 following exposure to 3 mJ/cm^2 of LP, MP, and PUV irradiation

Lamp Type	Average Log$_{10}$ Reduction		
	E. coli	Phage T4	Phage T7
LP	1.75	2.55	1.05
MP	-0.01^a	-0.07^a	0.06
PUV	4.26	4.29	2.72

Source: Adapted from Bohrerova, Z., H. Shemer, and R. Lantis, C.A. Impellitteri, and K.G. Lindene (2008). Water Res. 42: 2975–2982.
[a]Values are average of six replicates. The average value of N_0/N was less than 1.

no need to handle and store toxic chemicals, and small space requirement. Some disadvantages are no disinfectant residual in treated water, biofilm formation on UV lamp surface, lower disinfection in high turbidity effluents, or potential problems due to photoreactivation of UV-treated microbial pathogens.

3.5.2   Categories of UV Lamps

There are three categories of UV lamps (Mofidi et al., 2001):

- Continuous-wave emission lamps: they include low pressure (LP) (10^{-3} to 10^{-2} torr) mercury lamps with a peak emission at 253.7 nm, and medium pressure (MP) (10^2 to 10^4 torr) polychromatic mercury lamps which have a broader spectrum output of 185 nm to more than 400 nm.

- An alternative to mercury lamps is the pulsed xenon arc which generates pulsed UV radiation. An increase in the number of pulses increases the degree of inactivation of pathogens. For example, 10 pulses led to 4 log$_{10}$ reduction of poliovirus while 100 pulses were necessary for a 3 log$_{10}$ reduction of adenovirus (Lamont et al., 2007). Pulsed UV (PUV) radiation was found to be much more efficient than LP and MP mercury lamps as regard the inactivation of *E. coli* and phages (Table 3.2; Bohrerova et al., 2008).

- UV can also be generated by using light emitting diodes (LEDs) which do not contain mercury (i.e., no disposal problems), have low power requirements, fast start-up, and longer lifetime. Some disadvantages are low power output and higher cost.

UV disinfection systems generally use mercury lamps (LP or MP lamps) enclosed in quartz tubes. The tubes are immersed in flowing water in a tank and allow passage of UV radiation. However, transmission of UV by quartz decreases upon continuous use, due to fouling. The fouling rate depends on physical (e.g., temperature) and chemical (e.g., pH, hardness, alkalinity, iron concentration) characteristics of the water. Therefore, the quartz lamps must be periodically cleaned, using mechanical (on-line wiper systems), chemical (off-line acid cleaning), and ultrasonic cleaning methods. The recommended times for lamp replacement are approximately 12,000 hours for LP lamps and 5000 hours for MP lamps (Wright et al., 2006).

3.5.3 Mechanism of UV Damage

Ultraviolet radiation damages microbial DNA at a wavelength of approximately 260 nm. It causes thymine and cytosine dimerization which blocks DNA replication and effectively inactivates microorganisms. Thymine dimers are more prevalent due to the higher absorbance of thymine in the germicidal range. Thymine dimers were also detected by immunofluorescence microscopy, using a monoclonal antibody against cyclobutyl thymine, in *C. parvum* and *Cryptosporidium hominis* exposed to UV radiation (Al-Adhami et al., 2007). Studies with viruses have demonstrated that the initial site of UV damage is the viral genome, followed by structural damage to the virus coat (Nuanualsuwan and Cliver, 2003; Rodgers et al., 1985; Simonet and Gantzer, 2006). UV radiation damages are summarized in Figure 3.4 (Malato et al., 2009).

3.5.4 UV Damage Repair: Photoreactivation

The damage caused by UV radiation can be repaired by microorganisms, using two DNA repair processes, nucleotide excision repair (also called dark repair), and photoreactivation which requires visible light:

- *Dark repair*: In most bacteria, DNA damage can be repaired in the dark by the cell excision repair system which involves several enzymes. The UV-damaged DNA segment is excised and replaced by a newly synthesized segment. It appears that dark repair plays a smaller role in cell reactivation than photoreactivation, the other repair system. Viruses do not have a repair system but some can use the DNA repair system of the host cell (Hijnen et al., 2006; Sanz et al., 2007).
- *Photoreactivation* (i.e., photo repair). It may occur after exposure of UV-damaged microbial cells to visible light at wavelengths of 300–500 nm (Jagger, 1958). This repair of DNA dimers is activated by a protein called photolyase with a peak action response at 368 nm (Bohrerova and Linden, 2007). Photoreactivation increases with the visible light intensity and temperature, and decreases at higher UV fluences (Guo et al., 2012; Locas et al., 2008). Exposure to UV irradiation and subsequent photoreactivation do not seem to change bacterial characteristics. For example, plasmids carrying *amp* resistance or fluorescence did not affect the UV inactivation and photoreactivation patterns in *E. coli* (Guo et al., 2012).

Photoreactivation following UV irradiation has been demonstrated (Bohrerova and Linden, 2007; Carson and Petersen, 1975; Zimmer et al., 2003). Under laboratory conditions, it was shown that *E. coli* exposed to UV radiation at doses ranging from 3 to 10 mJ/cm^2, underwent photorepair when exposed to LP UV source, but less or no repair was detected following exposure to MP UV source (Locas et al., 2008;

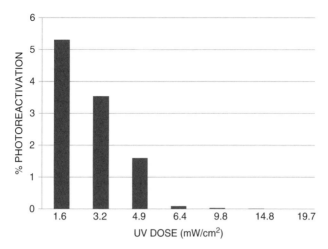

Figure 3.5 Inverse relationship between *E. coli* photoreactivation and UV dose. Adapted from Quek et al. (2006). Water Sci. Technol. 53 (6): 123–129.

Zimmer and Slawson, 2002), although no significant difference was found between LP and MP, especially at UV doses higher than 10 mJ/cm² (Guo et al., 2009). Others have shown that photoreactivation occurs when the microorganisms are exposed to MP UV lamps and an inverse relationship between the UV dose and the percent reactivation (Figure 3.5, Quek et al., 2006). Since both light intensity and lamp spectra significantly affect photorepair, it was suggested that photorepair experiments under laboratory conditions should be standardized (Bohrerova and Linden, 2007).

Figure 3.6 shows the photoreactivation of *E. coli* following exposure to UV at 10 mJ/cm² (Zimmer et al., 2003). Although photoreactivation and dark repair occurred in *Cryptosporidium* oocysts, their infectivity was not restored (Morita et al., 2002). This was later confirmed by others who showed that *C. parvum* and *C. hominis* oocysts damaged (i.e., formation of thymine dimers) by UV radiation failed to infect neonatal mice (Al-Adhami et al., 2007). Photoreactivation was also demonstrated in the cyanobacterium *Microcystis aeruginosa* (Sakai et al., 2011).

Photoreactivation was demonstrated in full-scale wastewater treatment plants using UV disinfection. Although total and fecal coliforms were photoreactivated, fecal streptococci showed no or slight photoreactivation (Baron, 1997; Guo et al., 2009; Harris et al., 1987; Locas et al., 2008; Whitby et al., 1984). (Figure 3.7).

Some suggested measures to reduce photoreactivation include the following (Guo et al., 2009; Locas et al., 2008; Martin and Gehr, 2007):

- A UV dose of 40 mJ/cm² has been recommended to avoid photoreactivation in effluents.
- Combine UV treatment (from LP lamps) with peracetic acid: It was found that the average photoreactivation following UV treatment was 1.2 logs as compared to 0.1 log following treatment with the combination of UV and peracetic acid.

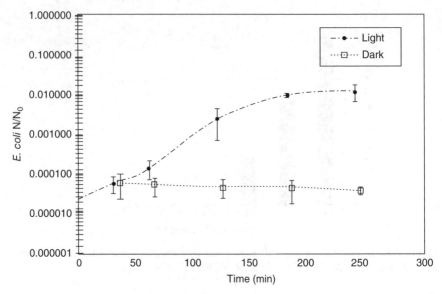

Figure 3.6 Photoreactivation of *Escherichia coli* following exposure to UV at 10 mJ/cm^2. *Source*: Zimmer et al. (2003). Water Res. 37: 3517–3523.

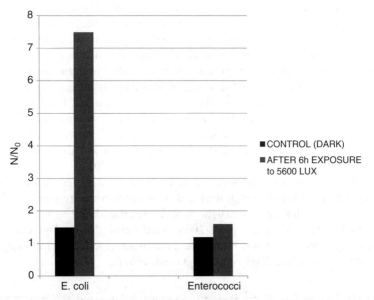

Figure 3.7 Comparison of photoreactivation levels for *E. coli* and enterococci from a UV-disinfected effluent (medium-pressure UV lamps) after 6h of exposure to visible light at 5600 lux. *Source*: Locas et al. (2008). Can. J. Microbiol. 54: 971–975.

- Keep the UV-treated water or wastewater sample in the dark for 3 hours before exposing to sunlight or artificial light. A practical suggested solution to photoreactivation is to conduct UV treatment at night (Sakai et al., 2011).
- Use alternative microbial indicators such as enterococci since they do not undergo photoreactivation.

More studies are needed to know whether the required fluence should be corrected for dark repair and photoreactivation.

3.5.5 Factors Controlling UV Action

Several factors affect the UV dose necessary for disinfection: organic and inorganic chemicals (e.g., humic substances, phenolic compounds, iron, manganese) interfere with UV transmission in water. Humic substances were first shown to protect bacteria from UV action (Bitton et al., 1972). The decay rate k of total coliforms exposed to UV decreased as the fulvic acid concentration increased from 0 to 10 mg/L (Alkan et al., 2007). Microorganisms are also partially protected from the harmful UV radiation when they are associated with or embedded within particulate matter (Madge and Jensen, 2006; Qualls et al., 1983). Furthermore, particle size affects the UV disinfection efficiency. Coliform inactivation tests with UV showed that the inactivation efficiency decreased in the presence of particles with size of 7 μm or larger were present in tertiary effluents (Jolis et al., 2001). More recent research showed that UV disinfection of fecal coliforms was slower when the bacteria were associated with particles over 20 μm (Madge and Jensen, 2006). Microbial aggregation, particularly in microbes with a hydrophobic surface, was also shown to reduce the inactivation of microorganisms by UV-A or UV-C (Bichai et al., 2011). The tailing (i.e., reduced inactivation) observed during the UV inactivation of MS2 phage could be explained by recombination within the aggregate, thus facilitating multiple infections of a host cells with several impaired viruses (Mattle and Kohn, 2012). Therefore, flocculation followed by filtration of effluents through sand or anthracite beds to remove interfering substances should improve UV disinfection efficiency (Templeton et al., 2007).

3.5.6 Pathogen and Protozoan Parasites Inactivation by UV

A UV dose of 40 mJ/cm^2 leads to a 4-log inactivation of bacteria, protozoan parasites, and most viruses (Yates et al., 2006). A survey of full-scale water treatment plants in the United States treating an impaired source water apply a 40 mJ/cm^2 dose (Dotson et al., 2012). Except for spore-forming bacteria, bacterial pathogens are generally quite sensitive to UV radiation (<20 mJ/cm^2 for a 3-log inactivation). A 3-log inactivation of enteric viruses (echoviruses, polioviruses, coxsackieviruses, caliciviruses, rotaviruses) can be achieved at UV doses of 12 to 44 mJ/cm^2 (Battigelli et al., 1993; Harris et al., 1987; Haynes et al., 2006; Hijnen et al., 2006; Meng and Gerba, 1996; Sommer et al., 1989; Yates et al., 2006) (Table 3.3). However, adenoviruses are much more resistant to UV than other viruses. This is because

TABLE 3.3   UV doses required for a 3-log inactivation of bacterial and
viral pathogens and protozoan cysts and oocysts

Microorganism	UV Dose (mW/cm²)
Bacteria	
Escherichia Coli	14
E. coli O157	2–14
Salmonella spp.	5–17
Shigella spp.	2–19
Legionella pneumophila	7–8
Vibrio cholerae	7
Helicobacter pylori[a]	8
Campylobacter jejuni	10
Yersinia enterocolitica	10
Bacillus subtilis spores	22–157
Streptococcus faecalis	23
Clostridium perfringens	145
Viruses	
Poliovirus	14–24
Coxsackievirus	20–27
Echovirus	16–25
Hepatitis A virus	12–17
Calicivirus	18–31
Rotavirus	23–44
Adenovirus	125–167
Protozoa	
Cryptosporidium parvum	3–9
Giardia lamblia	3–8
Acanthamoeba	119

Source: Adapted from data summarized by Hijnen et al. (2006) and Yates et al. (2006).
[a]4-log inactivation (Haynes et al., 2006).

these double-stranded DNA viruses are able to undergo repair within their host cells (Nwachcuku and Gerba, 2004; Roessler and Severin, 1996). Thus, an effective disinfectant for adenoviruses should be able to inhibit adsorption to and penetration into the host cell (Shannon et al., 2007). For example, adenovirus type 2 displayed a high resistance to UV, requiring a dose of 119 mJ/cm² for a 3-log inactivation (Gerba et al., 2002). This was confirmed in a pilot-scale study reporting adenovirus persistence in UV-treated wastewater tertiary effluents. A dose of approximately 170 mJ/cm² was necessary to achieve a 4-log inactivation of adenoviruses 2 and 15 (Thompson et al., 2003). These findings and others led the U.S. EPA to adopt the Long Term 2 Enhanced Surface Water Treatment Rule (LT2ESWTR) which states that a UV dose of 186 mJ/cm² was necessary for 4-log inactivation of viruses. Adenovirus inactivation can be enhanced by using polychromatic MP UV lamps (Eischeid et al., 2009; Linden et al., 2007). These lamps help achieve a 4-log inactivation at only 60 mJ/cm², potentially leading to lower cost for water utilities (Figure 3.8; Linden

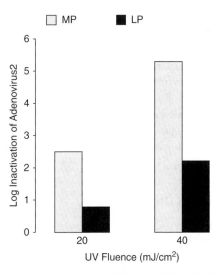

Figure 3.8 Inactivation of adenovirus 2 with UV light MP and LP UV (254 nm). Adapted from Linden et al. (2007). Appl. Environ. Microbiol. 73: 7571–7574.

et al., 2007). Another study with adenovirus 2 showed that a 4-log inactivation was achieved at approximately 80 mJ/cm^2 with MP UV lamps as compared to 180 mJ/cm^2 with LP UV lamps. Adenovirus 2 inactivation in a full-scale MP UV reactor showed that a 4-log inactivation can be achieved at approximately 100 mJ/cm^2 which is lower than the dose required by LT2ESWTR (Linden et al., 2009). The higher adenovirus inactivation observed for polychromatic MP UV lamps is due to their ability to cause both genome and protein damage as compared to only genome damage caused by LP UV-C lamps (Bosshard et al., 2013).

As regard protozoan parasites, they require relatively lower UV doses and their sensitivity is similar to that of bacterial pathogens. For example, a 3.9 log$_{10}$ inactivation was obtained for *Cryptosporidium* oocysts at a dose of 19 mJ/cm^2, using a MP UV lamp and animal infectivity assays (Bukhari et al., 1999). A UV dose of 11 mJ/cm^2 led to a 2-log inactivation of *C. parvum* (genotype 2 Iowa isolate) using pulsed UV light or MP UV lamps and cell culture infectivity assays (Mofidi et al., 2001). A 12 mJ/cm^2 dose is required by the U.S. EPA for a 3-log reduction in *Cryptosporidium* oocysts. Similar sensitivity to UV was observed for *Giardia lamblia* cysts (Mofidi et al., 2002). It was found that 10 and 20 mJ/cm^2 were required to obtain 2-log$_{10}$ and 3-log$_{10}$ inactivation, respectively (Campbell and Wallis, 2002). *Giardia* is quite sensitive to polychromatic emission from MP UV lamps. A 4-log inactivation was obtained in filtered drinking water at a UV fluence of 1 mJ/cm^2 (Shin et al., 2009). However, in full-scale wastewater treatment plants with UV at fluences varying from 18 to 57 mJ/cm^2, the reduction of cyst infectivity to gerbils was only 0.05 log. This low UV efficacy may have been due to particle shielding or to reactivation (Li et al., 2009). *Naegleria fowleri*, a free-living amoeboflagellate that causes primary meningoencephalitis (PAME), was found to be much more resistant to UV radiation than *Cryptosporidium* but less resistant than *Acanthamoeba* cysts (Sarkar and Gerba, 2012). *Encephalitozoon intestinalis* (microsporidia) spores, at a UV dose of 3 mJ/cm^2, showed an inactivation of 1.6–2.0 log$_{10}$ whereas at 6 mJ/cm^2 more than 3.6 log$_{10}$ of microsporidia were inactivated (Huffman et al., 2002).

Challenges microorganisms used in biodosimetry for UV irradiation include phage MS2, *Bacillus subtilis* spores for bacterial and viral inactivation, and phage T7 and T1 for UV-sensitive organisms such as protozoan parasites (Sommer et al., 2008).

Helminth parasites eggs appear to be the most resistant to UV radiation. A dose of 400 mJ/cm^2 resulted in 2.23-log inactivation (99.4%) of *Ascaris suum* eggs (Brownell and Nelson, 2006).

3.5.7 UV Disinfection of Drinking Water

UV disinfection is particularly useful for potable water (Wolfe, 1990). In 1998, Europe had more than 2000 drinking water facilities using UV disinfection as compared to 500 in North America (Parrotta and Bekdash, 1998). This disinfectant is particularly efficient against viruses that are major agents of waterborne diseases in groundwater. A drinking water plant using UV disinfection in London treats up to 14.5 MGD. In North America, several water treatment plants with a capacity of <1 to 2200 MGD have been built or are in the planning stage (Wright et al., 2006). In hospitals, chlorine is added to maintain a disinfectant residual after UV irradiation. Continuous UV irradiation rapidly inactivates *Legionella* in plumbing systems and in circulating hot tubs and whirlpools (Muraca et al., 1987). Smaller portable UV disinfection devices, capable of treating drinking water at a flow rate of about 4 gal/min, are being considered in small villages in rural areas of developing countries. These units provide a low cost alternative to chlorination which requires more expert supervision (see Chapter 8).

3.5.8 Coupling of UV Radiation with Other Technologies

UV disinfection can be coupled with the photocatalytic activity of TiO$_2$. A 3-log inactivation of *Cryptosporidium* was achieved at a UV dose of 25 mJ/cm^2 but was only 11 mJ/cm^2 when TiO$_2$ was used in combination with UV (Ryu et al., 2008). UV-C–TiO$_2$ combination reduced *E. coli* numbers by 6 logs in 60s, as compared to 90s for UVC alone. It caused structural damage to the cell DNA as well as cell membrane disruption after prolonged exposure (Kim et al., 2013a). Similarly, a synergistic effect of ozone and UV radiation on the disinfection of *B. subtilis* spores was also reported (Figure 3.9; Jung et al., 2008).

Upconversion of visible light to higher energy UV-C radiation is also being explored. Thus, yttrium orthosilicate crystals doped with praseodymium (Pr^{3+}) upconvert visible light to UV-C. Upconversion efficiency can be increased upon addition of gadolinium (Gd^{3+}) or lithium (Li$^+$). These materials, when coated on surfaces, completely inhibited the growth of *Pseudomonas aeruginosa* after 10-day incubation. The efficiency of the upconversion process must be increased when addressing drinking water disinfection (Cates et al., 2011; Webb, 2012).

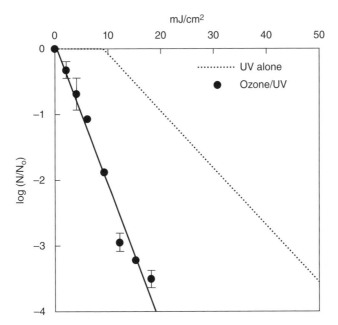

Figure 3.9 Synergistic effect of ozone and UV on the inactivation of *B. subtilis* spores. *Source*: Jung et al. (2008). Water Res. 42: 1613–1621.

3.6 USE OF PHOTOCATALYSTS IN WATER DISINFECTION

Photocatalysts (e.g., TiO_2) have been investigated extensively as regard their environmental applications such as oxidation of humic substances and hazardous organic materials and disinfection of water and wastewater (Beydoun et al., 1999). Photocatalysis requires activation by ultraviolet or visible light. Titanium dioxide (TiO_2), in combination with fluorescent light, sunlight, or near UV light, has been considered for the photocatalytic oxidation and, hence, inactivation of pathogens in drinking water and wastewater. TiO_2 photocatalysis produces reactive oxygen species (ROS) such as hydroxyl radicals (•OH), superoxide anions (•O_2^-), and hydrogen peroxide (H_2O_2) which have a high oxidizing power which leads to the destruction of the bacterial envelope (Cheng et al., 2007; Pigeot-Remy et al., 2012). TiO_2 photocatalysis not only results in the inactivation of *E. coli*, but also degrades the endotoxins released from the cells (Sunada et al., 1998). A study showed that TiO_2 immobilized on a glass surface led to an efficient inactivation of fecal and total coliforms when exposed to sunlight (Figure 3.10 (Gelover et al., 2006) and performed better than the SODIS process that will be discussed in Chapter 8. No bacterial regrowth was observed in the disinfected water, probably due to the oxidation of trace organics by TiO_2-sunlight.

Doping TiO_2 with nitrogen, carbon, or sulfur helps extend the absorbance of TiO_2 to the visible range (Li et al., 2007; Wong et al., 2006; Yu et al., 2005). The visible light-induced photocatalysis would be less costly than the one using UV. A

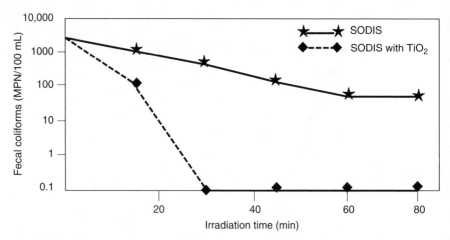

Figure 3.10   Fecal coliforms decrease during treatment SODIS and SODIS with TiO_2 disinfection (average radiation 1037 W m^{-2}). Adapted from Gelover et al. (2006). Water Res. 40: 3274–3280.

photocatalytic fiber made of palladium-modified N-doped TiO_2, in the presence of visible light (>400 nm; 40 mW/cm^2), reduced MS2 phage concentration (3 × 10^8 PFU/mL) by up to 3 $\log_{10}$. The phage decline was due to both adsorption to the fiber and photooxidation (Li et al., 2008b).

When considering the use of TiO_2 photocatalysis to disinfect surface waters, organic (e.g., humic acids) and, to a lesser extent, inorganic (sulfate, nitrate) constituents reduce the photocatalytic disinfection efficiency. A reduction of photocatalytic (use of immobilized TiO_2 and UV-A light at 370 nm) inactivation of *E. coli* was observed in surface water and in distilled water containing 7.0 mg/L total organic carbon (TOC) (Alrousan et al., 2009). Finally, the photocatalytic activity of 10-nm TiO_2 nanowires, as measured via oxidation of humic acid, was superior to the conventional P_{25} TiO_2. Another advantage of nano-sized TiO_2 is its relatively easy recovery by microfiltration without significant membrane fouling (Zhang et al., 2009b). Its application to disinfection of drinking water awaits further investigation.

A solar collector, in combination with a sensitizer consisting of Ru (III) complexes immobilized on a porous silicone, was used in water disinfection. The Ru (III) complexes absorb light in the visible range and inactivate microorganisms by producing singlet oxygen ($^1O_2$). This system was shown to inactivate up to 10^4 CFU/mL of *E. coli* and *Enterococcus faecalis* (Manjon et al., 2008).

3.7 PHYSICAL REMOVAL/INACTIVATION OF MICROBIAL PATHOGENS

Due to toxicological problems linked to the use of chemical disinfectants (e.g., formation of DBPs), disinfection using physical treatments is potentially attractive.

3.7.1 Membrane Filtration

Microfiltration, ultrafiltration, and nanofiltration remove efficiently the particulate matter in water and wastewater, and their small pore size also allows the physical removal of microbial pathogens and parasites. A set of three 0.4-μm pore size polyethylene membranes achieved up to a 7-log removal of fecal coliforms, 5-log removal of fecal streptococci, and 5.9-log removal of phage from settled wastewater. Phage removal efficiency increased as filtration resistance across the membrane increased. The increased resistance was due to biofilm accumulation on the membrane (Ueda and Horan, 2000). Ultrafiltration (pore size of 0.002–0.1 μm) helps achieve a good removal of protozoan cysts and oocysts from water and wastewater (Lonigro et al., 2006). This topic is covered in more details in Chapter 2.

3.7.2 Ultrasound

Ultrasound is known to cause damage to microbial cells (e.g., damage to membrane integrity, cell lysis). Most of the information on ultrasonic energy deals with wastewater disinfection. A study on disinfection of wastewater has shown that a 20-kHz ultrasound unit, operated at 700 W/L, causes a 4-log units inactivation of fecal coliforms within 6 minutes. Approximately half of the inactivation was due to thermal effects resulting from ultrasound treatment (Madge and Jensen, 2002). A 20-kHz ultrasound inactivated more than 90% *Cryptosporidium* oocysts within 1.5 minutes under laboratory conditions (Ashokkumar et al., 2003). At an ultrasonic intensity of 23–26 W/cm^2, the decimal reduction time (DRD) of gram-negative bacteria (*E. coli* and *P. aeruginosa*) varied between 5 and 7.3 minutes. A gram-positive bacterium, *Enterococcus avium*, required a higher DRD of 13.6–15.9 minutes. The DRD of sewage bacteria was 3.5- to 11-fold higher than that of pure cultures (Stamper et al., 2008). A synergistic effect was observed when ultrasound was combined with other disinfectants such as ozone, chlorine, or UV. The disinfecting power of ultrasound is also increased in the presence of TiO_2 (no UV irradiation was used). It is thought that hydroxyl radicals were responsible for the enhanced inactivation of *Legionella* (Dadjour et al. 2006).

3.7.3 Ultrahigh Hydrostatic Pressure

Ultrahigh hydrostatic pressure (UHP) is effective in inactivating pathogenic microorganisms. This approach is commercially used for processing foods such as cured ham or fruit juices and in the pharmaceutical industry (Demazeau and Rivalain, 2011; San Martin et al., 2002). However, sensitivity to UHP may vary among strains of the same species (e.g., *E. coli* O157:H7) (Malone et al., 2006).

3.7.4 Nanomaterials

Nanomaterials (e.g., chitosan, ZnO, TiO_2, Ce_2O_4, fullerenes, carbon nanotubes, silver nanoparticles) are beneficial in several applications in the biomedical, cosmetics,

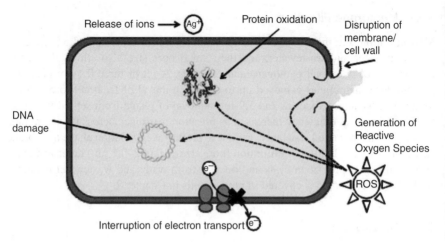

Figure 3.11 Mechanisms of antimicrobial activities exerted by nanomaterials. *Source*: Li et al. (2008). Water Res. 42: 4591–4602.

pharmaceuticals, and other industrial fields. Nanoparticles may serve as efficient nanosorbents owing to their high surface area, as nanocatalysts for the degradation of organic pollutants, redox active agents for metals, biosensors for pathogens, and are incorporated into membranes for water purification. Due to the formation of toxic by-products (DBPs) by conventional disinfectants (see Section 3.2.4), nanomaterials are being explored as alternative means of inactivating pathogens, and control of biofilms and membrane biofouling. They exhibit toxic properties toward microorganisms and this led to their consideration as potential chlorine-free disinfectants in the water treatment industry (Qu et al., 2013; Savage and Diallo, 2005; Stoimenov et al., 2002). The several proposed mechanisms for their antimicrobial activity are summarized in Figure 3.11 (Li et al., 2008a). We already discussed the use of TiO$_2$ as a disinfectant in combination with UV or sunlight (see Section 3.6).

The antimicrobial properties of silver have been known since the Antiquity. The medical applications of this metal included, among others, the treatment of burns and chronic wounds. Silver use in medicine declined following the introduction of antibiotics but is now regaining popularity as a result of the emergence of antibiotic resistance in bacteria (Rai et al., 2009). Silver nanoparticles are the most widely used antimicrobials in several consumer products and have been incorporated in membrane filters (Son et al., 2004) and activated carbon filters for point-of-use drinking water treatment (see Chapter 2). Silver nanoparticles–alginate composite beads were prepared by simultaneous Ag+ reduction and gelation of the alginate. A column packed with these beads inactivated over 5 logs of *E. coli* with 1-minute hydraulic retention time (Lin et al., 2013). Other composites, such as Ag-AgBr/TiO$_2$, were shown to exhibit *E. coli* inactivation under LED light illumination (Wang and Lim, 2013).

Chitosan, derived from chitin, is often used as a coagulant in wastewater treatment and is also an antimicrobial agent that could be useful as a disinfectant in water and wastewater treatment. The nanomaterials could be used as primary disinfectants or in combination with traditional ones (e.g., UV). For example, UV photosensitization of polyhydroxylated fullerene (fullerol) nanoparticles was shown to greatly enhance the inactivation of MS2 phage as compared to UV alone (Badireddy et al., 2007). The increased inactivation is due to the production of ROS such as singlet oxygen and superoxide.

Despite the use of some nanomaterials in point-of-use home water treatment systems, their implementation as disinfectants in large-scale water and wastewater treatment plants is still not feasible at the present time. Some potential challenges are separation and recycling problems, steady supply of nanomaterials, their incorporation into existing water treatment plants, their fate in the environment, and their potential impact on human and environmental health (Li et al., 2008a; Savage and Diallo, 2005).

Genetic tools employed in *toxicogenomics* will, in the future, greatly help shed light on the mechanisms of action of disinfectants used in hospitals and in the water and wastewater treatment industries. Hence, microarrays help in revealing the transcriptional responses of target pathogens, at the genomic level, upon exposure to disinfectants. A toxicogenomics approach was used to examine the response of *E. coli*, *P. aeruginosa*, and *S. aureus* to hydrogen peroxide or peracetic acid (Chang et al., 2005, 2006; Palma et al., 2004; Zheng et al., 2001). This approach will help identify the repressed or induced genes in response to a given disinfectant, which could lead to identify the optimal conditions for disinfectant use.

WEB RESOURCES

http://water.epa.gov/drink/contaminants/basicinformation/disinfectants.cfm
(disinfectants in drinking water: U.S. EPA)

http://www.epa.gov/enviro/html/icr/dbp.html
(disinfection by-products from U.S. EPA)

http://cfpub.epa.gov/ncea/cfm/recordisplay.cfm?deid=2829
(chloramines)

http://www.epa.gov/pesticides/factsheets/chemicals/chlorinedioxidefactsheet.htm
(disinfection with chlorine dioxide)

http://www.sodis.ch/index_EN
(Solar Disinfection (SODIS) process)

FURTHER READING

Hijnen, W.A.M., E.F. Beerendonk, and G.J. Medema. 2006. Inactivation credit of UV radiation for viruses, bacteria and protozoan (oo)cysts in water. *Water Res.* 40:3–22.

Hrudey, S.E. 2009. Chlorination disinfection by-products, public health risk tradeoffs and me. *Water Res.* 43:2057–2092.

Li, Q., S. Mahendra, D.Y. Lyon, L. Brunet, M.V. Liga, D. Li, and P.J.J. Alvarez. 2008. Antimicrobial nanomaterials for water disinfection and microbial control: potential applications and implications. *Water Res.* 42:4591–4602.

Qu, X., P.J.J. Alvarez, and Q. Li. 2013. Applications of nanotechnology in water and wastewater treatment. *Water Res.* 47:3931–3946.

Russel, A.D., J.R. Furr, and J.-Y. Maillard. 1997. Microbial susceptibility and resistance to biocides. *Am. Soc. Microbiol. News* 63:481–487.

Singer, P.C. 2006. DBPs in drinking water: additional scientific and policy considerations for public health protection. *J. Am. Water Wks. Assoc.* 98(10):73–79.

Sobsey, M.D. 1989. Inactivation of health-related microorganisms in water by disinfection processes. *Water Sci. Technol.* 21:179–195.

Templeton, M.R., R.C. Andrews, and R. Hofmann. 2008. Particle-associated viruses in water: impacts on disinfection processes. *Crit. Rev. Environ. Sci. Technol.* 38:137–164.

U.S. EPA. 2006. *Ultraviolet disinfection guidance manual for the final long term 2 enhanced surface water treatment rule.* EPA 815-R-06-007. U.S. Environmental Protection Agency, Office of Water, Washington, DC.

Webb, S. 2012. Harvesting photons to kill microbes: ES&T's top environmental technology article 2011. *Environ. Sci. Technol.* 46:3609–3810.

WHO. 2011. *Guidelines for Drinking-Water Quality*, 4th edition. World Health Organization, Geneva, Switzerland. http://www.who.int/water_sanitation_health/publications/2011/dwq_guidelines/en/.

4

DRINKING WATER DISTRIBUTION SYSTEMS: BIOFILM MICROBIOLOGY

4.1 INTRODUCTION

Following water treatment by processes mentioned in Chapters 2 and 3, water is distributed through a network of pipes to reach customers. The water distribution system (WDS), an essential component of drinking water treatment, is the "workhorse" that carries drinking water from the plant to the customers. The daily production in the United States is estimated 34 billion gallons of drinking water that flows through 1.8 million miles of distribution pipes. It is estimated that approximately two-third of the pipes are made of plastic, mostly polyvinyl chloride (PVC). Many waterborne outbreaks are attributed to the degradation of water quality in noncovered water reservoirs and in WDSs through cross-connections, main breaks (an estimated 240,000 water distribution main breaks per year in the United States; Lafrance, 2011), contamination during construction or repair, back siphonage (National Research Council, 2006), or negative pressure events (Gullick et al., 2004; NRC, 2006).

Drinking water regulations mostly concerns water quality at the end of the water treatment plant but the water quality can deteriorate during storage and transport through water distribution pipes and in indoor plumbing before reaching the consumer. More recently, US federal regulations addressed water quality in distribution systems through rules covered by the Safe Drinking Water Act (SDWA).

In the long history of water distribution which goes back to 2000 BC, pipes were made of a wide range of materials which include aqueducts, terracotta, wood, copper, lead, bronze, bamboo, and stone. Nowadays, piping materials include cast iron, ductile iron, stainless steel, concrete, asbestos-cement, or PVC, polyethylene, and medium density polyethylene (MDPE) (Bachmann and Edyvean, 2006; Berry et al., 2006; Lafrance, 2011; Walski, 2006; Percival et al., 2000). However, some of

Microbiology of Drinking Water Production and Distribution, First Edition. Gabriel Bitton.
© 2014 John Wiley & Sons, Inc. Published 2014 by John Wiley & Sons, Inc.

the piping materials lead to corrosion problems (e.g., cast iron and copper pipes) or to leaching of toxic materials (e.g., PVC pipes).

Some major problems concerning WDSs are the following:

- Microbial growth
- Pathogen survival and growth
- Nitrification problems
- Biocorrosion

Deterioration of water distributions systems due to leakage from cracks and joints or back siphonage may lead to drinking water contamination and disease outbreaks. In the United States, approximately 18% of the outbreaks reported for public water systems are due to contamination of the WDS by microbial pathogens and toxic chemicals (Craun, 2001). The contribution of WDSs to acute gastrointestinal illness (AGI) incidence was demonstrated by epidemiological studies which showed that the incidence was higher among consumers who drank from home taps than those who drank treated water bottled at the water treatment plant (Payment et al., 1991a, 1997). This shows that WDS and in-premise plumbing may contribute to disease in the community. The risk of viral AGI from nondisinfected groundwater-based drinking WDSs exceeded the U.S. EPA acceptable risk of 10^{-4} episodes/person-year. An extrapolation of the results to the United States as a whole shows an AGI risk of 470,000 to 1,100,000 episodes/year nationwide (Lambertini et al., 2012; U.S. EPA, 1989).

Drinking water deterioration in WDS has several causes (LeChevallier et al., 2011; Smith, 2002; Sobsey and Olson, 1983):

- Improperly built and operated storage reservoirs which should be covered to prevent airborne contamination and to exclude animals.
- Loss of disinfectant residual.
- Back siphonage and cross-contamination.

Some of the consequences are

- Public health problems due to the excessive growth and colonization of water distribution pipes by bacteria and other organisms, some of which are pathogenic. Microbial growth depends on the availability of nutrients (see Chapter 6)
- Microbial regrowth in storage reservoirs
- Protection of pathogens from disinfectant action
- Taste and odor problems due to the growth of algae, actinomycetes and fungi (see Chapter 5)
- Corrosion problems

Drinking water may contain humic and fulvic acids, as well as easily biodegradable natural organics such as carbohydrates, proteins, and lipids. The presence of dissolved organic compounds in finished drinking water is responsible for several problems, among which are taste and odors, enhanced chlorine demand, trihalomethane formation, and bacterial colonization of water distribution lines (see Chapter 6 for more details). Water utilities should maintain the distribution system by periodically flushing the lines to remove sediments, excessive bacterial growth, and encrustations due to corrosion.

4.2 BIOFILM DEVELOPMENT IN WDSs

4.2.1 Introduction

Biofilms develop at solid-water interfaces and are widespread in natural environments as well as in engineered systems. They are ubiquitous and are commonly found in trickling filters, rotating biological contactors, activated carbon beds, distribution pipe surfaces, groundwater aquifers, aquatic weeds, tooth surfaces (i.e., dental plaque), dialysis units, dental unit waterlines, and indwelling medical devices (IMDs) such as catheters, pacemakers, orthopedic implants, endotracheal tubes, prosthetic joints, and heart valves (Anwar and Costerton, 1992; Characklis, 1988; Hamilton, 1987; Schachter, 2003; Thomas et al., 2004; Walker and Marsh, 2007; van der Wende and Characklis, 1990). For example, in the United States, the risk of infection from bladder catheters ranges from 10% to 30% while the risk from cardiac pacemakers is 1–5% (Thomas et al., 2004).

Biofilms are relatively thin layers (up to a few hundred microns thick) of attached microorganisms that form microbial aggregates and grow on surfaces. Biofilms formed by prokaryotic microorganisms resemble in some ways tissues formed by eukaryotic cells (Costerton et al., 1995; Wingender and Flemming, 2011). The multicellular behavior is due to the fact that cells within the biofilm interact and coordinate their activities via direct cell-to-cell contact or by releasing signaling molecules (see quorum sensing in Section 4.2.5) (Branda and Kolter, 2004). Observations of living biofilms by scanning confocal laser microscopy have helped in our understanding of biofilm structure. Biofilm microorganisms form microcolonies separated by open water channels (Stoodley et al., 2002; Donlan, 2002). Liquid flowing through these channels allow the diffusion of nutrients, oxygen, and antimicrobial agents into the cells (Donlan, 2002). Some bacteria, such as *Staphylococcus epidermidis*, form honeycomb-like structures (Schaudinn et al., 2007) that provide mechanical stability to the biofilm, help lower the energy costs of individual cells and maximize nutrient absorption. Biofilms also include corrosion by-products, organic detritus, and inorganic particles such as silt and clay minerals. They contain heterogeneous assemblages of microorganisms (Figure 4.1), depending on the chemical composition of the surface, the chemistry of finished water, and oxido-reduction potential in the biofilm. They take days to weeks to develop, depending on nutrient availability and environmental conditions. Biofilm growth proceeds up to a critical thickness (approximately 100–200 μm) when nutrient diffusion across the biofilm becomes limiting. The decreased diffusion of oxygen is conducive to the development of facultative and anaerobic microorganisms in the deeper layers of the biofilm.

Figure 4.1 SEM pictures of biofilms.

4.2.2 Processes Involved in Biofilm Development

A number of processes contribute to biofilm development on surfaces exposed to water flow. The processes involved are the following (Bitton and Marshall, 1980; Gomez-Suarez et al., 2002; Olson et al., 1991). The sequence of events is described in Figures 4.2 and 4.3 (Gomez-Suarez et al., 2002; De Kievit, 2009):

4.2.2.1 Surface Conditioning. Surface conditioning is the first step in biofilm formation. Minutes to hours after exposure of a surface to water flow, a surface-conditioning layer, made of ions, proteins, glycoproteins, humic-like substances, and other dissolved or colloidal organic matter, initially adsorb to the surface. This results in a modified surface that is different from the original one (Schneider and Leis,

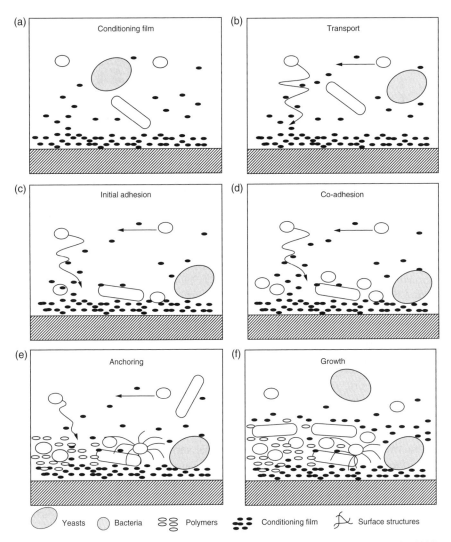

Figure 4.2 Sequence of events in biofilm formation *Source:* Gomez-Suarez et al. (2002). In: *Encyclopedia of Environmental Microbiology*, Gabriel Bitton, editor-in-chief, Wiley-Interscience, N.Y.

2002). The conditioning film is a source of nutrients for bacteria, particularly in oligotrophic environments such as drinking water or groundwater.

4.2.2.2 *Transport of Microorganisms to Conditioned Surfaces.*

Diffusion, Brownian motion, convection, and turbulent eddy transport are involved in turbulent flow regime. Chemotaxis may also enhance the rate of bacterial adsorption to surfaces under more quiescent flow.

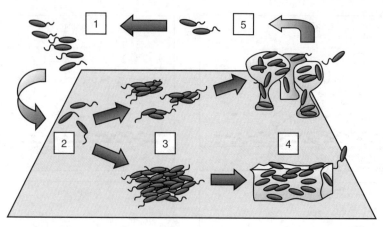

Figure 4.3   *Pseudomonas aeruginosa* biofilm development. Planktonic cells (stage 1) attach onto a solid surface (stage 2) and microcolonies are formed (stage 3). Under conditions that promote bacterial migration (e.g., succinate, glutamate), cells will spread over the substratum, ultimately developing into a flat, uniform mat (stage 4). Under motility-limiting conditions (e.g. glucose), the microcolonies proliferate forming stalk- and mushroom-like structures (stage 4). At various points throughout biofilm maturation, cells can detach and resume the planktonic mode of growth (stage 5). *Source:* De Kievit, T.R. (2009). Environ. Microbiol. 11: 279–288.

4.2.2.3 Adhesion of Microorganisms to Surfaces.   According to the thermodynamic theory, adhesion of a microorganism to a surface is favored when the free energy of adhesion is negative $(\Delta G_{adh} < 0)$ (Gomez-Suarez et al., 2002). Furthermore, according to the DLVO theory (Derjaguin, Landau, Verwey, and Overbeek theory), adhesion is a balance between Lifshitz-Van der Waals forces and repulsive or attractive electrostatic forces due to electrical surface charges on both microbial and surfaces. Hydrophobic interactions are also involved in microbial adhesion to surfaces.

4.2.2.4 Cell Anchoring to Surfaces.   Following adhesion to a given surface, microbial cells are anchored to the surface, using extracellular polymeric substances (EPSs). EPS, made of polysaccharides (e.g., mannans, glucans, uronic acids), proteins, nucleic acids, and lipids help support biofilm structure. EPS display hydrophilic and hydrophobic properties. They help in the adhesion of microorganisms to surfaces and their cohesion within biofilms, and play a structural role in biofilms (Bitton and Marshall, 1980; Characklis and Cooksey, 1983; Costerton and Geesey, 1979; Flemming and Wingender, 2002; 2010; Starkey et al., 2004; Zhu et al., 2010; Xue et al., 2013b). EPS also helps protect microorganisms from protozoan predation, chemical insult, antimicrobial agents, desiccation, and osmotic shock. Moreover, extracellular polysaccharides chelate heavy metals and reduce their toxicity to microorganisms (Bitton and Freihoffer, 1978).

Other attachment organelles include flagella, pili (fimbriae), stalks, or holdfasts (e.g., *Caulobacter*) (Bitton and Marshall, 1980; Olson et al., 1991).

4.2.2.5 Cell Growth and Biofilm Accumulation. Biofilms essentially consist of microbial aggregates embedded in EPSs attached to a solid surface (Rittmann, 1995b). It was estimated that biofilms harbor about 95% of the microbial communities in WDSs (Wingender and Flemming, 2004). However, bulk water bacteria may dominate in portions of the distribution system with no chlorine residual (Srinivasan et al., 2008).

4.2.3 Factors Involved in Biofilm Accumulation

The study of single-species biofilms has helped in our understanding of the steps involved in biofilm accumulation which depends on several factors:

4.2.3.1 Concentration of Assimilable Organic Carbon. The presence of even microgram levels of organic matter in distribution lines allows the growth and accumulation of biofilm microorganisms (Block et al., 1993; Ollos et al., 2003; Servais, 1996).

4.2.3.2 Limiting Nutrients. Microbial growth in drinking water can also be highly regulated by the availability of phosphorus, which can act as a limiting nutrient for microbial growth in drinking water (Charnock, and Kjønnø, 2000; Miettinen et al., 1997; Keinänen et al., 2002; Sathasivan and Ohgaki, 1999). For example, in most Finnish drinking waters, microbial growth was well correlated with the concentration of bioavailable phosphorus (Lehtola et al., 2002; Rubulis and Juhna, 2007). The presence of bioavailable phosphorus and iron in water distribution pipes also helps the survival of *Escherichia coli* (Appenzeller et al., 2005; Juhna et al., 2007). Thus, phosphorous removal by water treatment processes may decrease *E. coli* survival in this environment.

4.2.3.3 Trace Elements. Trace elements can stimulate (e.g., Fe, Mn, Zn) or inhibit (e.g., Cu, Ag) bacterial growth in distribution networks.

4.2.3.4 Disinfectant Concentration. Biofilm thickness and activity are inhibited by disinfection with chlorine. For example, the biofilm thickness ($\sim10^3$ pg ATP/cm^2) in the presence of a chlorine concentration of <0.05 mg/L was two orders of magnitude greater than the thickness (~10 pg ATP/cm^2) in the presence of a chlorine level of 0.30 mg/L (Hallam et al., 2001). Attached microorganisms are better controlled by monochloramine than by free chlorine (see also Chapter 3). In a bench-scale distribution system, it was found that a free chlorine residual of 0.5 mg/L or chloramine at 2 mg/L reduced biofilm microorganisms by several orders of magnitude (Ollos et al., 2003).

4.2.3.5 Type of Pipe Material. Biofilm thickness also depends on the type of pipe material. Plastic-based materials (polyethylene or PVC) support less attached biomass than iron (e.g., gray iron), cement-based materials (e.g., asbestos-cement, cemented cast iron) (Figure 4.4; Niquette et al., 2000), or stainless steel materials.

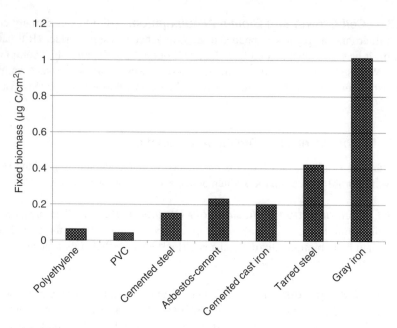

Figure 4.4 Effect of pipe material on biofilm accumulation in a drinking water distribution system. Adapted from Niquette et al. (2000). Water Res. 34: 1952–1956.

Copper pipes showed the lowest biofilm formation potential (Yu et al., 2010). Biofilm formation, as measured by heterotrophic plate counts and ATP content, was higher in galvanized steel pipes than in copper pipes, as the release of Cu from the pipes is toxic to microorganisms (Silhan et al., 2006). Iron corrosion, increased surface roughness, and porosity in iron pipes contribute to increased attached biomass. Iron, as a nutrient, is also a factor in increased biomass in the pipes. Pipe material type can also influence the community composition in the biofilms (Kalmbach et al., 2000). For example, 74% of the bacteria in PVC pipes were *Stenotrophomonas*, whereas iron pipes harbored a more diverse microbial populations which consisted mainly of *Nocardia*, *Acidovorax*, *Xanthobacter*, *Pseudomonas*, and *Stenotrophomonas* (Norton and LeChevallier, 2000). Thus, the use of pipe materials (e.g., stainless steel) that could reduce biofilm development is desirable (Percival et al., 2000).

4.2.3.6 *Flow Velocity and Regimen.* An increase in flow rate and a change from laminar to turbulent flow can enhance the mass transport of nutrients or biocides into the biofilm, leading to changes in biofilm thickness (see Bachmann and Edyvean (2006) for more details). Sudden changes in water flow leads to detachment of biofilm from the pipe surface with subsequent deterioration of water quality due to increase in suspended bacteria (Choi and Morgenroth, 2003). Hydraulic regime also affects the bacterial community structure in biofilms. A higher diversity was observed at highly varied flow in an experimental WDS (Douterelo et al., 2013).

4.2.3.7 Quorum Sensing. A bacterial communication system, which controls biofilm formation and function (Shrout and Nerenberg, 2012; see more details on this system in Section 4.2.5.

4.2.3.8 Other Factors. A number of other factors (pH, redox potential, TOC, water temperature, hardness) control the growth of microorganisms on pipe surfaces (Olson and Nagy, 1984). Temperature affects directly and indirectly the rate of biofilm formation. A study of biofilm growth in pipes made of different materials showed that biofilm accumulation was higher at 35°C than at 15°C (Silhan et al., 2006). Bacterial growth generally increases when temperature is above 15°C although psychtrophic iron bacteria can grow below this temperature.

4.2.4 Biofilm Ecology

Biofilms allow microorganisms to persist and grow in hostile environments such as WDSs under low nutrient conditions. In this protected environment, EPSs provide a diffusional barrier against disinfectants and other deleterious chemicals. EPS increase the resistance of biofilm and any detached biofilm clusters to chlorine (Xue et al., 2013b). Using model WDSs, it was found that during the initial stages of biofilm formation, the 16S RNA sequences in biofilms are similar to those found in the bulk solution. However, DNA- and RNA-based fingerprints of 20-year-old biofilms show a higher diversity in biofilm communities than in bulk water (Henne et al., 2012). The higher diversity in biofilms than in bulk water was also observed in an experimental WDS. Alphaproteobacteria dominated in the bulk water while beta- and gammaproteobacteria (which include many primary and opportunistic pathogens) were predominant in biofilms (Douterelo et al., 2013). Microcolonies are formed following attachment of single cells to the pipe surface. Species diversity increases thereafter, and biofilms may reach steady state and high ecological diversity, including both heterotrophs and autotrophs, only after 2–3 years (Berry et al., 2006; Martiny et al., 2003). Molecular techniques are now revealing the bacterial diversity in drinking water and biofilms (Kahlisch et al., 2010; Lee et al., 2010; Revetta et al., 2010; Schmeisser et al., 2003). Molecular probing showed the presence of gram-positive bacteria as well as alpha-, beta-, and gamma-proteobacteria. Bacteria generally found in drinking WDSs include *Pseudomonas, Acinetobacter, Flavobacterium, Alcaligenes, Aeromonas, Moraxella, Enterobacter, Citrobacter, Sphingomonas, Klebsiella, Burkholderia, Xanthomonas, Methylobacterium,* and *Bacillus* (Berry et al., 2006; Block et al., 1997). Examination of biofilms from household water meters showed that the major bacterial genera were *Sphingomonas* (alpha-proteobacteria), *Acidovorax, Methylophilus* (beta-proteobacteria), and *Lysobacter* (gamma-proteobacteria). Examination of the Ann Arbor drinking water by 16S rRNA gene sequencing also showed that most of the bacteria were proteobacteria and belonged to *Acidovorax, Variovorax, Sphingopyxis, Ralstonia,* and *Novosphingobium* (Figure 4.5; Lee et al., 2010). Another study in Ohio, using similar techniques, showed the presence of proteobacteria, cyanobacteria, actinobacteria, bacteroidetes, and planctomycetes but 57.6% of the sequences were hard to classify (Revetta et al., 2010).

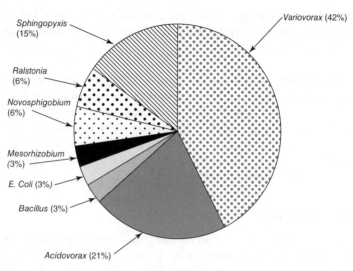

Figure 4.5 Bacterial diversity in drinking water. Pie chart shows the relative diversity of each genus identified by 16S rRNA gene sequencing. *Source:* Lee et al. (2010). Water Res. 44: 5050–5058.

Bacterial community composition changes following chlorine disinfection as alpha-, beta-, and gamma-proteobacteria decreased after chlorination (Poitelon et al., 2010). Community composition is also influenced by the type of disinfectant used (Hong et al., 2010; Poitelon et al., 2010; Williams et al., 2004; Yu et al., 2010). It is known that nutrient-depleted environments, such as marine waters and groundwater, harbor ultramicrocells (UMCs) that pass through 0.2 μm filters (Velimirov, 2001). Such ultramicrocells were detected in chlorinated drinking water and are part of the biofilm community. Analysis of 16S rRNA sequences showed that UMC belong to the Proteobacteria, Firmicules, and Actinobacteria. Some of the UMCs are potential opportunistic pathogens, such as *Stenotrophomonas maltophilia* or *Microbacterium* sp., that cause infections in immunocompromised patients in hospital setting (Silbaq, 2009).

4.2.5 Gene Exchange and Quorum Sensing in Biofilms

The proximity of microbial cells in biofilms offers opportunities for communication and exchange of metabolic products and genetic material (e.g., antibiotic resistance genes) between microorganisms (Cvitkovitch, 2004; Wuertz, 2002; Xi et al., 2009). The use of modern techniques such as confocal laser scanning microscopy and fluorescent-labeled (e.g., tagging with green fluorescent protein) plasmids has facilitated the study of gene transfer in biofilms and in the environment in general. These methods have generally shown that gene transfer rates were much higher than those determined by plating on selective media. In addition to gene transfer by conjugation and transformation, biofilms can also be a suitable environment for gene transfer via transduction. An example is the transfer of the Shiga toxin (Stx) gene to *E. coli* by an Stx-encoding bacteriophage (Solheim et al., 2013).

Much work is presently being done on the molecular mechanisms of biofilm development. For their survival and interactions in biofilms, bacteria are able to communicate with cells of the same species or with other prokaryotic and eukaryotic species. This cell-to-cell communication mechanism, also called "bacterial twitter," is quorum sensing which requires a threshold cell density for its triggering. It helps microorganisms in biofilm development and function and improves their survival by increasing their access to nutrients and defense against competing microorganisms (Boyle et al., 2013; de Kievit, 2009; Shrout and Nerenberg, 2012; Williams et al., 2007). Several quorum sensing bacteria has been found in water and wastewater treatment systems. These include, among others, *Aeromonas, Arcobacter, Legionella, Pseudomonas, Sphingomonas, Vibrio*, and *Nitrobacter* (Shrout and Nerenberg, 2012).

Quorum sensing diffusible signaling molecules (e.g., *N*-acyl homoserine lactone), called autoinducers, play an important role in density-dependent gene expression and biofilm differentiation. Bacteria undergo changes when the signaling molecule has reached a threshold concentration and when the bacterial population has reached a certain density. There are three major classes of signaling molecules: acyl homoserine lactones (AHLs), peptides, and LuxS/autoinducer 2. Antibiotics can also function as signaling molecules at relatively low concentrations (Fajardo and Martínez, 2008). AHL signals are emitted by gram-negative bacteria and their activity was demonstrated in single-species biofilms (e.g., *Pseudomonas aeruginosa*) and in naturally occurring biofilms (McLean et al., 1997). Peptides are the signaling molecules in gram-positive bacteria. Autoinducer-2 (AI-2) is another cell-to-cell signaling molecule that may function in cell communication in some bacteria (Winzer et al., 2002). Some other functions regulated by the AI-2 signal are biofilm formation in *Bacillus cereus*, motility in *Helicobacter pylori*, virulence factor production in *P. aeruginosa*, or susceptibility to antibiotics in *Streptococcus anginosus* (Pereira et al., 2013). In addition to biofilm formation, quorum sensing in both gram-negative and gram-positive bacteria is also known to be involved in the regulation of several activities, such as virulence, motility, EPS production, sporulation, production of secondary metabolites such as antibiotics, conjugation, symbiosis, and interspecies competition and other survival strategies (Miller and Bassler, 2001; Withers et al., 2001).

4.2.6 Biofilm Detachment from Surfaces

Biofilm detachment from surfaces releases microorganisms, including pathogens, which may colonize other surfaces. It affects the occurrence of waterborne diseases resulting from the transport of opportunistic pathogens to the consumer's faucet. Detachment occurs following *erosion* (detachment of single cells), *sloughing* (detachment of large pieces of biofilm, sometimes extending all the way to the substratum), *scouring* following abrasion or scraping and grazing by predators (Percival et al., 2000; Rittmann, 2004; Rittmann and Laspidou, 2002).

Biological processes also involved in biofilm detachment, include, among other factors, a decrease in EPS production, rhamnolipids, anti-biofilm poylysaccharides, and predation on biofilm microorganisms. Detachment events are genetically regulated (Rendueles et al., 2013; Schooling et al., 2004; Stoodley et al., 2002).

The net accumulation of biofilms on surfaces is described by the following equation (Rittmann and Laspidou, 2002).

$$X_f L_f = \frac{YJ}{b + b_{\text{det}}} \tag{4.1}$$

where;

X_f = biomass density (mg biomass/cm^3)
L_f = biofilm thickness (cm)
Y = true yield of biomass (mg biomass/mg substrate)
J = substrate flux (mg substrate/cm^2-day)
b = specific decay rate of biofilm microorganisms (d^{-1})
b_{det} = specific detachment rate (d^{-1})

Biofilm accumulation increases when the substrate is increased or when detachment is decreased. Under steady state conditions, the average specific growth rate (μ_{ave}) is equal to the specific detachment rate (b_{det}).

Slow growing microorganisms (e.g., methanogens) are protected from detachment because they live in the deeper layers of the biofilm.

There are several factors affecting biofilm detachment (Pedersen, 1990; Rittmann and Laspidou, 2002). Detachment is:

- increased by tangential flow shear stress which is related to the detachment rate (at least for smooth surfaces);
- increased by axial force on the biofilm, due to abrasion and to pressure changes caused by turbulent changes and expressed by the Reynolds number Re;
- decreased by surface roughness;
- affected by physiological parameters such as EPS content (higher detachment at low EPS) or biofilm density (lower detachment for dense biofilms).

4.2.7   Some Methods Used in Biofilm Study

Several physical, chemical, biochemical, molecular, and microbiological methods are available for biofilms study. Some of these methods, including molecular techniques, are summarized in Table 4.1 (Broschat et al., 2005; Characklis et al., 1982; Lazarova and Manem, 1995; Neu and Lawrence, 2002; Percival et al., 2000; Schaule et al., 2000; Schmidt et al., 2004).

In the cultivation-based method, a low nutrient medium, designated as R2A agar, is recommended to determine the heterotrophic plate count (HPC) in environmental samples such as drinking water or groundwater. However, this approach detects only from 0.1% to 10% of the total bacterial counts in most aquatic environments (Pickup, 1991). Microbial biomass in biofilms can be measured via determination of specific

TABLE 4.1 Some methods used for biofilm study

Type	Analytical Method
Microscopy	Use of light microcopy, fluorescence microscopy, scanning confocal laser microscopy, scanning electron microscopy, environmental scanning electron microscopy, transmission electron microscopy, atomic force microscopy.
Image analysis	Image structure analyzer, time lapse imaging
Direct measurement of biofilm quantity	Biofilm thickness (using optical methods, image analysis, thermal resistance, quartz crystal microbalance)
	Biofilm mass (total cell count via staining with Acridine Orange or DAPI, crystal violet assay)
Indirect measurement of biofilm quantity	Extracellular polymeric substances (EPS)
	Specific biofilm constituents (polysaccharides, proteins, nucleic acids)
	Total organic carbon (TOC)
	Total proteins
	Peptidoglycans
	Lipid biomarkers
Microbial activity within biofilms	Viable cell count (plate counts, preferably in low nutrient medium such as R2A agar, Live/Dead BacLight™)
	Active bacteria: direct viable count (DVC)
	DVC-FISH
	ATP
	Lipopolysaccharides
	Substrate removal rate
	Dehydrogenase activity (e.g., TTC, INT, or CTC dyes) or esterases such as carboxyfluorescein diacetate (CFDA).
	Oxygen uptake rate (OUR)
	DNA
	rRNA
	mRNA
Microbial observation and identification	Immunofluorescence (monoclonal or polyclonal antibodies)
	FISH (fluorescent *in situ* hybridization)
	mRNA amplified by polymerase chain reaction
	Green fluorescent protein (GFP)
	Use of commercial fluorochromes such as DAPI, acridine orange, SYTOX Green, PicoGreen and propidium iodide, or FUN-1 (for fungi)
Microenvironment	Fluor conjugates for determining diffusion and permeability in biofilms
	Use of microelectrodes for determining pH, temperature, O_2, NH_4, NO_3, NO_2, H_2S, and CH_4 within the biofilm
Other methods	Mass spectroscopy
	Infrared spectroscopy
	X-ray spectroscopy

Source: Adapted from Broschat et al. (2005); Characklis et al. (1982); Lazarova and Manem (1995); McLean et al. (2004); Mezule et al. (2013); Neu and Lawrence (2002); Percival et al. (2000); Schaule et al. (2000); Schmidt et al. (2004).

cell biochemical constituents such as ATP, DNA, RNA, proteins, lipid biomarkers, bacterial cell wall components, or photosynthetic pigments (Sutton, 2002).

Total cell counts in biofilms are difficult due to the presence of aggregated cells. This task is made easier by using modern techniques such as scanning confocal microscopy (SCLM) which gives information about the biofilm structure. The non-destructive SCLM in conjunction with 16S rRNA-targeted oligonucleotide probes help to show the heterogeneous structure of biofilms and their complex biodiversity by giving information on the distribution and activity of specific groups of microorganisms in biofilms (Costerton et al., 1995; Schramm et al., 1996). rRNA-targeted, fluorescently labeled oligonucleotide probes are useful in identifying biofilm microorganisms and community composition (Kalmbach et al., 2000). Other nondestructive microscopic techniques include epifluorescence microscopy using fluorogenic dyes (e.g., acridine orange or 6-diamidino-2-phenylindol known as DAPI), transmission electron microscopy (TEM), scanning electron microscopy (SEM), and environmental SEM (ESEM) which allows the observation of the specimen in its own hydrated state.

Microelectrode probes give information on the physicochemical properties (pH, temperature, O_2, NH_4, NO_3, NO_2, H_2S, CH_4) of the biofilm microenvironment (Burlage, 1997). These microelectrodes have helped shed light on the distribution of nitrifiers and sulfate reducers along the depth of biofilms. Microelectrodes for biofilm study will be replaced in the future with fiber-optic microsensors (Beyenal et al., 2000).

Microbial activity within biofilms can also be assessed by measuring enzyme activity (e.g., esterase or dehydrogenase activity), reduction of tetrazolium salts such as cyanoditolyl tetrazolium chloride (CTC), 2-p-(iodophenyl)-3-(p-nitrophenyl)-5-tetrazolium chloride (INT), triphenyl tetrazolium chloride (TTC), or resorufin by biofilm microorganisms. Fluorescein diacetate (FDA) or carboxyfluorescein diacetate (CFDA) are used to measure esterase activity and the active fluorescent cells are counted under a fluorescent microscope. Cell viability can also be determined by the LIVE/DEAD BacLight kit, or special dyes which measure membrane integrity (e.g., SYBR Green propidium iodide). The fluorescence intensity can be determined via flow cytometry (Berney et al., 2007, 2008; Cerca et al., 2011).

Several techniques have been proposed for gaining information about biofilm formation and thickness. Some of the methods are (Broschat et al., 2005; Milferstedt et al., 2006; Nivens et al., 1995; Reipa et al., 2006):

- *Quartz crystal microbalance* (QCM). This nondestructive technique measures biomass changes on a quartz crystal. Both the quartz oscillator frequency (f) and resistance (R) are monitored during biofilm formation and the data are expressed as $\Delta R/\Delta f$ which reflects changes in the viscoelastic properties of the biofilms. QCM was used in the long-term monitoring of biofilm formation by *Pseudomonas cepacia* and *P. aeruginosa*.

- *Optical methods*: Early on, biofilm thickness was estimated by focusing a light microscope on the top and bottom layers of a biofilm. A more recent approach consists of exposing the biofilm to a desktop scanner and then following with image analysis. The gray level of the images is well correlated with the biofilm biomass. A reflectance assay was considered to determine biofilm formation of 14 *Enterococcus* isolates on opaque and nonopaque surfaces

- *Crystal violet assay:* This method consists of staining dry biofilms with 0.1% crystal violet, washing with buffer to remove excess dye, releasing the dye from the biofilm with 95% ethanol and measuring the absorbance at 595 nm.

4.3 GROWH OF PATHOGENS AND OTHER MICROORGANISMS IN WDSs

Pathogen accumulation in biofilms of WDSs is an important public health issue as it may contribute to the spread of waterborne diseases. Pathogens enter the WDS through insufficient water treatment or through leaks. They grow in WDSs and colonize pipe inner walls, connections, tubercles, and dead ends. However, biofilm growth over surfaces is not continuous as noncolonized spots are observed. Bacterial accumulation preferably occurs in the colonized area and increases with biofilm age (Paris et al., 2009).

4.3.1 Earlier Studies on the Microbiology of Distribution Systems

The presence of a wide range of microorganisms (eubacteria, filamentous bacteria, actinomycetes, diatoms) was demonstrated in distribution pipes by SEM. Some of the bacteria were seen attached to the surfaces by means of extracellular fribrillar materials (Ridgway and Olson, 1981; Ridgway et al., 1981). Turbercles provide a high surface area for microbial growth and protect microorganisms from the lethal action of disinfectants. Turberculated cast iron pipe sections from the WDS in Columbus, Ohio, metropolitan area harbored high numbers of aerobic and anaerobic bacteria (e.g., sulfate reducing bacteria). Some samples contained up to 3.1×10^7 bacteria per gram of tubercle material (Tuovinen and Hsu, 1982). Several investigators have also reported the proliferation of iron and manganese bacteria that grow attached to pipe surfaces and, subsequently, cause a deterioration of water quality (e.g., pipe clogging and color problems). Attached iron bacteria such as *Gallionella* have *stalks* that are partially covered with iron hydroxide. This was confirmed by X-ray energy-dispersive microanalysis (Ridgway et al., 1981).

4.3.2 Opportunistic Bacterial Pathogens

As mentioned in Chapter 1, this group includes genera such as *Pseudomonas, Aeromonas, Klebsiella, Flavobacterium, Enterobacter, Citrobacter, Serratia, Acinetobacter, Proteus, Providencia, L. pneumophila, S. maltophilia*, and nontubercular mycobacteria (NTM) (Sobsey and Olson, 1983; van der Wielen and van der Kooij,

2013). They are of health concern to the young, the elderly, and immunocompromised consumers. The routine occurrence of these opportunistic pathogens in distribution systems have been documented in the literature (Keevil, 2002). For example, most *Aeromonas* isolates from 13 Swedish drinking WDSs produced virulence factors such as cytotoxins, suggesting potential public health problems (Kuhn et al., 1997). *Sphingomonas* spp. can also be found in water distribution biofilms. They are often overlooked due to their slower growth than other members of the HPC (Koskinen et al., 2000). They have been isolated from activated carbon and sand filters, and are potential reservoirs of antibiotic-resistant bacteria in drinking water (Vaz-Moreira et al., 2011). Biofilms can also be a potential reservoir for *H. pylori* which can survive in the viable but nonculturable (VBNC) state (Linke et al., 2010; Percival and Thomas, 2009). The VBNC state is a bacterial response to stress (e.g., nutritional stress, exposure to disinfectants and other toxicants such as heavy metals). A wide range of bacteria are now known to enter in the VBNC state when exposed to environmental stress (Oliver, 2005). Biofilms may also harbor antibiotic-resistant bacteria (Schwartz et al., 2003; see also Chapter 1).

4.3.3 Nontubercular Mycobacteria (NTM)

The genus *Mycobacterium* includes pathogenic species such as *M. tuberculosum* and *M. leprae*, and NTM which are found in soil and aquatic environments and are considered as opportunistic pathogens (Falkinham, 2002; Falkinham et al., 2001). NTM were detected around the world in 21% to more than 80% of samples from WDSs (see review by Vaerewijck et al., 2005). A more recent survey in two chloraminated WDSs in Florida and Virginia showed that the percent occurrence of *Mycobacterium* spp. was 93.7% and 94.4%, respectively. The percent occurrence of *M. avium* was 10% and 8.9%, respectively (Wang et al., 2012b). In Germany and France, mycobacterial species were found in 90% of biofilm samples taken from water treatment plants (Schulze-Robbecke et al., 1992). A survey of seven water treatment plants in the Netherlands showed that NTM were detected in all distributed drinking water samples and their levels were much higher in distributed water than in treated water (van der Wielen and van der Kooij, 2013). Approximately one-third of the 100 known mycobacterial species have been detected in WDSs. In the Netherlands, several mycobacterial species (*M. peregrinum, M. salmoniphilum, M. llatzerense, and M. septicum*) were identified in distributed water and in shower water (van Ingen et al., 2010). The primary *Mycobacterium* species detected in biofilms of WDSs in China (Guangzhou and Beijing) were *M. arupense* and *M. gordonae* (Liu et al., 2012a). *Mycobacterium avium* subsp. *paratuberculosis* (MAP), a possible causative agent of Crohn's autoimmune disease in humans, was detected, via PCR, in finished drinking water (Aboagye and Rowe, 2011; Beumer et al., 2010). This opportunistic pathogen persists for several weeks in biofilms and its survival is much higher than that of *E. coli* (Lehtola et al., 2007; Torvinen et al., 2007). The growth of *M. avium* in WDSs depends on the level of available organic carbon (Falkinham et al., 2001; LeChevallier, 2004).

4.3.4 *Legionella* in Hot-Water Tanks and Distribution Systems

Due to airborne transmission through showering, much work has been carried out on the survival and growth of Legionellae in potable WDSs and plumbing in hospitals and homes (Colbourne et al., 1988; Felfoldi et al., 2010; Muraca et al., 1988). A recent survey showed that the percent occurrence of *Legionella* spp. was 67.5% in a Florida WDS and 30% in a Virginia system (Wang et al., 2012b). Furthermore, legionellae appear to be more resistant to chlorine than *E. coli* (States et al., 1989), and small numbers may survive in distribution systems that have been judged to be microbiologically safe. Under high shear turbulent-flow conditions, survival of *L. pneumophila* and *Mycobacterium avium* is much higher than that of *E. coli* which is traditionally used as an indicator (Lehtola et al., 2007). Legionellae may also grow to detectable levels inside hot-water tanks in hospitals and homes and thus pause a health threat (Stout et al., 1985; Witherell et al., 1988). The *Legionella* isolates responsible for nosocomial Legionnaires' disease corresponded to isolates from potable water (Garcia-Nunez et al., 2008). *Legionella* survives well at 50°C and may even grow and multiply in tap water at 32–42°C (Dennis et al., 1984; Yee and Wadowsky, 1982). The enhanced survival and growth in these systems have also been linked to stagnation (Ciesielski et al., 1984), stimulation by rubber fittings in the plumbing system (Colbourne et al., 1988), and trace concentrations of metals such as Fe, Zn, and K (States et al., 1985, 1989). It was found that sediments found in WDSs and tanks (i.e., scale and organic particulates) and the natural microflora significantly improved the survival of *Legionella pneumophila*. Sediments indirectly stimulate *Legionella* growth by promoting the growth of commensalistic microorganisms (Stout et al., 1985; Wadowsky and Yee, 1985). For example, *Mycobacterium chelonae*, an opportunistic bacterial pathogen, can have a positive effect on the cultivability of *L. pneumophila* and *H. pylori* in biofilms (Gião et al., 2011).

Protozoa, mostly amoebas and ciliates, play a role in the survival of *Legionella* in water distribution pipes. Biofilms and the presence of protozoa (e.g., *Acanthamoeba* spp.) play an essential role in the survival, proliferation, and pathogenicity of *Legionella* in drinking water (Corsaro et al., 2010; Lau and Ashbolt, 2009). Biofilms growing in in-premise plumbing harbor *Legionella* cells as well as protozoan hosts which help in the propagation of this pathogen. The biofilm-associated *Legionella* is detached and aerosolized during showering and can be subsequently inhaled.

Several methods have been considered for controlling *Legionella* in water: superchlorination (chlorine concentration of 2–6 mg/L), chloramination, maintaining the hot-water tanks at temperatures above shock treatment at 70°C for 30 minutes, ultraviolet irradiation (Antopol and Ellner, 1979; Knudson, 1985; Kool et al., 2000; Muraca et al., 1987; Stout et al., 1986), biocides (Fliermans and Harvey, 1984; Grace et al., 1981; Soracco and Pope, 1983), and alkaline treatment (States et al., 1987). There are less *Legionella* nosocomial infection outbreaks in hospitals that use monochloramine than those using free chlorine (Flannery et al., 2006; Heffelfinger et al., 2003; Kool et al., 1999). *Legionella* can, however, become heat-resistant after repeated heat treatments (Allegra et al., 2011). The control of legionellae in potable water systems may also be achieved via the control of protozoa which may harbor these pathogens (States et al., 1990).

4.3.5 Antibiotic-Resistant Bacteria and Antibiotic Resistance Genes

Antibiotic-resistant bacteria (ARB) and antibiotic resistance genes (ARG) to several antibiotics (amoxilline, chloramphenicol, ciprofloxacin, gentamicin, rifampin, sulfisoxasole, tetracycline) were found in several finished and tap water samples. The ARB levels were often higher in tap water than in finished water, indicating that WDSs can serve as a reservoir for ARB and ARG (Xi et al., 2009). Antibiotic resistance genes have been detected in biofilms from drinking WDSs. Enterobacterial *amp*C β-lactam-resistance and enterococcal *van*A vancomycin-resistance genes were detected in drinking water biofilms although no enterobacteriaceae or enterococci were detected in those samples. However, the Staphylococcal *mec*A methicillin-resistance gene was not found in drinking water biofilms (Schwartz et al., 2003).

Several studies have shown that chlorination of water and wastewater induces selection for ARB (Murray et al., 1984; Shrivastava et al., 2004). The mechanism of this selection is not well known.

However, at the present time, we do not know if increased resistance to antibiotic present a significant risk to consumers, especially the young, the elderly, and immunocompromised people.

Table 4.2 lists some human pathogens and opportunistic pathogens detected in biofilms of domestic plumbing systems (Eboigbodin et al., 2008).

4.3.6 Protozoan Parasites

Biofilms can also serve as reservoirs for protozoan parasites cysts and oocysts. The trapping and concentration of *C. parvum* oocysts and their survival in biofilms was demonstrated under laboratory conditions (Rodgers and Keevil, 1995). Free-living

TABLE 4.2 Some human pathogens detected in biofilms from domestic plumbing systems

Pathogen	Comments
Legionella pneumophila	Causes Legionnaires' disease
Legionella bozemanii	Causes human pneumonia
Helicobacter pylori	Causes gastric diseases including cancer
Pseudomonas sp.	Some species (e.g., *P. aeruginosa, P. maltophilia*) are human pathogens
Sphingomonas sp.	Some species cause nosocomial infections in humans
Micrococcus kristinae	Found in catheter-associated bacteremia
Corynebacterium sp.	Part of human skin flora. Some species are pathogenic
Acanthamoeba keratitis	Can cause infection in the cornea

Source: Adapted from Eboigbodin et al. (2008). Amer. Water Works Assoc. J. 100 (10): 131–137.

pathogenic protozoa, such as *Naegleria fowleri*, may also find refuge in biofilms, particularly in stagnant areas of the distribution network. There is experimental evidence that *N. fowleri* can colonize and persist for months in pipe biofilms, although none was detected in two full-scale distribution networks (Biyela et al., 2012).

The cysts and oocysts can subsequently be resuspended in WDSs by biofilm sloughing following an increase in water flow (Helmi et al., 2008; Puzon et al., 2009; Searcy et al., 2006).

4.3.7 Enteric Viruses

Enteric viruses are occasionally detected in treated drinking water (Bitton et al., 1986). Laboratory studies have shown that, following their entry into the WDS, viruses may adsorb to biofilms where they may survive for relatively long periods. A similar phenomenon was observed in wastewater systems where viruses (noroviruses, enteroviruses, FRNA phages) were found to attach and persist in biofilms for longer periods than in wastewater (Skraber et al., 2009). However, only one of three field studies of virus monitoring in biofilms showed the presence of infectious enteroviruses (Vanden Bossche, 1994). Similarly to other pathogens, pathogenic viruses may contaminate drinking water through sloughing of biofilm pieces.

4.4 SOME ADVANTAGES AND DISADVANTAGES OF BIOFILMS IN DRINKING WATER TREATMENT AND DISTRIBUTION

The development of biofilms on surfaces can be beneficial or detrimental to processes in water and wastewater treatment plants. Trickling filters and rotating biological contactors are examples of processes that rely on microbial activity in biofilms to treat wastewater.

4.4.1 Advantages

Some advantages offered by biofilms are (Rittmann, 1995b; Shrout and Nerenberg, 2012)

- The fact that anytime a solid surface is exposed to water is followed by biofilm formation demonstrates well the benefits of biofilms to microorganisms.
- Biofilms act as biocatalysts in natural systems for the self-purification of surface waters and groundwater, and in engineered fixed-film processes (e.g., trickling filters, rotating biological contactors, biofilters in water purification, riverbank filtration, membrane-based biofilm reactors) for the treatment of water and wastewater.
- They are desirable whenever microorganisms must be retained (i.e., need for a high mean cell residence time) in a system with a short hydraulic retention time,

as is the case for biological treatment of drinking water which involves the processing of large volumes of water with very low concentrations of biodegradable compounds.

- Many of the processes used in drinking water treatment allow biofilm formation and include, among others, granular activated carbon, slow and rapid sand filters, and biological water treatment.

4.4.2 Disadvantages

However, biofilms can cause problems in water treatment and distribution systems (Rittmann, 1995b, 2004; Smith, 2002; van der Wende and Characklis, 1990):

- Because of mass-transport resistance, bacteria are exposed to lower concentrations of substrates than in the bulk liquid.
- Biofilm accumulation increases fluid frictional resistance in distribution pipelines (Figure 4.6; Bryers and Characklis, 1981), leading to an increase in pressure drop and to reduced water flow if the pressure drop is held constant. The microbial community structure of the biofilm also affects frictional resistance. A predominantly filamentous biofilm appears to increase frictional resistance (Trulear and Characklis, 1982).
- Biofouling decreases flow rates in membrane-based water treatment processes. The biofouling is due to the overproduction of EPSs mainly composed of polysaccharides.

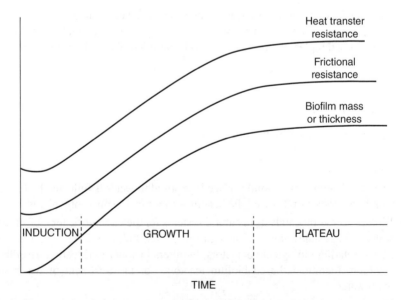

Figure 4.6   Increase in heat transfer resistance and frictional resistance resulting from biofilm growth. *Source:* Bryers and Characklis (1981). Water Res. 15: 483–491.

- Anaerobic conditions lead to the production of H_2S, a toxic gas characterized by a rotten-egg odor.
- Biofilms have an impact on public health and are the cause of persistent infections (Costerton, 1999). They are conducive to the accumulation of pathogens and parasites in distribution pipes biofilms which are habitats for opportunistic pathogens such as *Legionella* and *Mycobacterium* and lead to infections related to implants and dental plaques. Biofilms are involved in 80% of chronic inflammatory and infectious diseases due to pathogenic bacteria (Sauer et al., 2007). Biofilms developing on implanted medical devices harbors microbial pathogens such as *Candida albicans*, *Staphylococcus aureus*, *P. aeruginosa*, *Kelbsiella pneumoniae*, and *Enterococcus* spp., and *E. coli*. Most of the circulatory and urinary tract infections are due to biofouled implanted devices. These infections increase patient morbidity and mortality and the cost of medical care. Biofilms associated with medical implants are often resistant to antibacterial and antifungal drugs (Anwar and Costerton, 1992; Donlan, 2002; Thomas et al., 2004).
- Corrosion problems: Biocorrosion of pipes is associated with biofilm growth. Biofilm accumulation is associated with corrosion of iron pipes where Fe^{2*} serves as an electron donor. Complaints about red and black waters, due to the activity of iron- and manganese-oxidizing bacteria (e.g., *Gallionella*, *Hyphomicrobium*) (Rittmann, 2004).
- Resistance of biofilm microorganisms to biocides: Biofilm microorganisms can be up to 1500 times more resistant to biocides and antibiotics than freely suspended bacteria (Sauer et al., 2007). Increase in resistance to chlorine disinfection contributes to the regrowth of indicator and pathogenic bacteria in distribution systems. It was suggested that this increased resistance may be due to diffusional resistance of the biofilm matrix as measured with a chlorine microelectrode (de Beer et al., 1994), the protective effect by extracellular polymeric materials, the presence of efflux pumps that export biocides, antibiotics, and other toxic substances to the bacterial surrounding medium (Marquez, 2005), selection of biofilm microorganisms with enhanced resistance to the disinfectant, attachment of bacteria to biological (e.g., macroinvertebrates and algal surfaces) and nonbiological surfaces (particles, activated carbon) or cell surface hydrophobicity (Steed and Falkinham, 2006). Furthermore, multispecies biofilms appear to be more resistant to disinfection than single-species ones (Simoes et al., 2010). As shown for activated carbon, biofilm bacteria appear to be more resistant to residual chlorine and monochloramine than suspended bacteria (Berry et al., 2009; LeChevallier et al., 1988). The protective effect of biofilms toward pathogens exposed to chlorine was also demonstrated for *Mycobacterium avium* and *Mycobacterium intracellulare* (Steed and Falkinham, 2006). Particulates, including particulate organic matter, protect pathogenic microorganisms from disinfectant action (Hoff, 1978). It was found that *Enterobacter cloacae*, in the presence of particulates from WDSs, is protected from chlorine action as a result of its attachment to the particles (Herson et al., 1987). Chloramines appear to

be more efficient in biofilm control than free chlorine (HOCl or OCl⁻) (see Chapter 3). This may be due to the lower affinity of chloramines for bacterial polysaccharides (LeChevallier et al., 1990; van der Wende and Characklis, 1990). It was proposed that, for biofilm control, free chlorine should be used as a primary disinfectant but the residual should be converted to chloramine (LeChevallier et al., 1990). The increased resistance of biofilm microorganisms to chlorine applies to other antibacterial agents such as antibiotics. The free chlorine residual decreases as the water flows through the distribution system. In a laboratory study, silver (100 µg/L) used as a biocide did not have any significant effect on preventing biofilm formation on PVC and stainless steel surfaces (Sylvestry-Rodriguez et al., 2008).

4.5 BIOFILM CONTROL AND PREVENTION

Biofilms can be controlled or removed by mechanical, chemical, or enzymatic means (Brisou, 1995). The drinking water industry has a few options for controlling biofilm growth in WDSs (Camper et al., 2003; Rittmann, 2004):

- Mechanical cleaning: It is combined with the use of biocides when the biofouling is severe.
- Maintain a proper level (1 mg/L as Cl_2) of chlorine residual throughout the WDS to reduce biofilm accumulation. The chlorine level cannot exceed 4 mg/L as Cl_2. Maintaining a chlorine residual is, however, difficult to achieve in large systems and cannot prevent biofilm growth in the presence of high levels of electron donors.
- Obtain a biostable water by removing or drastically reducing the level of organic and inorganic electron donors in treated water. This can be accomplished by using membrane filtration, enhanced coagulation, or biological filtration (see Chapter 6).
- Consider the effect of the pipe material: Replace deteriorating iron pipes with other materials (e.g., PVC or CPVC pipes). However, one should be concerned about the leaching and accumulation of the monomer, vinyl chloride, a proven human carcinogen (Walter et al., 2011).
- Potential "jamming" of quorum sensing in biofilms (Barraud et al., 2009; Hentzer et al., 2004; Shrout and Nerenberg, 2012): Attempts to interrupt quorum sensing in biofilms are potentially a useful strategy in the control of infectious diseases. As a result of the emergence of antibiotic resistance in bacteria, there is presently a quest for quorum sensing inhibitors to attenuate and control bacterial infections (Hentzer et al., 2003). Laboratory studies have shown that several chemicals can control biofilms by interrupting quorum sensing. Some examples are 2,4-dinitrophenol, a furanone produced by red algae, nitric oxide to disperse single- or multi-species biofilms, or protoanemonin, a natural compound which inhibits quorum sensing in *P. aeruginosa*. The application of this research

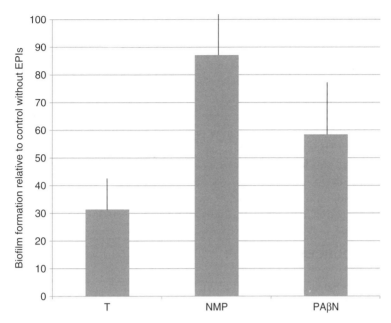

Figure 4.7 Effect of efflux pump inhibitors (T, NMP and PAβN) on biofilm formation in *Staphylococcus aureus*. T = Thioridazine; NMP = 1-naphthylmethyl piperazine; PaβN = Phe-Arg-β Naphthylamide. Adapted from Kvist et al. (2008). Appl. Environ. Microbiol. 74: 7376–7382.

to WDSs under field conditions awaits further investigation. Quorum sensing interruption by chemicals could be used in combination with other control measures to improve biofilm control. Some microorganisms make endogenous nitric oxide (NO) which is generated by NO-synthase under the control of the *nos* gene. The presence of NO helped decrease the ability of *P. putida* to form biofilms (Bobadilla Fazzini et al., 2013; Liu et al., 2012b; Njoroge and Sperandio, 2009). There are also quorum enzymes or bacteria that release enzymes that inhibit signaling in biofilms, leading to control of biofouling in membrane bioreactors (Jiang et al., 2013; Kim et al., 2013b). As an example, a recombinant *E. coli* that produces *N*-acyl homoserine lactonase, inhibited biofouling in a membrane bioreactor (Oh et al., 2012; Yeon et al., 2009a, 2009b). The feasibility of this approach for controlling biofilms is yet to be demonstrated.

- As regard the formation of biofilms on medical devices, great efforts are being made to prevent their formation and to develop drugs to treat existing biofilms. One original control method is to block quorum-sensing pathways in order to prevent the formation of biofilms on medical devices. Bacteria also have several chromosome-encoded efflux pumps that help export antibiotics, antiseptics, heavy metals, solvents, detergents, and other biocides to the surrounding medium (Martinez et al., 2009). These efflux pumps would lead to increased resistance of biofilms to toxic chemicals, including disinfectants. The addition of efflux pumps inhibitors (e.g., thioridazine, phe-arg β-naphthylamide, 1-naphthylmethyl piperazine) helped reduce biofilm formation by *Staphylococcus aureus* (Figure 4.7; Kvist et al., 2008). A wide range of chemicals are being screened by biotechnological companies to find a way to control biofilms. Some of these chemicals are the halogenated furanones produced by marine algae (Schachter, 2003).

- Antibiofilm polysaccharides: Some bacteria are able to produce antibiofilm polysaccharides which can inhibit biofilm formation or destabilize preformed biofilms. They potentially have useful applications in industrial and medical fields and can help reduce the incidence of infections caused by biofilms on medical devices (Rendueles et al., 2013).
- Other biofilm control approaches are surface modification or application of an electric current (Chiang et al., 2009; Wellman et al., 1996).

WEB RESOURCES

http://www.personal.psu.edu/faculty/j/e/jel5/biofilms/
(biofilm online manual, Amer. Soc. Microbiology)

http://www.personal.psu.edu/faculty/j/e/jel5/biofilms/primer.html
(biofilm primer)

http://www.erc.montana.edu/
(Center for Biofilm Engineering, Montana State University, Bozeman)

http://www.nesc.wvu.edu/ndwc/articles/OT/SP01/History_Distribution.html
(history of drinking water distribution)

http://www.nottingham.ac.uk/quorum/
(general information about quorum sensing)

http://www.mrc-lmb.cam.ac.uk/genomes/awuster/wecb/
(gives many references on quorum sensing)

http://www.epa.gov/safewater/dwh/index.html
(Drinking water: EPA consumer information)

http://www.epa.gov/safewater/sdwa/index.html
(SDWA: Safe Drinking Water Act)

http://www.epa.gov/safewater/standards.html
(General information on drinking water)

http://www.cdc.gov/safewater/
(Safe Water System, Center for Disease Control and Prevention)

FURTHER READING

Bachmann, R.T. and R.G.J. Edyvean. 2006. Biofouling: an historic and contemporary review of its causes, consequences and control in drinking water distribution systems. *Biofilms* 2:197–227.

Berry, D., C. Xi, and L. Raskin. 2006. Microbial ecology of drinking water distribution systems. *Curr. Opin. Biotechnol.* 17:297–302.

Costerton, J.W., Z. Lewandowski, D.E. Caldwell, D.R. Korber, and H.M. Lappin-Scott. 1995. Microbial biofilms. *Annu. Rev. Microbiol.* 49:711–745.

Donlan, R.M. 2002. Biofilms: microbial life on surfaces. *Emerging Infect. Dis.* 8:880–890.

Eboigbodin, K.E., A. Seth, and C.A. Biggs. 2008. A review of biofilms in domestic plumbing. *Am Water Works Assoc. J.* 100(10):131–137.

Flemming, H.-C. and J. Wingender. 2010. The biofilm matrix. *Nat. Rev. Microbiol.* 8:623–633.

Ghannoum, M. and G.A. O'Toole (Eds). 2004. *Microbial Biofilms.* ASM Press, Washington, DC, 426 p.

Keevil, C.W. 2002. Pathogens in environmental biofilms. In *Encyclopedia of Environmental Microbiology,* G. Bitton (editor-in-chief). Wiley-Interscience, New York, pp. 2339–2356.

LeChevallier, M.W., M.-C. Besner, M. Friedman, and V.L. Speight. 2011. Microbiological quality control in distribution systems. In *Water Quality and Treatment: A Handbook on Drinking Water,* 6th edition, J.K. Edzwald (Ed.). AWWA, Denver, CO, pp. 21.1–21.84.

Miller, M.B. and B.L. Bassler. 2001. Quorum sensing in bacteria. *Annu. Rev. Microbiol.* 55:165–199.

National Research Council. 2006. *Drinking Water Distribution Systems: Assessing and Reducing Risks.* NCR, Washington, DC, 400 pp. http://www.nap.edu/catalog/11728.html

Peirera, C.S., J.A. Thompson, and K.B. Xavier. 2013. AI-2-mediated signaling in bacteria. *FEMS Microbiol. Rev.* 37:156–181.

Percival, S.L., J.T. Walker, and P.R. Hunter. 2000. *Microbiological Aspects of Biofilms and Drinking Water.* CRC Press, Boca Raton, FL, 229 pp.

Rendueles, O., J.B. Kaplan, and J.-M. Ghigo. 2013. Antibiofilm polysaccharides. *Environ. Microbiol.* 15:334–346.

Rittmann, B.E. 2004. Biofilms in the water industry. In *Microbial Biofilms,* M. Ghannoum and G.A. O'Toole (Eds). ASM Press, Washington, DC, pp. 359–378.

Rittmann, B.E. and C.S. Laspidou. 2002. Biofilm detachment. In *Encyclopedia of Environmental Microbiology,* G. Bitton (editor-in-chief). Wiley-Interscience, New York, pp. 544–550.

Sauer, K., A.H. Rickard, and D.G. Davies. 2007. Biofilms and diversity. *Microbe* 2(7):347–353.

Shrout, J.D. and R. Nerenberg. 2012. Monitoring bacterial twitter: Does quorum sensing determine the behavior of water and wastewater treatment biofilms? *Environ. Sci. Technol.* 46:1995–2005.

Simpson, D.R. 2008. Biofilm processes in biologically active carbon water purification. *Water Res.* 42:2839–2848.

Stoodley, P., K. Sauer, D.G. Davies, and J.W. Costerton. 2002. Biofilms as complex differentiated communities. *Annu. Rev. Microbiol.* 56:187–209.

Walski, T.M. 2006. A history of water distribution. *Amer. Water Works Assoc. J.* 98(3):110–121.

5

ESTHETIC AND OTHER CONCERNS ASSOCIATED WITH DRINKING WATER TREATMENT AND DISTRIBUTION

5.1 INTRODUCTION

In Chapter 4, we focused on the formation of biofilms in water distribution systems and on the public health concerns mostly associated with bacterial pathogens and viruses. In this chapter, we focus on the impact of other microorganisms (fungi, actinomycetes, cyanobacteria, algae, protozoa) and invertebrates on the quality of treated water as it flows through the distribution system. Some of the concerns address the esthetic quality of drinking water (e.g., taste and odor problems caused by a wide range of microorganisms or color problems resulting from the activity of iron- and manganese-oxidizing bacteria). Others address the public health aspects of distributed water (e.g., toxicity problems due to cyanotoxins and the growth of nitrifying bacteria in distribution lines, or protozoa and invertebrates as reservoirs for pathogens and parasites).

5.2 TASTE AND ODOR PROBLEMS IN DRINKING WATER TREATMENT PLANTS

Taste and odors in drinking water are a major cause of consumers' complaints to water utilities. Several surveys have shown that water treatment plants, especially those using surface waters, have taste and odor problems. One such survey in the United

Microbiology of Drinking Water Production and Distribution, First Edition. Gabriel Bitton.
© 2014 John Wiley & Sons, Inc. Published 2014 by John Wiley & Sons, Inc.

States and Canada showed that 43% of the 800 plants surveyed had experienced these problems, with an average cost of 4.5% of the utilities total budget (Khiari and Watson, 2007; Suffet et al., 1996). This preoccupies consumers who often base their perception of safe water on objectionable tastes and odors in the drinking water. A 2001 survey in South Korea revealed that 30% of the consumers did not drink tap water because of offensive odors as compared to 37% fearing the presence of health-related microorganisms (Bae et al., 2007).

5.2.1 Sources

The sources of tastes and odors in water may be natural or anthropogenic.

1. Anthropogenic sources include phenol, chlorinated phenols (e.g., 2-chlorophenol, 2,4-dichlorophenol, 2,6-dichlorophenol), hydrocarbons, and halogenated compounds such as chloroform. Methyl tertiary butyl ether (MTBE), a gasoline additive, is also a source of odor problems in drinking water.

   Approximately half of all pipes used by the water industry are plastic pipes made predominantly of polyethylene (PE) or polyvinyl chloride (PVC) (Raynaud, 2004). These materials release odorous organic compounds that may affect the consumer's perception of drinking water quality (Heim and Dietrich, 2007; Rigal and Danjou, 1999; Skjevrak et al., 2003; Tomboulian et al., 2004). The extent of leaching of organic compounds (ketones, phenols, hydrocarbons), measured as TOC concentration, from high density PE (HDPE) is illustrated in Figure 5.1 (Heim and Dietrich, 2007). Both chlorinated PVC and HDPE plastic pipes release more TOC than glass pipes and the release is not affected by chlorination or chloramination. Moreover, HDPE pipes release more TOC than chlorinated PVC (cPVC) pipes. Although the cPVC pipes did not release significant odors, water from HDPE pipes had a weak to moderate waxy/plastic/citrus odor. No trihalomethanes (THMs) were detected in chlorinated water exposed to pipe materials (Heim and Dietrich, 2007). Levels of total VOCs migrating from HDPE pipes did not significantly change during three successive tests (Skjevrak et al., 2003). As for crossbonded PE pipes, the major VOC migrating into water was MTBE, and the threshold odor number (TON) was above 5. No significant odor was detected in water from PVC pipes (Skjevrak et al., 2003).

2. Among the natural sources, the major taste- and odor-causing compounds are geosmin and 2-methyl isoborneol (MIB) which are products of actinomycete metabolism (e.g., *Streptomyces* and *Nocardia* species), and cyanobacteria (e.g., *Oscillatoria, Anabaena, Lyngbia, Phormidium, Schizothrix*) (Izaguirre et al., 1982; Lalezary-Craig et al., 1988; Zaitlin and Watson, 2006). Geosmin, 2-MIB, sesquiterpenes, β-cyclocitral, 3-methyl-1-butanol, and others have been isolated from water supplies and implicated as the cause of earthy-musty odor of drinking water, leading to complaints from consumers who perceive the water as unsafe to drink. Geosmin and 2-MIB have very low threshold odor concentrations in the nanograms per liter (4 and 9 ng/L, respectively (Cees

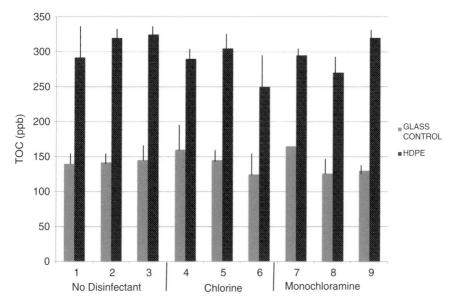

Figure 5.1   Average TOC concentrations leaching from HDPE in the absence of disinfectant and in the presence of 2 mg/l free Cl_2 or 4 mg/l monochloramine as Cl_2. *Source:* Heim, T.H., and A.M. Dietrich (2007). Water Res. 41: 757–764.

et al., 1974; Persson, 1979). The biosynthetic mechanisms for geosmin and MIB synthesis in cyanobacteria were recently elucidated (Ciglio et al., 2008, 2011). Other microbial metabolites are 2-isopropyl-3-methoxypyrazine and 2-isobutyl-3 methoxypyrazine (Lalezary-Craig et al., 1988). In addition to actinomycetes and cyanobacteria, more microorganisms should be investigated for their contribution to taste and odor problems in drinking water (Khiari and Watson, 2007).

5.2.2   Control Approaches

Several approaches are used to control taste- and odor-causing compounds in water treatment plants (Namkung and Rittmann, 1987):

- *Adsorption to solids*. It is the most popular treatment for removing these odorous compounds. Concentrations as low as 5 mg/L can successfully reduce geosmin and MIB to acceptable levels (Lalezary-Craig et al., 1988; Quinlivan et al., 2005). Adsorption depends on the type of carbon used, the nature and concentration of natural organic matter (NOM) which competes for the same sites on the activated carbon, powdered activated carbon (PAC) concentration, and contact time (Cook et al., 2001; Drikas et al., 2009; Ndiongue et al., 2006; Summers et al., 2013). The NOM interference can be reduced by the traditional coagulation–flocculation–filtration, enhanced coagulation, or by pretreatment with MIEX, a strong base anion exchange resin with a magnetic iron oxide core (Drikas et al., 2009, 2011; Ho et al., 2012b). Another issue is the reaction of chlorine with activated carbon, leading to the reduction of adsorption sites on the carbon surface. Among many carbon types tested, bituminous activated carbon was found to display the highest adsorption capacity toward MIB (Chen et al., 1997). Zeolites have also been suggested for taste and odor removal (Ellis and Korth, 1993).

• *Oxidation processes (chlorination, chloramines, ozonation, K permanganate, UV)*. Regular chlorination removes little geosmin and MIB. Chlorine dioxide is one of the most effective oxidant for removing taste- and odor-causing compounds but is not effective in eliminating odors caused by hydrocarbons (Walker et al., 1986). Ozone, at a concentration of 62.5 μmol/L, removed 95% of MIB (0.0744–0.2976 μmol/L) in 30 minutes. The MIB removal is due to the oxidizing action of ozone and hydroxyl radicals. Ozonation followed by biofiltration is effective in removing MIB and geosmin (Nerenberg et al., 2000). However, ozonation of MIB results in the formation of by-products, mainly *d*-camphor and aldehydes (e.g., formaldehyde, acetaldehyde) which are taste and odor compounds (Qi et al., 2009). Peroxone, a mixture of ozone and hydrogen peroxide, appears to be more effective for the oxidation of MIB and geosmin than ozone alone. The optimum $H_2O_2:O_3$ ratio for oxidation of taste and odor compounds is 0.1 to >0.3, depending on the water being treated (Ferguson et al., 1990).

UV light at 254 nm wavelength is not efficient in the destruction of geosmin and MIB. However, UV displays higher efficiency when combined with vacuum UV (VUV) at a wavelength of 185 nm. The higher removal observed with VUV is due to the generation of hydroxyl radicals. The combination of UV and H_2O_2, an advanced oxidation process, is also efficient for the removal of taste and odor compounds such as MIB, geosmin, and MTBE (Cater et al., 2000; Jo et al., 2011; Rosenfeldt et al., 2005).

• *Biodegradation*. Various laboratory bacterial cultures have been isolated, with a capacity to degrade taste and odor compounds (Ho et al., 2012a). For example, *Bacillus cereus* can degrade geosmin in water while MIB was found to be degraded by *Bacillus fusiformis* and *Bacillus sphaericus*, as shown by microscopy and 16S rRNA probes (Lauderdale et al., 2004). In drinking water biofilters, many bacterial isolates (e.g., *Flavobacterium* spp., *B. cereus*, *Enterobacter* spp., *Pseudomonas* spp.) were shown to biodegrade geosmin and MIB (Elhadi et al., 2006). A novel gram-negative bacterium isolated from the biofilm of a sand filter was able to degrade geosmin at a rate of 0.010 to 0.029 per hour. This bacterium belonged to the *Sphingomonadaceae* and was 97.2% similar to *Sphingopyxis alaskensis* (Hoefel et al., 2009). Seeding of sand filters with a bacterial consortium (*Sphingopyxis* sp., *Novosphingobium* sp., and *Pseudomonas* sp.) enhanced geosmin biodegradation and removal (McDowall et al., 2009). *Rhodococcus*, *Variovorax*, and *Comamonas*, isolated from the digestion basin of an aquaculture unit, were found to use MIB and geosmin as a sole source of carbon and energy (Guttman and van Rijn, 2011). The taste- and odor-causing compounds can also be used co-metabolically by biofilms grown on fulvic acid which serves as a primary substrate (Namkung and Rittmann, 1987).

5.3 ALGAE AND CYANOBACTERIA

Eutrophication, an enhanced biological production in surface waters, is due to excessive concentrations of nitrogen and phosphorus from wastewater treatment plants discharges and fertilizers, and generally results in algal and cyanobacterial blooms with dire consequences on water quality (e.g., fish kill due to drop in dissolved oxygen) and potential health implications.

5.3.1 Impact on Water Treatment Plants

Algae and cyanobacteria may cause the following problems in water treatment plants:

- Clogging of sand filters, particularly during the warm season.
- Taste and odor problems: We have seen that cyanobacteria (e.g., *Oscillatoria, Anabaena, Microcystis*) can also cause taste and odor problems in drinking water. For example, the chrysophite, *Synura petersenii* is responsible for the cucumber odor of some drinking waters. This odor is caused by a chemical identified as 2,6-nonadienal (Hayes et al., 1989). Another diatom species, *Sinura uvella* in parts a cod-liver oil odor to drinking water.
- Release of algal-derived organic compounds (e.g., hydrophilic and hydrophobic proteins) may also result in taste and odor compounds (e.g., 2-methylisoborneol) and increase in disinfection by-products formation upon chlorination of drinking water (Li et al., 2012; Lui et al., 2011). Several cyanobacteria (*Anabaena, Anacystis*), green algae (*Chlamydomonas, Selenastrum, Scenedesmus*), and diatoms (*Navicula*) were implicated in chloroform production. Organic matter (both extracellular and intracellular organic matter) from *Microcystis aeruginosa*, following chlorination, can potentially lead to the formation of chloroform, chloroacetic acid, and nitrosodimethylamine (Li et al., 2012). Ozonation of *Anabaena flos-aquae* and *M. aeruginosa* cells also led to the formation of DBPs (THM and haloacetic acid (HAA)) as a result of release of organic carbon from the cells. The DBP concentration increased with the pH of the water and the ozone concentration (Coral et al., 2013).
- Cyanobacteria are hardy microorganisms which can adapt to life in extreme environments, giving them a competitive advantage over eukaryotic algae. Their proliferation in water has a negative impact on water quality due to the production of secondary metabolites, namely the cyanotoxins (Chorus and Bartram, 1999).

5.3.2 Cyanotoxins

The proliferation of cyanobacteria (e.g., *M. aeruginosa, Anabaena flos-aquae, Oscillatoria, Schizothrix, Lyngbya, Phormidium, Cylindrospermopsis, Nostoc, Nodularia*) in surface waters result in the production of allergenic, hepatoxic, neurotoxic, and possibly tumor-promoting toxins (Carmichael, 1989; Falconer and Buckley, 1989; Chorus and Bartram, 1999). Based on their chemical structure, these toxins are divided into cyclic peptides, alkaloids, and lipopolysaccharides (Sivonen and Borner, 2008; Chorus and Bartram, 1999). There are four categories of cyanotoxins: hepatoxins (e.g., microcystin, nodularin), neurotoxins (e.g., anatoxin-a, saxitoxin), cytotoxins (e.g., cylindrospermopsin), and dermatotoxins (e.g., lyngbyatoxin A, aplysiatoxin). The toxic effects of cyanotoxins are summarized in Table 5.1 (Dittmann et al., 2013).

The toxicity of cyanobacteria to humans and animals was recorded for decades around the world. Cyanotoxins that may contaminate drinking water cause livestock poisoning and human health effects. The two cyanotoxins of concern are the microcystins and the cylindrospermopsins (Falconer, 2005; Falconer and Humpage, 2006; Ho et al., 2011a). Microcystins are cyanobacterial hepatoxins that have been isolated from freshwater environments and drinking water supplies, and cause liver tumors in humans and animals. Among over 80 types of microcystins, microcystin-LR (MC-LR) is the most common as well as the most toxic cyanotoxin with an $LD_{50} = 50 \, \mu g/L$ for mice. The mechanism of toxicity is the inhibition of two protein phosphatases. MC-LR is produced by cyanobacteria such as *Microcystis*, *Anabaena*, *Nostoc*, and *Planktothrix*. The use of microcystin-contaminated water in hemodialysis treatment systems has led to the death of 88 patients in Brazil, due to hepatic failure. Moreover, high liver cancer rates have been correlated with high toxin concentrations in drinking water sources in China. There is also evidence that exposure to *Microcystis* may induce vitellogenin production in zebrafish, pointing to an estrogenic effect of this cyanobacterium. There is concern that the estrogenic effect may also occur in birds and mammals (Rogers et al., 2011). The estrogenic effect was observed in cultured human cells (breast carcinoma cell line) exposed to the microcystin MC-LR (Oziol and Bouaïcha, 2010).

Cylindrospermopsin is a tricyclic alkaloid produced by cyanobacteria such as *Cylindrospermopsis*, *Anabaena*, and *Aphanizomenon*. It is hepatoxic, cytotoxic, genotoxic, and possibly carcinogenic. This toxin is released in the water following cell death. Cylindrospermopsin is released by at least five cyanobacteria in drinking water reservoirs and reach concentrations of $1–100 \, \mu g/L$, and a tentative guideline of $1 \, \mu g/L$ has been recommended for this biotoxin. Using solid-phase extraction followed by liquid chromatography–mass spectrometry, cylindrospermopsin was detected in Taiwan tap water at a concentration of $8.6 \, \mu g/L$ (Yen et al., 2011).

Anatoxin is another cyanotoxin produced by *Anabaena*, *Planktothrix*, *Microcystis*, *Nostoc*, *Phormidium*, *Oscillatoria*, and *Aphanizomenon* (Osswald et al., 2007). This alkaloid is neurotoxic and causes cramps followed by paralysis and may cause death if respiratory muscles are affected. Some of the long-term chronic effects of these by-products on human health are not well known but may include damage to the liver, kidneys, and other organs, and may lead to genotoxicity and carcinogenicity.

5.3.3 Water Treatment Options for Algae, Cyanobacteria, and Cyanotoxins

Several treatments have been suggested for removing algal and cyanobacterial cells and/or their toxins from water. These treatments include coagulation–flocculation, sand filtration, adsorption to activated carbon (GAC or PAC), treatment with oxidizing

Cyanobacterial Toxin	Producing Genera/Species	Toxic Mechanism	Biosynthetic Genes	GenBank Accession Numbers
HEPATOTOXINS				
Microcystin	*Microcystis* sp. *Planktothrix* sp. *Anabaena* sp. *Nostoc* sp. *Hapalosiphon* sp. *Phormidium* sp.	Hepatotoxic; inhibition of eukaryotic protein phosphatases of type 1 and 2A	*mcyA-J*	AF183408 AJ441056 AJ536156
Nodularin	*Nodularia spumigena*	Hepatotoxic; inhibition of eukaryotic protein phosphatases	*ndaA-I*	AY210783
CYTOTOXINS				
Cylindrospermopsin	*Cylindrospermopsis raciborskii* *Aphanizomenon ovalisporum* *Umezakia natans* *Raphidiopsis curvata* *Anabaena* sp. *Oscillatoria* sp.	Hepatotoxic, cytotoxic, neurotoxic; inhibition of glutathione Synthesis, protein synthesis and cytochrome P_{450}	*cyrA-O* *aoaA-C*	EU140798 AF395828 FJ418586
NEUROTOXINS				
Anatoxin-a Homoanatoxin-a	*Aphanizomenon flos-aquae* *Oscillatoria* sp. *Aphanizomenon* sp.	Neurotoxic, mimics the neurotransmitter acetylcholine	*anaA-H*	FJ477836 JF803645
Saxitoxin	*Anabaena circinalis* *Aphanizomenon* sp. *Aphanizomenon gracile* *Cylindrospermopsis raciborskii* *Lyngbya wollei*	Neurotoxic, blocks voltage-gated Na+ channels	*sxtA-Z*	DQ787200
DERMATOTOXINS				
Lyngbyatoxin	*Lyngbya majuscula* (*Moorea producens*)	Tumor promoting, binds to protein kinase C (PKC)	*ltxA-D*	AY588942
Aplysiatoxin	*L. majuscula* (*M. producens*)	Tumor promoting, binds to protein kinase C (PKC)	*Unknown*	–

Source: Adapted from Dittmann et al. (2013). FEMS Microbiol. Rev. 37: 23–43.

agents such as chlorine, chlorine dioxide, chloramine, permanganate or ozone, and UV-based advanced oxidation processes.

Water utilities can opt for another source water if available or consider the treatment of source water at the intake with oxidants, such as potassium permanganate ($KMnO_4$), which does not lyse the cells (Cheng et al., 2009). Other treatment options are bank filtration, microfiltration, and ultrafiltration (Gijsbertsen-Abrahamse et al., 2006) or lime precipitation (Lam et al., 1995). In the plant, algal and cyanobacterial cells are effectively removed by coagulation/flocculation using Al- and Fe-based compounds. However, this process does not remove well cyanobacterial toxins. Due to the low density of these flocs, dissolved air flotation (DAF) is effective in removing the cells, particularly cyanobacteria which have gas vacuoles which serve as flotation devices. Filtration through coarse filters does not effectively remove algal and cyanobacterial cells (Westrick et al., 2010) while fine filters are prone to clogging. The treatment efficiency depends on cyanobacterial type. Although *Microcystis*, *Anabaena*, and *Pseudanabaena* were removed by coagulation and filtration processes, *Aphanizomenon* was poorly removed by these processes (Zamyadi et al., 2013).

Under laboratory conditions, depending on the frequency, ultrasound energy can be used to rupture and inactivate cyanobacterial cells. Under low frequency (20 kHz), *M. aeruginosa* cells were ruptured via acoustic cavitation whereas under higher frequency (580 kHz) the cells and chlorophyll function were inactivated via production of free radicals (Wu et al., 2012).

Cyanobacterial cells naturally release cyanotoxins following senescence, death, and lysis. This release can also be induced by mechanical stress (e.g., addition of coagulants) or chemical stress (e.g., addition of oxidizing agents or copper). As the monitoring of cyanotoxins is time-consuming and costly, some have suggested the use of fluorescent cyanopigments, phycocyanin (blue fluorescence), and phycoerythrin (red fluorescence) as indicators of the removal of cyanotoxins by water treatment processes (e.g., coagulation, filtration, disinfection). However, this indicator function is lost when chlorine dioxide is used as a disinfectant (Schmidt et al., 2009).

Cyanotoxins are generally not well removed by conventional coagulation/ flocculation–filtration–chlorination treatment train, although the addition of powdered or granular activated carbon (GAC) or ozonation to the treatment sequence increases their removal from water (Himberg et al., 1989; Yoo et al., 1995). However, other studies have shown that 1–3 log removal of microcystins was obtained following conventional water treatment, and in all drinking water samples, the toxin concentration was below the WHO microcystin guideline of 1 μg/L (Drikas et al., 2001). Graphene oxide was found to be a better adsorbent toward microcystin-LR (MC-LR) and microcystin-RR (MC-RR) than activated carbon. For example, the adsorption capacity of graphene oxide toward MC-RR was 1878 μg/g, as compared to 1034 μg/g for activated carbon (Pavagadhi et al., 2013).

Chemical oxidants (chlorine, chlorine dioxide, chloramines, ozone and potassium permanganate) react with cyanotoxins, leading to their removal from drinking water. Chlorine helps release and subsequently oxidize the free and cell-bound toxins (Zamyadi et al., 2012) but may, at high concentrations, lead to the formation of THMs and HAAs. These disinfection by-products (DBPs) are produced through the interaction of chlorine with cyanobacterial cells (e.g., *Anabaena*, *Microcystis*) and released extracellular organic matter. For example, the yield of DBP formation by *Anabaena* ranges from 2 to 11 μmol/mmol C for total THMs, and between 2 and 17 μmol/mmol C for total HAAs (Huang et al., 2009). Chloramines and chlorine dioxide are not effective in cyanotoxin inactivation (Westrick et al., 2010). Ozonation is quite effective in removing dissolved microcystins (Rapala et al., 2006). Using water from a eutrophic lake (Lake Greifensee, Switzerland), the effects of three oxidants (ozone, chlorine, K permanganate) on three cyanotoxins (microcystin LR, cylindrospermopsin, and anatoxin) were compared. Ozone was the most efficient in eliminating the three toxins while K permanganate can effectively reduce anatoxin, a neurotoxin (Figure 5.2; Rodriguez et al., 2007a). Using a protein phosphatase inhibition assay, Rodriguez et al. (2008) showed that the oxidation products of MC-LR and MC-RR with chlorine and K permanganate were nontoxic. Advanced oxidation processes affect cyanotoxins via production of hydroxyl radicals. The reaction with hydroxyl radicals proceeds according to the following order: microcystins > cylindrospermopsin > anatoxin-a (Onstad et al., 2007).

Several studies have been carried out on the biodegradation of microcystins (see review of Ho et al., 2012a). The removal of microcystins following passage through biologically active filters (activated carbon and sand filters) is partly due to adsorption as well as biodegradation. Activated carbon is generally used in water treatment to remove taste and odor compounds, NOM, and synthetic compounds. Adsorption of microcystins to activated carbon is controlled by various factors such as the type of activated carbon, the size and conformation of microcystin molecule, and the presence of NOM. Figure 5.3 (Ho et al., 2011) illustrates the extent of removal of total microcystins and cylindrospermopsin by two types of PAC in two water supplies in Australia.

Biodegradation of microcystins is carried out by bacteria such as *Sphingomonas* spp. and *Sphingopyxis* (Ho et al., 2006, 2007). A bacterial strain (*Sphingosinicella* sp.) isolated from a lake was able to biodegrade microcystin LR and nodularin in 2 hours under laboratory conditions. Strains of *Arthrobacter* sp., *Rhodococcus* sp., and *Brevibacterium* sp., isolated from surface water, were found to biodegrade from 96% to 100% of MC-LR in 3 days (Manage et al., 2009). A wide range of microcystins and nodularin were biodegraded by actinobacteria such as *Brevibacterium* sp., *Rhodococcus* sp., *Arthrobacter* sp., and *Paucibacter toxinivorans* (Lawton et al., 2011). Similarly, *Sphingomonas* sp. (strain MJ-PV) was used, under laboratory conditions, to bioremediate water and slow sand filters contaminated with microcystin-LR (Bourne et al., 2006). Microorganisms implicated in the biodegradation of nodularin include *Arthrobacter* sp., *Brevibacterium* sp., *P. toxinivorans*, *Rhodococcus* sp., and *Sphingomonas* sp. Less is known about the biodegradation of cylindrospermopsin and anatoxin (Ho et al., 2012a). The large-scale application of this approach to bioremediate water supplies awaits further investigations.

Photodegradation of cyanotoxins is possible only at UV doses several orders of magnitude higher than the doses required for pathogen and protozoan parasites inactivation (up to 40 mJ/cm^2). Therefore, UV was not recommended for cyanotoxin removal (Westrick et al., 2010). Advanced oxidation processes produce hydroxyl

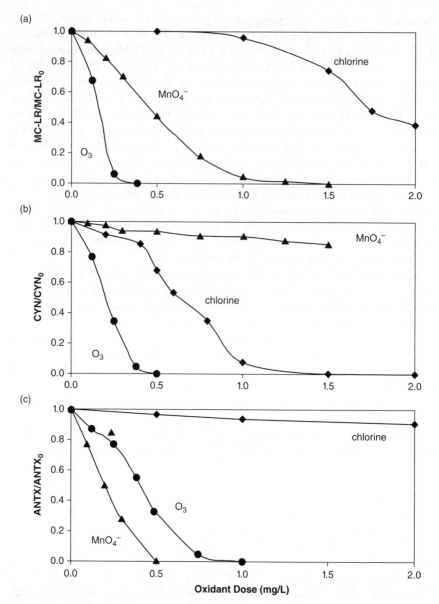

Figure 5.2 Effect of oxidants on cyanotoxins elimination. *Source:* Rodriguez et al. (2007a). Water Res. 41: 3381–3393

radicals that are effective in the destruction of a wide range of trace organic contaminants such as pharmaceuticals, pesticides, taste and odor compounds, or cyanotoxins. For example, the combination of low pressure UV (dose of 200 mJ/cm^2) and H$_2$O$_2$ (30 mg/L) helped destroy 70% of anatoxin-a. Hydroxyl radicals are responsible for anatoxin oxidation (Afzal et al., 2010). A 94% destruction of MC-LR (1 μM

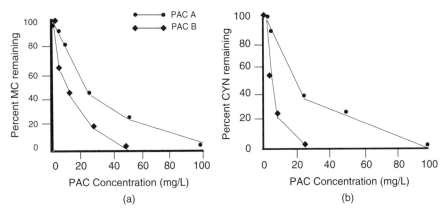

Figure 5.3 Comparison in the removal of: (a) total microcystins (MC); and (b) cylindrospermopsin (CYN) in Warragamba water by PAC-A and PAC-B after a contact time of 30 min. Adapted from Ho et al. (2011). Water Res. 45: 2954–2964.

concentration) was achieved at a UV fluence of 80 mJ/cm^2 in combination with 882 μM of H$_2$O$_2$. The presence of both alkalinity and NOM led to a decrease in the efficiency of the UV-C/H$_2$O$_2$ process (He et al., 2012).

Other advanced oxidation technologies, using TiO$_2$, have also been used for the detoxifixation of cyanobacterial toxins, using UV or solar radiation as the energy source. A more sustainable solution is the use of nitrogen-doped TiO$_2$ and solar radiation for the photocatalytic degradation of microcystin-LR (Choi et al., 2007).

Membrane filtration was also considered for the removal of algal and cyanobacterial cells and their toxins. Microfiltration and ultrafiltration are adequate for cell removal but not cyanotoxins. However, nanofiltration was found to be quite effective for the removal of anatoxin and microcystins regardless of variation in water quality (Ribau Teixeira and Rosa, 2005, 2006). Over 96% of microcystins and anatoxin was removed by nanofiltration (Gijsbertsen-Abrahamse et al., 2006). The average retention of microcystin-LR and microcystin-RR by reverse osmosis was 96.799.6% (Neumann and Weckesser (1998).

5.4 FUNGI

Fungi have been isolated from drinking water distribution systems around the world (Arvanitidou et al., 1999; Doggett, 2000; Göttlich et al., 2002; Hageskal et al., 2006, 2007, 2009; Hinzelin and Block, 1985; Rosenszweig and Pipes, 1989; van der Wielen and van der Kooij, 2013) and are part of the biofilms forming on pipe surfaces. A few methods are available (e.g., staining with Calcofluor White, fluorescent *in situ* hybridization) for their *in situ* detection in biofilms (Siqueira et al., 2013). Filamentous fungi occur in drinking water at levels up to approximately 10^2 CFU/100 mL (Olson and Nagy, 1984). Genera such as *Penicillium, Verticillium, Fusarium, Alternaria, Trichoderma, Mucor, Cephalosporium, Cladosporium, Aspergillus, Aureobasidium,*

Phoma, Rhizoctonia, Stachybotrys, Cladosporium, Mucor, Epicoccum, Phialophora, candida, and *Rhodotorula* are encountered in drinking water and biofilms because of their ability to survive chlorination and to grow on the surface of drinking water reservoirs and water distribution pipes (Doggett, 2000; Nagy and Olson, 1982; Rosenszweig et al., 1983). Fungi were found in biofilms of water distribution pipes at concentrations varying from 4.0 to 25.2 CFU/cm^2, with *Aspergillus* and *Penicillium* being the dominant fungi. Yeasts were found at lower concentrations (Doggett, 2000).

Monitoring of mesophilic fungi in tap waters from several municipalities in Finland showed that they occur at concentrations up to approximately 100 per liter of tap water (Niemi et al., 1982). Approximately 50% of potable water samples taken from small municipal water distribution systems contain fungi at levels of 1–6 fungal propagules/50 mL. The predominant genera found were *Aspergillus, Alternaria, Cladosporium,* and *Penicillium* (Rosenszweig et al., 1986). A Norwegian survey of fungi in drinking water revealed that *Penicillium, Aspergillus,* and *Trichoderma* were the predominant genera found in treated drinking water as well as in tap and shower water (Hageskal et al., 2006). High concentrations of *Fusarium* (up to 10^5 CFU/L) were detected in a hospital water distribution system in Dijon, France. The predominant species detected were *F. oxysporum* and *F. dimerum* (Sautour et al., 2012). Using quantitative PCR, fungi were detected in all unchlorinated distributed water samples in eight water treatment plants in the Netherlands (Figure 5.4). The 18S RNA gene copies were higher in the summer than in the winter season and were also generally

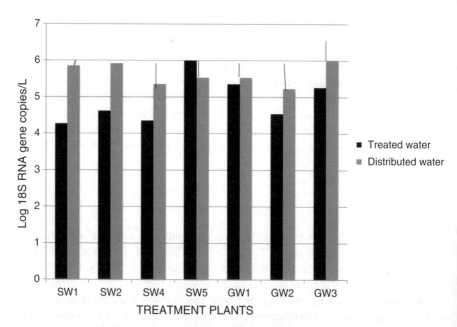

Figure 5.4   Geometric means of the 18S rRNA gene copies of fungi in drinking water from the distribution systems of seven water treatment plants sampled. Adapted from van der Wielen, and D. van der Kooij (2013). Appl. Environ. Microbiol. 79: 825–834.

higher in the distributed water than in treated water (van der Wilen and van der Kooij, 2013).

Water treatment processes can remove up to 2 logs of fungal numbers. However, these microorganisms appear to be more resistant to chlorine and ozone than bacteria or viruses (Haufele and Sprockhoff, 1973; Pereira et al., 2013; Rosenszweig et al., 1983).

Fungi may cause the following problems in water distribution systems:

- They exert a chlorine demand and may protect bacterial pathogens from inactivation by chlorine (Rosenszweig and Pipes, 1989; Rosenszweig et al., 1983).
- They colonize sand filter and may affect water treatment.
- They may degrade gaskets and some of the jointing compounds used in distribution systems.
- Some of them form humic-like substances that may act as precursors of THMs (Day and Felbeck, 1974).
- Some fungal species may cause taste and odor problems. A survey of 450 waterworks showed that 10% of the plants had taste and odor problems caused by fungi or actinomycetes (Miettinen et al., 2007).
- Fungi are generally transmitted via the airborne route but drinking water may also be a transmission route. Some fungi may be pathogenic (e.g., *A. flavus, A. fumigatus, Fusarium*) and some of them may cause respiratory illnesses (e.g., hypersensitivity pneumonitis, pulmonary hemorrhage) and allergic reactions (e.g., allergic asthma, hay fever) in sensitive persons, particularly in hospital settings where 9% of the nosocomial infections are caused by fungal pathogens which affect mostly immunocompromised patients. *Fusarium* species (e.g., *F. solani, F. oxysporum*) are the second most frequent cause of fungal infections (Rosenszweig et al., 1986). Although the public health significance of their presence in drinking water is little known, it is suspected that they may be of concern following mold aerosolization during showering (Anaissie et al., 2001, 2002).

Advances in the study of fungi in drinking water will benefit from the development of standard methods for enumeration and identification of these water contaminants (Hageskal et al., 2009).

5.5 ACTINOMYCETES

Actinomycetes are gram-positive filamentous bacteria characterized by mycelial growth with branching filaments with a diameter similar in size to bacteria. Most of them are strict aerobes and can be grown on starch media. They produce clinically important antibiotics such as streptomycin, erythromycin, tetracycline, and chloramphenicol. Soil runoff is the most likely source of actinomycetes in water. Actinomycetes are found in drinking water following water treatment (Niemi et al.,

1982), and the most widely found actinomycetes in water distribution systems belong to the genus *Streptomyces*. Other genera found are *Nocardia* and *Micromonospora*. These organisms are found at levels up to 10^3 CFU/100 mL. Actinomycetes were the most frequently detected microorganisms in chlorinated distribution water from a coastal community in Oregon (LeChevallier et al., 1980). Actinomycetes are important because, as mentioned in Section 5.2, they produce volatile metabolites such as geosmin and 2- MIB which are the sources of taste and odors in drinking water.

Chemical coagulation followed by slow sand filtration and disinfection remove actinomycetes from water. However, routine chlorination or chloramination have little effect on actinomycetes and they are not very effective in taste and odor reduction (Jensen et al., 1994; Sykes and Skinner, 1973).

5.6 PROTOZOA

Protozoa are eukaryotic unicellular organisms which play an important role in wastewater and drinking water treatment plants. They form cysts or oocysts that resist adverse environmental conditions and allow their transmission in the environment. These structures confer resistance to desiccation, starvation, high temperatures, lack of oxygen, and chemical insult, specifically disinfection in drinking water and wastewater treatment plants. Their classification has been traditionally based on means of locomotion. Hence, they include the amoebas (movement via pseudopods), flagellates (movement via flagella), ciliates (movement via cilia), and sporozoa with no means of locomotion (Bitton, 2011).

5.6.1 Occurrence of Protozoa in Water Treatment Plants

Protozoa are found in treated drinking water (e.g., *Hartmannella*) (Otterholt and Charnock, 2011), and some are part of the microflora of biofilms developing on the surface of water reservoirs (e.g., *Bodo*, *Vorticella*, *Euplotes*). Although much effort has been focused on parasitic protozoa (e.g., *Giardia lamblia*, *Cryptosporidium*, *Entamoeba histolytica*, *Acanthamoeba*, *Balantidium coli*, *Naegleria fowleri*), work is being carried out about the ecology of these microorganisms in reservoirs, water treatment plants, and distribution networks. Molecular techniques (e.g., terminal restriction fragment length polymorphism, analysis of 18S rRNA fragments) have shown highly diverse protozoan communities in distribution networks (Loret and Greub, 2010; Valster et al., 2009).

Protozoa were found at average concentrations of 10^5 cells/L in distribution systems and 10^3 cells/cm^2 in biofilms. Flagellates are predominant in water whereas biofilms harbor mostly ciliates and amoebae. Free-living amoebas (FLAs) enter water distribution systems where they are isolated with a frequency greater than 10%. This environment allows their persistence and replication. FLA also persist and replicate in treated drinking water reservoirs and are isolated at a higher frequency than in distribution systems (Thomas and Ashbolt, 2011).

5.6.2 FLAs Grazing Role in Water Treatment Plants

Protozoa graze on biofilm bacteria but the grazing rates appear to be lower than those found in other aquatic environments (Sibille et al., 1998). In biofilms developing in drinking water flow chambers, grazing by amoebae reduced the bacterial population by 70%. The consumption rate by amoeba was estimated at 1.2×10^4 to 2.3×10^4 bacterial cells/cm^2 h. Since amoebae preferably graze on isolated cells, the biofilm structure changes from single cells to aggregated cells and microcolonies (Paris et al., 2007). Some may reach bacterial microcolonies deep in the biofilm by swimming through the water channels in the biofilms. Some of the predatory amoebas are human parasites (e.g., *Acanthamoeba, Hartmannella*) (Keevil, 2002). A study of the distribution of FLA in a water treatment plant in France revealed the presence in the plant of 31 FLA isolates, four of which potentially harboring human pathogens. Amoeba colonization was detected mostly in the biofilms developing on sand and GAC filters (Thomas et al., 2008). FLA were also detected in 27% of samples taken from three water treatment plants in Spain. The amoebas were mainly *Naegleria* and *Acanthamoeba* species and *Hartmannella vermiformis* (Corsaro et al., 2010). A molecular identification of FLA in reservoirs and water treatment plants in Spain showed the presence of *Acanthamoeba* spp., *H. vermiformis, Naegleria* spp., and *Vanella* spp. (Garcia et al., 2013). Biofilm formation in waterlines in dental units also favors the FLA proliferation. The presence of *Acanthamoeba* spp. and *Naegleria* in water from dental units was detected in 24% and 3% of the samples, respectively. *Hartmannella vermiformis* was found to internalize *Candida* species, allowing the proliferation of the yeast in tap water from dental units. This may pose a health risk to patients and dentists following aerosolization (Barbot et al., 2012; Leduc et al., 2012).

Protozoa (e.g., ciliates, amoebas) also graze on *Cryptosporidium* oocysts which gain protection from environmental stresses and, probably, from chlorination (Stott et al., 2003).

5.6.3 FLAs as Reservoirs of Pathogens

Amoebas can serve as reservoirs for viruses and amoeba-resistant bacteria (ARB) some of which are human pathogens (e.g., *Legionella, Mycobacterium* spp., *Pseudomonas* spp., *Listeria monocytogenes, Francisella, Helicobacter pylori, Escherichia coli* 0157) which are protected from environmental adverse conditions and from disinfection in drinking water treatment systems (Bichai et al., 2008; Garcia et al., 2013; LeChevallier et al., 2011; Le Dantec et al., 2002a, 2002b; Loret and Greub, 2010; Thomas et al., 2006, 2008, 2010; Thomas and Ashbolt, 2011). Sixteen percent of FLA isolated from various stages of a water treatment plant supplied with river water were infected by ARB (Thomas et al., 2008). In Spanish reservoirs and water treatment plants, *L. pneumophila, Pseudomonas* spp., and *Mycobacterium* spp. were detected in 13.9%, 32.6%, and 41.9%, respectively, inside FLA (Garcia, et al., 2013). These ARMs are able to survive and sometimes multiply inside the amoebae and are subsequently released in the water following amoeba lysis. These released

TABLE 5.2   Diversity of environmental FLA naturally infected with pathogenic ARM isolated from treated drinking water systems

FLA Genus	ARM	Drinking Water Sampling Location
Environmental FLA	*Flavobacterium indologenes* *Pseudomonas stutzeri* *Stenotrophomonas maltophilia*	Water treatment plant or distribution system
Acanthamoeba sp.	*Pseudomonas aeruginosa* *Legionella* sp. *Neochlamydia* sp.	In-premise (tap) Distribution system Reservoir sediment
Echinamoeba	*Mycobacterium mucogenicum* *Legionella londiniensis* *Legionella donaldsonii*	Water treatment plant
Hartmannella (*H. vermiformis*)	*Neochlamydia* sp. *Legionella pneumophila*	Water treatment plant In-premise (tap)
Naegleria	*Flavobacterium johnsoniae* *Legionella micadei* *Legionella anisa*	Water treatment plant
Vannella	*Microsporidia*-like organism	In-premise (tap)

Source: Adapted from Thomas and Ashbolt (2011). Environ. Sci. Technol. 45: 860–869.
FLA, free-living amoeba; ARM, amoeba-resisting microorganisms.

pathogens are of concern because their virulence may be enhanced following their passage through amoebas (Cirillo et al., 1997). It was demonstrated that two strains of *Mycobacterium avium* survived in several FLA species (*Naegleria clarki, H. vermiformis, Acanthamoeba castellanii, A. polyphaga, A. lenticulata*) but high replication rates were shown only for *A. lenticulata* (Ovrutsky et al., 2013). A list of microbial pathogens harbored by FLAs is shown in Table 5.2 (Thomas and Ashbolt, 2011).

To address protozoan cyst and oocysts removal from water, amendments of the Safe Drinking Water Act (PL 93–523) mandates the U.S. EPA to require filtration for all surface water supplies.

5.7 INVERTEBRATES

5.7.1 Occurrence of Invertebrates in Water Distribution Systems

This group includes nematodes, rotifers, oligochetes (annelids), amphipods, crustaceans, isopods, flatworms, water mites, and insect larvae (e.g., chiromonid larvae) (van Lieverloo et al., 2002, 2004). A study of the occurrence of invertebrates in water distribution systems in Denmark showed that they were present in 94% of the samples. Some invertebrates (e.g., the water lice *Asellus aquaticus*) were present in higher numbers in iron pipes than in plastic pipes (Christensen et al., 2011). A survey of 34 plants in the Netherlands showed that invertebrates found in drinking water were mostly rotifers and nematodes whereas water flushed from mains contained the larger invertebrates such as Cladocera, Copepoda, Oligochaeta, and Asellidae (van Lieverloo et al., 2012).

5.7.2 Invertebrates as Reservoir of Pathogens and Parasites

A survey of nematodes in distribution pipes showed the presence of 12–17 nematodes per liter. The nematodes were shown to harbor bacteria (0–643 CFU per nematode) (Locas et al., 2007). Their presence in water treatment plants and in distribution lines triggers complaints from consumers. There are several reports of bacterial ingestion and excretion as viable cells by nematodes, leading to their consideration as potential vectors of pathogenic microorganisms (see review by Bichai et al., 2008). They may also be of public health significance because these invertebrates protect pathogens from disinfectant action and environmental insult (Caldwell et al., 2003; Tang et al., 2011). Nematodes such as *Caenorhabditis elegans* protected *E. coli* and *B. subtilis* from UV-A and UV-C at a fluence of 40 mJ/cm^2 (Bichai et al., 2009, 2011). It was found that about 24% of the UVA fluence (5.6 J/m^2) reached *E. coli* cells inside the nematode (Bichai et al., 2011). *Caenorhabditis elegans* can also serve as a vector for *Cryptosporidium* oocysts. The intact and viable oocysts can be released in the gut following ingestion by the nematode (Huamanchay et al., 2004). The use of high concentrations of prey in the ingestion experiments has been criticized because it does not represent the typical situation in a water treatment plant where the prey load is not as high. The risk to public health from ingested bacterial pathogens with high infectious doses is lower than that posed by ingested viruses or protozoan parasites with low infectious doses (Bichai et al., 2008). However, some investigators question whether the protective effect of invertebrates toward microbial pathogens has a significant impact on drinking water safety (van Lieverloo et al., 2012).

In controlled laboratory experiments, the isopod, *A. aquaticus*, did not significantly affect the survival of *E. coli*, *Klebsiella pneumoniae*, and *Campylobacter jejuni* but the growth rate of heterotrophic plate count bacteria increased in the presence of dead *A. aquaticus* (Christensen et al., 2012). As regard the grazing of pathogens and protozoan parasites by rotifers and crustaceans, their survival inside these microinvertebrates, and protection from disinfectants, most of the studies have dealt with natural waters but less is known about water distribution systems.

5.7.3 Control of Microinvertebrates

Invertebrates can be controlled in water distribution lines by chemical (copper sulfate, chlorine) and physical methods (air scouring, pipe and hydrant flushing, filtration of source waters) or a combination of both approaches. Invertebrate growth in distribution line sediments can be prevented by reducing the food supply. This can be achieved by controlling the formation of biofilms which, through sloughing, are the main contributors to sediment accumulation inside the pipes (Gray, 1994; Levy, 1990; Levy et al., 1986; van Lieverloo et al., 2002).

5.8 ENDOTOXINS

These heat-stable lipopolysaccharides are structural components of the outer membrane of gram-negative bacteria and some cyanobacteria, and may be detected in surface and ground waters and in drinking water produced from reclaimed wastewater (Burger et al., 1989). High concentrations of endotoxin-producing cyanobacteria (e.g., *Schizothrix calcicola*) have been found in uncovered reservoirs containing finished water (Sykora et al., 1980). Raw (untreated) water may contain endotoxin concentrations ranging from 1 to 1049 ng/mL (approximately 10–10,000 endotoxin units (EU)/mL) (Anderson et al., 2002; Rapala et al., 2002, 2006).

Traditional water treatment processes (e.g., coagulation, flocculation, sedimentation, and sand filtration) may remove from 78% to 98% of endotoxins (Rapala et al., 2002, 2006). Endotoxins can also be destroyed by oxidants such as chlorine, monochloramine, or potassium permanganate (Anderson et al., 2003). Water distribution systems may contain from 2 to 32 EU/mL (Korsholm and Søgaard, 1988; Rapala et al., 2002). Although endotoxins have been associated with human health problems (fever, diarrhea, vomiting, and allergies), the full implications of their presence in drinking water are still unclear.

5.9 IRON, MANGANESE, AND SULFUR BACTERIA

Iron and manganese bacteria are autotrophic microorganisms that derive energy from the oxidation of inorganic compounds and use CO_2 as a carbon source. They attach to and grow on pipe surfaces and subsequently cause a deterioration of water quality (e.g., pipe clogging and color problems leading to laundry and sink staining), triggering consumer complaints. They form red (iron oxidizing bacteria) or black (manganese oxidizing bacteria) deposits on pipe surfaces in the water distribution system. In the case of a ground water source, it is sometimes necessary to install an iron-removing filter to solve this problem. Bacterial genera implicated in biological iron oxidation include *Crenothrix, Leptothrix, Toxothrix, Sphaerotilus, Gallionella,* and nonspore forming *Acidithiobacillius* (e.g., *Acidithiobacillus ferrooxidans*). Manganese oxidation is carried out by microorganisms such as *Metallogenium, Leptothrix, Hyphomicrobium,* or *Rhodococcus.* There is a competition between chemical and biological oxidation of iron and manganese. Biological oxidation is generally favored by low pH and low oxygen level. However, it was recently reported that *Gallionella* can thrive at slightly alkaline pH and full aeration (de Vet et al., 2011).

Sulfur bacteria play a major role in biocorrosion or microbially influenced corrosion (MIC) of distribution pipes in drinking water and wastewater treatment plants (Beech and Sunner, 2004; Coetser and Cloete, 2005). Under anaerobic conditions, sulfate-reducing bacteria, such as *Desulfovibrio desulfuricans*, produce hydrogen sulfide (H_2S) which has a rotten-egg smell and plays a role in pitting corrosion of pipes. Under aerobic conditions, sulfur-oxidizing bacteria (e.g., *Thiobacillus concretivorus, Acidithiobacillus thiooxidans*) and iron-oxidizing bacteria (e.g., *Acidithiobacillus ferrooxidans*, *Gallionella*) can also cause pipe and pump corrosion and tubercle formation in distribution pipes (Hamilton, 1985).

5.10 NITRIFYING BACTERIA IN WATER DISTRIBUTION SYSTEMS

Nitrification is a microbial-driven process that converts ammonia to nitrite and nitrate. It is carried out by two categories of bacteria:

- Conversion of ammonia to nitrite by autotrophic ammonia-oxidizing bacteria such as *Nitrosomonas*, *Nitrosospira*, *Nitrosococcus*, *Nitrosolobus*, and *Nitrosovibrio* (Focht and Verstraete, 1977):

$$NH_3 + 1.5O_2 \rightarrow NO_2^- + H^+ + H_2O \tag{5.1}$$

- Oxidation of nitrite to nitrate by chemolithotrophic bacteria such as *Nitrobacter*, *Nitrospira*, and *Nitrococcus* (Wolfe and Lieu, 2002):

$$NO_2^- + 0.5O_2 \rightarrow NO_3 \tag{5.2}$$

Early indicators of the occurrence of nitrification in water distribution systems and premise plumbing are nitrite and nitrate production, difficulties in maintaining a chlorine residual, decrease in pH and alkalinity, increase in heterotrophic plate counts, and decrease in dissolved oxygen due to its consumption by nitrifiers (Zhang et al., 2009c). Excess in nitrate and nitrite concentrations in drinking water can lead to toxicological problems such as methemoglobinemia and formation of carcinogenic nitrosamines. Methemoglobinemia or "blue babies syndrome" is due to the formation of methemoglobin which is incapable of binding O_2 in the blood. It occurs in infants and certain susceptible segments of the adult population (e.g., Navajo Indians, Inuits, and people with genetic deficiency of methemoglobin reductase). Under the Safe Drinking Water Act, the maximum contaminant levels for nitrate-N and nitrite-N in drinking water were set at 10 mg/L and 1 mg/L, respectively.

5.10.1 Occurrence of Nitrification in Chloraminated Water

Many water treatment plants around the world now use inorganic chloramines as disinfectants to reduce the formation of DBPs which are of health concern (see Chapter 3). However, surveys have shown that a large percentage (about two-thirds) of the plants using chloramination experience nitrification problems. Excess free ammonia in water serves as a substrate for nitrifiers which proliferate in water distribution systems. The predominant nitrifiers are *Nitrosomonas* and *Nitrosospira* (ammonium oxidizing bacteria), *Nitrospira* and *Nitrobacter* (nitrite-oxidizing bacteria), and Archaea that can also carry out nitrification in distribution systems (Berry et al., 2006; Francis et al., 2005). A molecular study of nitrifiers diversity in full-scale chloramination plants showed that the ammonia-oxidizing bacteria (AOB) and nitrite-oxidizing bacteria (NOB) communities are dominated by *Nitrosomonas oligotropha* and *Nitrospira*, respectively (Regan et al., 2003).

Ammonia oxidation by AOB results in an increase in the concentration of nitrite which has a significant chlorine demand and increases the breakdown of chloramine residual. In the absence of nitrification chloramine is more stable than chlorine while the reverse situation is observed in the presence of nitrification (Zhang and Edwards, 2009). Nitrification also results in the proliferation of heterotrophic bacteria in water distribution systems and premise plumbing. This was also demonstrated for opportunistic pathogens such as *Legionella*, mycobacteria, *Pseudomonas aeruginosa,* and *Acanthamoeba* in a simulated distribution system (Wang et al., 2012a). Indicators of nitrification include decrease in chloramine residual, elevated bacterial count, decrease in pH and dissolved oxygen concentration, increase in nitrite (NO_2-N = 15–100 µg/L during nitrification episodes) or nitrate levels, and release of metals such as copper and lead. The occurrence of nitrification is influenced by several factors which include increased water temperature, chloramine residual, possible synergistic relationship between AOB and heterotrophic bacteria, as well as operational practices such long detention times leading to the growth of AOB (Wolfe et al., 1988; Wolfe and Lieu, 2002; Zhang and Edwards, 2009; Zhang et al., 2009c).

5.10.2 Nitrification Control in Water Distribution Systems

The main preventive/control measures for nitrification in water distribution systems include the following (AWWA, 2006; Harrington et al., 2002; Pintar and Slawson, 2003; Wolfe and Lieu, 2002; Zhang et al., 2009e):

- Maintenance of an adequate chloramine residual ($\geq$2mg/L)
- Increasing the chlorine-to-ammonia ratio: Traditionally, this ratio is 3:1 to optimize monochloramine formation. Nitrification control occurs when the ratio is raised to 4–5.
- Improving the removal of NOM which is responsible for the depletion of chloramines with subsequent release of ammonia. This can be accomplished by enhanced coagulation (i.e., coagulation at a pH of approximately 6).
- Although a short-term measure, periodic flushing of the distribution system helps to remove biofilms harboring nitrifying bacteria.
- Switching to free chlorine residual to control nitrification in water distribution systems: A 2007 survey showed that 23% of US water utilities switched to free chlorine residual to control nitrification (AWWA Committee Report, 2008). Periodic switching from chloramines to free chlorine results in ammonia oxidation, which ultimately leads to N_2 gas if breakpoint chlorination is used. Some states (e.g., North Carolina) require an annual switching to free chlorine for 1 month. This strategy was successful in eliminating nitrification. However, one drawback is the increase in total THMs levels (Carrico et al., 2008).

- A demonstration study in the Glendale, CA, distribution system showed that chlorite was effective in preventing nitrification due to its toxicity toward AOB. However, chlorite is not as effective in controlling nitrification and may increase corrosion in copper pipes (Rahman et al., 2011). Since chlorite is a regulated contaminant, it must be used at levels below the maximum contaminant level goal (MCLG) of 0.8 mg/L (McGuire et al., 2009).
- Biological denitrification: Drinking water suppliers may also consider biological denitrification (microbial conversion of nitrate to N_2) to remove nitrate from drinking water. This is possibly the most economical measure for removing nitrate.
- Other used methods include ion exchange, electrodialysis, reverse osmosis, and membrane bioreactors (Mateju et al., 1992; Nuhoglu et al., 2002; Ovez et al., 2006).

WEB RESOURCES

Algae

http://www.dipbot.unict.it/sistematica/Index.html
(good pictures of algae and cyanobacteria from Italy)

http://www.psaalgae.org/
(Phycological Society of America)

http://www.fotosearch.com/photos-images/algae.html
(algae images)

http://www.microscopy-uk.org.uk/mag/indexmag.html?http://www.microscopy-uk.org.uk/mag/wimsmall/diadr.html
(diatoms images)

http://www.micrographia.com/specbiol/alg/diato/diat0200.htm
(diatoms images)

Cyanobacteria

http://www.cyanosite.bio.purdue.edu/
(Excellent cyanobacteria and general pictures from Purdue University)

http://www.who.int/water_sanitation_health/dwq/cyanobactox/en/
(WHO documents on cyanotoxins)

Fungi

http://www.microbemagazine.org/index.php?option=com_content&view=article&
id=1433:fungi-from-the-faucet&catid=377:animalcules-and-forum&Itemid=453
(Fungi from the faucet : Microbe magazine_2009)

Protozoa

http://www.uga.edu/~protozoa/
(Pictures and information from the International Society of Protistologists)

http://www.microscopy-uk.org.uk/micropolitan/index.html
(excellent pictures of freshwater and marine microorganisms_UK)

http://www.slideshare.net/SujeeshSebastian/pathgenic-free-living-amoeba
(pathogenic free-living amoebas: slide show)

http://www.patient.co.uk/doctor/Acanthamoeba-and-Balamuthia.htm
(pathogenic free-living amoebas)

http://www.google.com/search?q=Free+living+amoeba&tbm=isch&tbo=u&
source=univ&sa=X&ei=P-Y2UuLBCeTd2QWwz4DYBA&ved=0CFwQsAQ&biw
=1215&bih=727&dpr=1
(images of free-living amoebas)

Nitrification in Drinking Water

http://www.google.com/url?sa=t&rct=j&q=&esrc=s&source=web&cd=20&ved=
0CGgQFjAJOAo&url=http%3A%2F%2Fwww.techstreet.com%2Fproducts%2F
preview%2F1858817&ei=M_g2UoisIOTEigLy6ICgCw&usg=AFQjCNFWiyg8oY
Xo_z_ctc9C5EBqXaicNQ&sig2=XHV-4UooJYJJIFA4RgCo5Q
(nitrification prevention and control in drinking water: AWWA Manual_2013)

FURTHER READING

AWWA. 2006. *Fundamentals and Control of Nitrification in Chloraminated Drinking Water Distribution Systems*, 1st edition. American Water Works Association, Denver, CO.

Falconer, I.R. 2005. *Cyanobacterial Toxins of Drinking Water Supplies: Cylindrospermopsins and Microcystins*. CRC Press, Boca Raton, FL.

LeChevallier, M.W., M.-C. Besner, M. Friedman, and V.L. Speight. 2011. Microbiological quality control in distribution systems. In *Water Quality and Treatment: A Handbook on Drinking Water*, 6th edition, J.K. Edzwald (Ed.). AWWA, Denver, CO, pp. 21.1–21.84.

van Lieverloo, J.H.M., D. van der Kooij, and W. Hoogenboezem. 2002. Invertebrates and protozoa (free living) in drinking water distribution systems. In *Encyclopedia of Environmental Microbiology*, G. Bitton (editor-in-chief). Wiley-Interscience, New York, pp. 1718–1733.

Post, G.B., T.B. Atherholt, and P.D. Cohn. 2011. Health and aesthetic aspects of drinking water. In *Water Quality and Treatment: A Handbook on Drinking Water*, 6th edition, J.K. Edzwald (Ed.). AWWA, Denver, CO.

Thomas, V., G. McDonnell, S.P. Denyer, and J.-Y. Maillard. 2010. Free-living amoebae and their intracellular pathogenic microorganisms: risks for water quality. *FEMS Microbiol. Rev.* 34:231–259.

Vaerewijck, M.J.M., G. Huys, J.C. Palomino, J. Swings, and F. Portaels. 2005. Mycobacteria in drinking water distribution systems: ecology and significance for human health. *FEMS Microbiol. Rev.* 29:911–934.

WHO. 1999. *Toxic Cyanobacteria in Water: A Guide to Their Public Health Consequences.* World Health Organization, Geneva, Switzerland. ISBN 0-419-23930-8.

Wolfe, R.L., and N.I. Lieu. 2002. Nitrifying bacteria in drinking water. In *Encyclopedia of Environmental Microbiology*, G. Bitton (editor-in-chief). Wiley-Interscience, New York, pp. 2167–2176.

Zhang, Y., N. Love, and M. Edwards. 2009. Nitrification in drinking water systems. *Crit. Rev. Environ. Sci. Technol.* 39:153–208.

6

BIOLOGICAL TREATMENT AND BIOSTABILITY OF DRINKING WATER

6.1 INTRODUCTION

In Chapters 2 and 3, we discussed the impact of physical and chemical processes on the fate of pathogens and parasites in drinking water treatment plants. However, the treated water may still contain trace amounts of bioavailable organic matter and other electron donors that may favor microbial growth in water distribution systems. In this chapter, we will cover the biological treatment of treated water and the methodology to determine the assimilable organic matter (AOC) in drinking water.

6.2 BIOLOGICAL TREATMENT OF DRINKING WATER

Key factors controlling microbial growth in water distribution systems were discussed in Chapter 4.

Biological treatment of drinking water has been practiced in Europe since the early twentieth century. Water containing organic compounds as well as ammonia nitrogen is biologically instable. A biologically stable water is produced by limiting bacterial growth in water distribution systems, and consequently reducing the disinfectant concentration to control regrowth. The presence of organic (biodegradable organic matter or BOM) and inorganic (e.g., NH_4, Fe^{2+}, Mn^{2+}) electron donors in water causes significant problems such as trihalomethane formation following disinfection with chlorine, taste and odor problems, regrowth of bacteria in distribution systems, and reduced bed life of granular activated carbon (GAC) columns (see Chapters 4 and 5; Hozalski et al., 1992). Several European countries (e.g., France, Germany, Netherlands) and Japan include biological processes in the water treatment train in

Microbiology of Drinking Water Production and Distribution, First Edition. Gabriel Bitton.
© 2014 John Wiley & Sons, Inc. Published 2014 by John Wiley & Sons, Inc.

an effort to remove/decrease BOM and nutrient levels (e.g., NH_4, NO_2, Fe^{2+}, Mn^{2+}) in the treated water and thus obtain biologically stable water which limits the growth of microorganisms in distribution pipes and reservoirs (van der Kooij, 1995). To prevent or reduce biofilm formation in water distribution systems, it was recommended that the assimilable organic carbon (AOC) concentration of drinking water should be <10 µg of acetate-C equivalents per liter (eq/L) (van der Kooij, 1990, 1992, 2000) and the biodegradable dissolved organic carbon (BDOC) less than 0.15 mg C/L (Servais et al., 1995). However, certain bacteria, such as *Aeromonas*, grow in biofilms even at AOC concentrations below 10 µg of acetate-C eq/L. Hence, a biofilm formation rate (BFR) measurement was proposed to better assess the biostability of drinking water in distribution systems. A BFR value below 10 pg ATP/cm^2 day indicates biostable water (van der Kooij, 2002). Certain practices (e.g., chloramination, ozonation) in water treatment plants can reduce biostability (Rittmann, 2004). Furthermore, biostability can also be affected by the materials in contact with drinking water (type of pipe material), flow rate, temperature, biofilms, sediments, or contamination of source water (Lautenschlager et al., 2013; Prèvost et al., 2005). This led to the determination of the biofilm formation potential which varies from less than 10 to 3000 pg ATP/cm^2, depending on the material being used (van der Kooij and Veenendaal, 1994).

Biological treatment is carried out in slow sand filters, fixed-bed and fluidized-bed systems, membrane bioreactors, river bank filtration, or in second-stage biologically active GAC filters. In North America, single-stage biofiltration removes both BOM and turbidity (Urfer et al., 1997). Biofiltration is based on aerobic biofilm processes that offer several advantages over physicochemical processes in regard to drinking water treatment. Activated carbon provides a large surface area (up to 1000 m^2/g carbon) for the accumulation of microorganisms as a biofilm (see Chapter 2). Three important factors affecting BOM removal by biofilters are the presence of chlorine in the backwash water, media type (e.g., GAC-sand vs. anthracite-sand filters), empty bed contact time (EBCT), and temperature (Liu and Slawson, 2001; Urfer et al., 1997).

The advantages provided by biofiltration are the following (Andersson et al., 2001; Carraro et al., 2000; Halle et al., 2009; LeChevallier et al., 1992; Hozalski and Bouwer, 1998; Hozalski et al., 1992; Manem and Rittmann, 1992; Nerenberg et al., 2000; Rittmann, 1987, 1989, 1995a; Rittmann and Snoeyink, 1984):

- Biotreatment removes organic compounds (total and AOC), thus reducing bacterial growth in water distribution systems, and producing biostable water.
- Removal of taste and odor compounds (e.g., geosmin and 2-methyl isoborneol). Ozonation followed by biofiltration provide an effective treatment for these compounds.
- Biotreatment, using GAC filtration, can also remove ammonia from water and reduces chlorine demand (theoretical consumption of 7.6 mg of Cl_2 per mg of N-NH_4). The chlorine demand may be reduced by 25–50% leading to a reduction in disinfection costs.

- Removal of trihalomethane precursors and other disinfection by-products (see Chapters 2 and 3). A comparison of chemical and biological treatments shows that the latter produces drinking water with lower mutagenic activity.
- Removal of iron and manganese.
- Potential reduction of mutagenic/carcinogenic compound levels.
- Biodegradation and removal of xenobiotics (e.g., petroleum hydrocarbons, halogenated hydrocarbons, pesticides, perchorate, chlorinated phenols, and benzenes).
- Finally, biological filtration can serve as a coagulant-free pretreatment prior to ultrafiltration to reduce fouling of the membranes by polysaccharides, proteins, and particulate matter.

Backwash of biological filters with water partially removes the attached biomass but maintains the ability of the filter to remove biodegradable organic carbon. Preozonation enhances the biodegradability of organic compounds in biofilms by oxidizing large molecular-weight compounds to smaller ones which are more easily biodegraded. Thus, ozonation increases the AOC in plant effluents and distribution systems (Escobar and Randall, 2001). Ozonation followed by biofiltration should be able to remove AOC and thus reduce bacterial regrowth in distribution systems.

To address the problem of bacterial regrowth in distribution systems, a standardized algorithm was developed to generate biostability curves for a specific water treatment plant. Bacterial regrowth is controlled by two main factors: substrate (organic or inorganic substrate) concentration and disinfectant concentration. Regrowth is prevented when the decay rate resulting from disinfectant addition is equal or higher than the growth rate of biofilm microorganisms which is controlled by the substrate concentration according to Monod's kinetics (Srinivasan and Harrington, 2007). This approach has been used to control nitrifiers in distribution systems (Fleming et al., 2005).

6.3 ASSESSMENT OF BIOSTABILITY OF DRINKING WATER

6.3.1 Introduction

Treated water flowing in distribution pipes is an oligotrophic environment where heterotrophic bacteria must be able to survive and grow in the presence of very low (μg C/L) carbon concentrations. Natural organic matter (NOM) is made of two fractions: BOM and nonbiodegradable fraction (Figure 6.1; Volk and LeChevallier, 2000). BOM is contributed by the raw water entering in the water treatment plant, sediments accumulated in the water distribution system, and leachates from pipes and other construction materials used in the plant. Most of the total organic carbon (TOC) pool occurs as dissolved organic carbon (DOC), and it is estimated that heterotrophs use only 0.5–5% of the DOC pool because the bulk of this pool is made of microparticulate and colloidal carbon that cannot be used by bacteria as

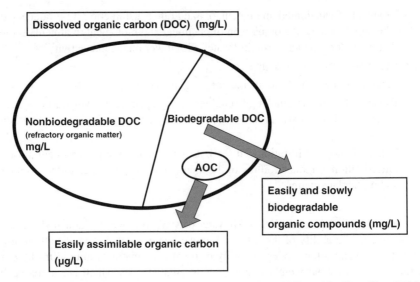

Figure 6.1   Fractionation of organic matter. Adapted from Volk and LeChevallier (2000). Amer. Water Works Assoc. J. 92 (5): 64–76.

source of carbon and energy (Egli, 2010). Others estimated that this "assimilable organic carbon" (AOC) in drinking water is 0.1–9% of the TOC (van der Kooij and Hijnen, 1985; van der Kooij et al., 1982a; LeChevallier et al., 1991a). The development of novel technologies such as flow cytometry (FCM) in combination with staining techniques, microautoradiography, or direct viable counts (see review in Bitton, 2011; Khan et al., 2010; Vital et al., 2012; Wang et al., 2010) to study substrate utilization in oligotrophic environments has helped in our understanding of bacterial growth in such dilute environments. Bacteria have developed strategies for coping with such low amounts of carbon. Some investigators invoke the concepts of obligate oligotrophy and "viable but not culturable" (VBNC) bacterial lifestyles. Pure culture studies suggest a third strategy, the so-called "mixed substrate growth," which allows bacteria to simultaneously assimilate several carbon substrates (Egli, 1995, 2010). Bacterial growth in distribution systems is influenced by several factors, among them the level of trace biodegradable organic materials found in finished water, water temperature, nature of pipe surfaces, disinfectant residual concentration, and detention time within the distribution system (Joret et al., 1988). As discussed previously, drinking water harbors bacteria (e.g., *Flavobacterium, Pseudomonas*) that can grow at extremely low substrate concentrations (van der Kooij and Hijnen, 1981). *Pseudomonas aeruginosa* (strain P1525) and *Pseudomonas fluorescens* (strain P17) are able to grow in treated water at relatively low concentrations (μg/L) of low molecular weight organic substrates such as sugars, organic acids (e.g., acetate, lactate, succinate), and amino acids. The K_s values (K_s represents the affinity of microorganisms for the growth substrate) for bacteria isolated from drinking water may vary from 2 to 100 μg/L in the presence of substrates such as glucose, acetate,

TABLE 6.1   Growth kinetics of bacteria isolated from drinking water

Microorganism	Substrate	Temperature (°C)	K_s (µg C/L)	V_{max} (h^{-1})
Aeromonas hydrophila	Acetate	15	11	0.15
A. hydrophila	Glucose	15	16	0.28
A. hydrophila	Amylase	15	93	0.26
A. hydrophila	Oleate	15	2.1	0.23
Flavobacterium sp.	Glucose	15	3.3–109	0.15–0.21
Flavobacterium sp.	Maltose	15	23.7	0.37
Flavobacterium sp.	Amylose	15	26	0.50
Flavobacterium sp.	Amylopectin	15	11	0.48
Klebsiella pneumoniae	Maltose	15	51	0.49
K. pneumoniae	Maltopentaose	15	92	0.41
Pseudomonas fluorescens	Acetate	15	4	0.18
Pseudomonas fluorescens	Glucose	15	57	0.22
Spirillum sp.	Oxalate	15	15.2	0.24
Citrobacter freundii	Glucose	15	95	0.17
Enterobacter sp.	Glucose	15	60	0.21
Escherichia coli	Lactose	25	142	0.37

Source: Adapted from van der Kooij (2002). In: Encyclopedia of Environmental Microbiology, Gabriel Bitton, editor-in-chief, Wiley-Interscience, N.Y., pp. 312–327.

oleate, or starch (van der Kooij, 1995). The growth kinetic parameters of some bacteria isolated from drinking water are shown in Table 6.1 (van der Kooij, 2002). A strain of *Klebsiella pneumoniae* is able to use organic compounds at concentrations of a few micrograms per liter. This bacterium grows on maltose with a yield $Y = 4.1 \times 10^6$ CFU per 1 µg C (van der Kooij and Hijnen, 1988).

6.3.2   Bioassays for Determination of Biostability of Drinking Water

Bioassays have been developed to assess the biological stability of drinking water by determining the potential of drinking water to support microbial growth, to exert a chlorine demand, or to lead to the formation of disinfection by-products. These bioassays essentially include (Huck, 1990; Volk and LeChevallier, 2000): (1) determining the bacterial biomass following consumption of AOC; and (2) DOC-based methods, the goal of which is to assess chlorine demand and disinfection by-products formation by measuring the concentration of BDOC.

6.3.2.1   *Biomass-Based Methods*

• *Methodology*

These methods are based on the measurement of the growth potential of pure bacterial cultures or indigenous microorganisms following incubation in the water samples for several days. Colony-forming unit counts, ATP, FCM, or bioluminescence are used to assess bacterial numbers or biomass.

A pasteurized or filtered water sample is seeded with pure bacterial cultures of *P. fluorescens* (strain P17), *Spirillum* (strain NOX), or *Flavobacterium* (strain

S12). *Pseudomonas fluorescens* (strain P17) utilizes a wide range of low molecular weight substrates whereas *Spirillum* (strain NOX) utilizes only carboxylic acids. *Flavobacterium* sp. (strain S12) specializes in the utilization of carbohydrates. The growth of these reference bacteria in a given water sample is compared with that obtained in a sample spiked with substrates such as acetate or oxalate. The water sample is inoculated with the bacteria and incubated at 15°C for 9 days, using nutrient agar plates. Bacterial CFUs are counted after 7, 8, and 9 days, and an average colony count is recorded.

The AOC is calculated using the following formula (van der Kooij, 1990, 2002):

$$AOC(\mu gC/L) = \frac{N_{max} \times 1000}{Y} \tag{6.1}$$

where N_{max} = maximum colony count (CFU/mL); Y = yield coefficient (CFU/mg of carbon); AOC concentration expressed as µg of acetate-C eq/L

When using *P. fluorescens* strain P17, AOC is calculated by using the yield factor Y for acetate (Y = 4.1 × 10^6 CFU/µg C) (van der Kooij and Hijnen, 1985; van der Kooij et al., 1982b). For *Spirillum* sp., strain NOX, $Y_{acetate}$ = 1.2 × 10^7 CFU/µg C. *Flavobacterium* sp. *(strain S12)* has a yield of 2–2.3 × 10^7 CFU/µg C (van der Kooij and Hijnen, 1984). AOC concentration in the sample is expressed as µg acetate C equivalents/L. This approach necessitates a relatively long incubation period (van der Kooij et al., 1982a; van der Kooij, 1995; van der Kooij and Hijnen, 1984, 1985). P17 and NOX strains can also be used in combination, sequentially or simultaneously, to measure AOC in environmental samples. Efforts have been made to simplify and standardize this method (Kaplan et al., 1993). Other bacterial mixtures (e.g., mixture of *P. fluorescens*, *Curtobacterium* sp., *Corynebacterium* sp.) have also been proposed to estimate AOC. Calibration curves help estimate the AOC from the bacterial colony count (Kemmy et al., 1989).

Biomass production can also be determined via ATP measurement (Stanfield and Jago, 1987) or turbidimetry (Werner, 1985). ATP has been used to determine the active microbial biomass in drinking water and distribution systems and thus is a useful indicator of microbial activity in drinking water (Vital et al., 2012). The average ATP concentration in the distribution systems of six water treatment plants varied between 0.8 and 12.1 ng ATP/L (Figure 6.2; van der Wielen and van der Kooij, 2010). Therefore, ATP measurement has been considered for the rapid determination of AOC (LeChevallier et al., 1993).

AOC was also conveniently determined with a bioluminescent *P. fluorescens* P17 containing the luciferase operon (Noble et al., 1994) as well as with the bioluminescent marine bacterium *Vibrio harveyi* or mutagenized *P. fluorescens* P17 and *Spirillum* sp. NOX for freshwaters (Weinrich et al., 2009, 2011).

Another approach to AOC determination consists of staining the nucleic acids with SYBR green stain followed by rapid counting (about 15 minutes) by flow cytometry method (FCM) instead of using the time-consuming and labor-intensive heterotrophic plate counts (HPCs). FCM was found to be well correlated with ATP but not as well with HPC (only 0.04–5.99% of intact cells detected by FCM are cultivable). This

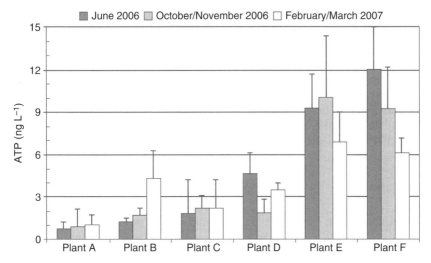

Figure 6.2 Average ATP concentration and standard deviations in drinking water sampled at 27 different locations in the distribution system of six different water treatment plants. *Source:* van der Wielen and van der Kooij (2010). Water Res. 44: 4860–4867.

kinetics approach method uses a higher incubation temperature (30°C) and natural microbial consortia but could be used with pure cultures of bacteria such as *P. fluorescens* P17 (Hammes and Egli, 2005; Siebel et al., 2008; Vital et al., 2012).

- *AOC levels in various environmental samples*
  AOC levels vary from less than 1 μg C/L to >3 mg/L and account for 0.03–27% of the total DOC. The average AOC values of drinking water in distribution systems of five plants in China varied between 177 and 235 μg acetate-C eq/L (van der Kooij et al., 1982b; Liu et al., 2002). Another study in China showed AOC values from 100 to more than 200 μg/L (Lu and Zhang, 2005). A US survey of 95 treatment plants revealed that AOC levels varied from 18 to 214 μg/L (Volk and LeChevallier, 2000). AOC of drinking water from 20 water treatment plants in the Netherlands varied between 1 and 57 μg acetate-C eq/L. Most of the AOC was utilized by *Spirillum*, indicating that AOC was composed mostly of carboxylic acids. AOC represented less than 1.7% of DOC. A significant correlation was observed between AOC of water and counts of heterotrophic bacteria in distribution systems (Escobar et al., 2001; van der Kooij, 1992). AOC of five Norwegian brands of bottled water varied between 13.5 and 115 μg/L (Ottherholt and Charnock, 2011).

  This biomass-based technique is useful for determining the effect of water treatment processes on AOC and the release of biodegradable materials from pipe surfaces (van der Kooij et al., 1982a). Coagulation, sedimentation, sand filtration, and GAC treatment generally reduce the AOC concentration whereas ozonation, UV/H_2O_2 treatment, and to a lesser extent chlorination increase the AOC levels as a result of production of low molecular weight compounds mostly consisting of organic acids, aldehydes, and ketones (Bazri et al., 2012; Hammes et al., 2006; van der Kooij and Hijnen, 1985; LeChevallier et al., 1992; Servais et al., 1987, 1991). In water treatment plants using surface water as the source water, ozonation of algae may increase the AOC and DOC concentrations (Hammes et al., 2007). It was recommended that the AOC concentration of drinking water in distribution systems should be <10 μg acetate-C eq/L to limit regrowth potential of heterotrophic bacteria (van der Kooij, 1990, 1992, 2000). In unchlorinated drinking water in the Netherlands, *Legionella pneumophila* was detected more often when the AOC concentration was above 10 μg C/L (van der Wielen and van der Kooij, 2013). AOC concentration is also an important parameter to consider in regard to the biological clogging of sand beds which occurs as a result of infiltration of pretreated surface water in recharge wells. AOC level should be <10 μg acetate-C eq/L to prevent biological clogging (Hijnen and van der Kooij, 1992).

Determination of AOC level, based on the growth potential of *P. fluorescens*, does not appear to be a good indicator of the growth potential of coliforms in water distribution systems; coliform monitoring appears to be more suitable (McFeters and Camper, 1988). A *coliform growth response* test, using *Enterobacter cloacae* as the test organism, was developed by the U.S. EPA to measure specifically the growth potential of coliform bacteria in water. Some investigators recommend the use of environmental coliform isolates found in drinking water (e.g., *Enterobacter aerogenes*, *K. pneumoniae*, *Escherichia coli*) to determine AOC in distribution systems (Camper et al., 1991). These isolates appear to be more adapted to oligotrophic conditions than their clinical counterparts (McFeters and Camper, 1988; Rice et al., 1988).

6.3.2.2 DOC-Based Methods.

These assays are based on the measurement of DOC before and after incubation of the sample in the presence of an indigenous bacterial inoculum (e.g., river water or sand filter bacteria). It was argued that indigenous bacterial populations are more suitable than pure cultures for testing the biodegradation of natural organic compounds (Block et al., 1992; Neilson et al., 1985). The biodegradable dissolved organic carbon, *BDOC*, is given by the following formula:

$$BDOC \ (mg/L) = initial \ DOC - final \ DOC \qquad (6.2)$$

The general approach is as follows: A water sample is sterilized by filtration through a 0.2-μm pore size filter, inoculated with indigenous microorganisms, and incubated in the dark at 20°C for 10–30 days, until DOC reaches a constant level. BDOC is the difference between the initial and final DOC values (Servais et al., 1987). Another approach consists of seeding the water sample (300 mL) with prewashed biologically active sand with mixed populations of attached indigenous bacteria. A European collaborative study has demonstrated that the source of the indigenous bacterial inoculum is not a major source of variance in the BDOC procedure (Block et al., 1992). BDOC is estimated by monitoring the decrease in DOC (Joret et al., 1988). The experimental setup for this method is displayed in Figure 6.3 (Joret and Levi, 1986; Joret et al., 1988). The incubation period at 20°C is 3–5 days.

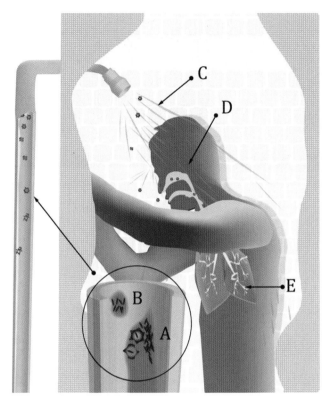

Figure 1.2 Conceptual model for Legionella exposure from inhalation of shower aerosols containing Legionella derived from the in-premise plumbing of buildings. First, Legionella multiply within the premise plumbing biofilm, potentially by colonizing the biofilm or within a protozoan host (A). The biofilm-associated Legionella detaches from the biofilm during a showering event (B), is transported to the shower head, and is aerosolized (C). Finally, the aerosolized Legionella is inhaled (D) and a fraction of that inhaled dose is deposited in the alveolar region of the lungs (E). *Source:* Schoen and Ashbolt (2011). Water Res. 45: 5826–5836.

Microbiology of Drinking Water Production and Distribution, First Edition. Gabriel Bitton.
© 2014 John Wiley & Sons, Inc. Published 2014 by John Wiley & Sons, Inc.

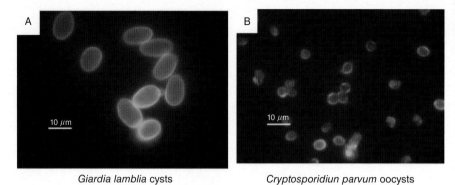

Giardia lamblia cysts   Cryptosporidiun parvum oocysts

Figure 1.6 Immunofluorescence images of *Giardia lamblia* cysts (A) and *Cryptosporidium parvum* oocysts (B). Courtesy of H.D.A. Lindquist, U.S. EPA.

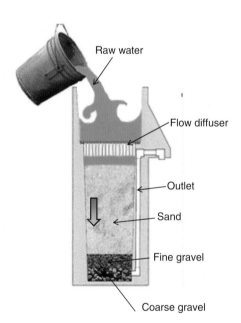

Figure 8.4 Biosand filter. Adapted from http://www.waterforcambodia.org/projects/biosand-filters.

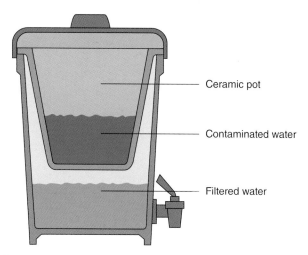

Figure 8.5 Ceramic filtration. *Source*: Van Hallem, et al. (2009). Physics and Chem. of the Earth 34 (1-2): 36–42.

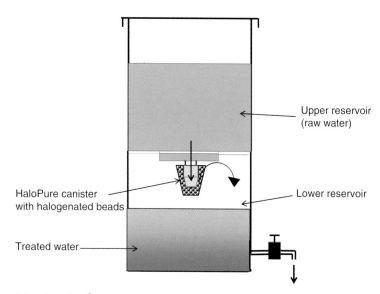

Figure 8.8 AquaSure® system: Arrows represent the flow of water in the system. Adapted from Coulliette et al. (2010). Amer. J. Trop. Med. Hygiene 82: 279–288.

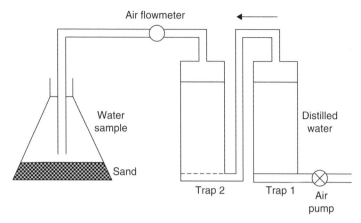

Figure 6.3 Experimental set-up for BDOC determination. Adapted from Joret and Levi (1986). Tribune CEBEDEAU 510: 3–9.

One advantage of this method is the use of biofilm microorganisms as inoculum, thus simulating situations occurring in water distribution systems. The monitoring of water treatment plants in Paris, France, suburbs showed that treated drinking water contains enough BDOC to sustain bacterial growth in distribution systems (Joret et al., 1988). Substrate availability, as measured by this technique, appears to correlate well with the regrowth potential of bacteria (Joret et al., 1990). A survey of treatment plants in the United States showed that BDOC levels varied between 0.03 and 1.03 mg/L (geometric mean of 0.32 mg/L for all sites) (Volk and LeChevallier, 2000).

Another approach consists of passing water continuously through one or two glass columns filled with sand or sintered porous glass and conditioned to obtain the development of a biofilm on the supports provided. BDOC is the difference between the inlet of the first column and the outlet of the second column (Ribas et al., 1991; Frias et al., 1992) (Figure 6.4). BDOC is obtained in hours to days, depending of the type of water being assayed. Immobilized cell bioreactors have also been considered for determining BDOC in drinking water. The pre-aerated drinking water is passed through the bioreactor with a hydraulic retention time of 3 hours. This rapid procedure does not require a start-up period. AOC fraction represents the easily degradable compounds whereas BDOC fraction represents a wider range of compounds, some of which are slowly biodegradable. It was estimated that AOC values in drinking water represent 15–22% of the BDOC concentrations. This may be mainly due to the fact that, in the BDOC method, several drinking water microorganisms are involved in the biodegradation of the wide range of compounds found in drinking water. The BDOC approach also necessitates a longer incubation (up to 4 weeks) than the AOC method (1 week) (van der Kooij, 1995).

6.3.2.3 Determination of Biofilm Growth to Assess Biological Stability of Water. Some contend that AOC and BDOC methods for measuring the biostability of water have some limitations, such as the use of pure cultures of nonindigenous

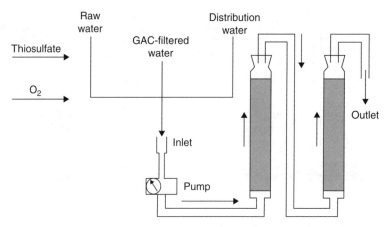

Figure 6.4   Apparatus for BDOC determination. Adapted from Ribas et al. (1991). J. Appl. Bacteriol. 71: 371–378.

bacteria in AOC method or the use of highly acclimated bacteria in the BDOC method. They proposed to add biofilm growth parameters to make this assessment. A biofilm annular reactor (BAR) was used to obtain data on biofilm growth (Sharp et al., 2001). The reactor consists of a rotating (60 rpm) drum supporting 20 polycarbonate slides which support biofilm development, and which can be sampled to assess biofilm growth. It allows the measurement of both organic matter biodegradability parameters (AOC and BDOC) and biofilm-associated parameters such as protein content. Both of these parameters can be used to obtain a biostability factor for a given aquatic sample.

Schaule et al. (2007) proposed a fiber optical sensor to detect biofilm growth resulting from changes in biodegradable organic carbon. This optical sensor could be installed online in water treatment plants.

6.3.2.4 *Multiparametric Approach to Determine Biostability.*   So far, the methods used (AOC, BDOC and HPCs) to assess biostability of drinking water are based on organic carbon data. However, some argue that biological instability may also be due to the growth of autotrophic bacteria such as nitrifiers and some denitrifiers. Thus, a multiparametric approach was proposed and was applied to the nonchlorinated water distribution system of Zürich water treatment plant (Lautenschlager et al., 2013). This approach is based on determination in seven stations along the distribution system (hydraulic retention times of 6–52 hours) of total cell counts via FCM, ATP concentration, and microbial community structure, using two molecular techniques. Except for the last two stations, this study demonstrated a biostable drinking water and confirmed the results of conventional methods.

6.3.2.5 *Assessment of the Impact of Leachates from Polymers in Pipes and Other Construction Materials on the Biostability of Drinking Water.*
In addition to the biofilm growth potential bioassays, it is also important to assess

the contribution of organic materials released from polymers used in distribution and storage of drinking water. Thus, it is necessary to carry out leaching or migration tests (three to seven migration cycles) to determine quantitatively and, sometimes, qualitatively the organic compounds leaching from the polymeric materials into the water. A comprehensive protocol was proposed and consists of migration tests followed by determination of AOC in the leachates (Bucheli-Witschel et al., 2012).

WEB RESOURCES

http://www.nae.edu/Publications/Bridge/ExpandingFrontiersofEngineering/BiologicalTreatmentsofDrinkingWater.aspx
(biological treatments of drinking water: National Academy of Engineering)

http://www.iwapublishing.com/template.cfm?name=w21prodnews080211b
(construction of a biological treatment plant to remove perchlorate)

FURTHER READING

van der Kooij, D. 1995. Significance and assessment of the biological stability of drinking water. In *Water Pollution: Quality and Treatment of Drinking Water*. Springer, New York, pp. 89–102.

van der Kooij, D. 2002. Assimilable organic carbon (AOC) in treated water: determination and significance. In *Encyclopedia of Environmental Microbiology*, G. Bitton (editor-in-chief). Wiley-Interscience, New York, pp. 312–327.

Urfer, D., P.M. Huck, S.D. Booth, and B.M. Bradley. 1997. Biological filtration for BOM and particle removal: a critical review. *Am. Water Works Assoc. J.* 89(12):83–96.

Volk, C.J. and M.W. LeChevallier. 2000. Assessing biodegradable organic matter. *Am. Water Works Assoc. J.* 92(5):64–76.

7

BIOTERRORISM AND DRINKING WATER SAFETY

7.1 INTRODUCTION

Drinking water experts are witnessing some challenges in their quest for safe drinking water. Threats to water safety may emanate from the accidental or deliberate contamination of drinking water with the 60,000 biological and chemical agents or the 760 radionuclides found in the commerce. Hopefully these numbers were reduced to 21 priority agents following rigorous screening (Porco, 2010). This chapter will deal mostly with biological agents that pose many challenges which may include, among others, emerging pathogens resistant to drinking water treatments including disinfection, climate change with subsequent floods which may contribute to contamination of freshwater resources, and bioterrorism (Brettar and Höfle, 2008). Bioterrorism involves the deliberate use of microbial pathogens or microbe-derived products to cause harm to humans, livestock, or agricultural crops.

A biological weapon is a four-component system made of the payload (biological agent), munition (container ensuring the maintenance of the potency of the biological agent), delivery system (e.g., missile, vehicle), and a dispersion system (e.g., sprayer) (Hawley and Eitzen, 2001). The production of biological weapons and their delivery are relatively inexpensive as compared to conventional weapons. The cost of inflicting civilian casualties is much lower for biological weapons than for conventional weapons.

In this chapter, we will introduce the main biowarfare (BW) agents and biotoxins and we will discuss their potential impact on the safety of drinking water supplies.

7.2 EARLY HISTORY OF BIOLOGICAL WARFARE

Biological agents were used for hundreds of years to cause harm to humans, animals, and agricultural crops (Christopher et al., 1997; Hawley and Eitzen, 2001; Sheehan,

Microbiology of Drinking Water Production and Distribution, First Edition. Gabriel Bitton.
© 2014 John Wiley & Sons, Inc. Published 2014 by John Wiley & Sons, Inc.

2002). Some notorious examples illustrating the use of biological agents to cause harm to the enemy in war times are

Fourteenth century: the Tartar army catapulted the diseased bodies of dead soldiers into territory held by the enemy.

1754–1767: smallpox-contaminated blankets were sent to native American Indians who experienced an outbreak of smallpox.

1914–1918: During World War I, the Germans used *Bacillus anthracis* and *Burkholderia mallei* to infect livestock, horses, and mules of the Allied Forces.

1932–1945: Japan subjected prisoners to infectious agents such as *B. anthracis*, *Shigella*, *Salmonella*, *Yersinia pestis*, and smallpox virus.

World War II (1939–1945): the Germans experimented on prisoners in concentration camps by exposing them to infectious agents.

1943: the United States started a biological weapons program at Fort Detrick, MD, involving *B. anthracis* and other BW agents. Studies were carried out on weaponization systems as well as the survival of BW microbial agents in the environment. The Allied Forces dropped anthrax bombs on Gruinard Island off the coast of Scotland. The island was decontaminated some 50 years later. Eventually, the US military terminated its biological program in 1969.

1962–1968 Vietnam War: the Viet Kong used spear traps and spikes contaminated with human and animal excreta.

1970s: The former Soviet Union also maintained an extensive biological warfare arsenal. An outbreak of inhalation anthrax accidentally occurred in a production facility in 1979, resulting in a death toll of 66 persons. The program was terminated following the fall of the Soviet Union.

1991 Gulf War: Iraq admitted to having engaged in the development of a biological warfare program. After the war, Iraq claimed that it has destroyed its biological agents.

7.3 BW MICROBIAL AGENTS AND BIOTOXINS

7.3.1 Some Important Features of BW Agents

Some characteristics of the BW agents must be taken into account when assessing the threat posed by the pathogens and when comparing chemical and biological weapons.

- As compared to chemical agents, microbial agents are infectious and can multiply within the host. BW agents can be transmitted via person to person or via the waterborne, foodborne, airborne, and fomite routes. Person-to-person transmission may be quite easy for certain agents such as smallpox virus.
- The microbial dose necessary for causing infection (i.e., minimum infectious dose) may vary from a few to thousands of microbial cells (Bitton, 2011). The minimum infectious doses of some inhaled or ingested BW agents are given in Table 7.1 (Burrows and Renner, 1999; Sheehan 2002).

TABLE 7.1   Characteristics of some biological warfare agents

Disease Agent	Potential for Weaponization	Estimated Infectious Dose	Equivalent Drinking Water Concentration (number/L)		Water Threat
			5 L/day[a]	15 L/day[a]	
Anthrax (*Bacillus anthracis*)	Yes	6000 spores (inh)	171	57	Yes
Brucellosis (*Brucella*)	Yes	10–100 cells	300	100	Probable
Cholera (*Vibrio cholerae*)	Unknown	10^3 (ing)	30	10	Yes
Plague (*Yersinia pestis*)	Probable	10^2–10^3 (inh) 70 (ing)	2	<1	Yes
Glanders (*Burkholderia mallei*)	Probable	3.2×10^6	9×10^4	3×10^4	Unlikely
Tularemia (*Francisella tularensis*)	Yes	10^8 cells (ing) 10–50 cells (inh)	3×10^6	10^6	Yes
Q fever (*Coxiella burnetii*)	Yes	25 (unsp)	<1	<1	Possible
Smallpox virus	Possible	10–100 particles (inh)			?
Cryptosporidiosis (*Cryptosporidium*)	Unknown	132 oocysts (ing)	3	1	Yes

Source: Adapted from Burrows and Renner (1999). Env. Hlth. Persp. 107: 975–984; Bitton (1999). Wastewater Microbiology, 2nd. Ed., Wiley
inh, inhalation route; ing, ingestion route; unsp, unspecified route.
[a]The equivalent drinking water concentration is obtained by dividing by 7 (the maximum number of days for accumulation of the infectious organism without any clearance) and by 5 or 15 to account for water consumption of 5 L/day and 15 L/day, respectively.

- The production of biological agents may be less complex and less costly that for chemical poisons.
- Stability of the BW agents in the environment (water, soil, groundwater, sediments, water or wastewater treatment plants, air, surfaces, etc.)
- Impact of water treatment processes on BW agents, including disinfection, on their removal and/or inactivation
- Biological agents are colorless and odorless, and the onset of symptoms varies from days to weeks for these agents, as compared to minutes or hours for chemical agents.
- In the United States, the baseline incidence of diseases reported to the National Notifiable Disease Surveillance System is generally low except possibly for tularemia and brucellosis.

7.3.2　Categories of BW Agents

The Centers for Disease Control and Prevention (CDC) in the United States has categorized BW agents as follows (Table 7.2; Rotz et al., 2002; Atlas, 2002):

Category A. These agents pose the highest threat and are of most concern to public health. They include variola virus (smallpox), *B. anthracis*, *Y. pestis*, *Clostridium botulinum* toxin, *Francisella tularensis*, and viruses causing hemorrhagic fevers (e.g., Ebola, Marburg, Lassa viruses).

Category B. These agents cause lower morbidity and mortality than category A agents. They include *Brucella* (brucellosis) *Coxiella burnetii* (Q fever), *B. mallei* (glanders), alphaviruses, Venezuelan encephalomyelitis, Eastern and Western equine encephalomyelitis, ricin toxin from *Ricinus communis* (castor beans), and *Staphylococcus* enterotoxin. A subset of category B agents includes pathogens that are foodborne or waterborne. These pathogens include *Salmonella* spp., *Shigella* spp., *Escherichia coli* O157:H7, *Vibrio cholerae*,

TABLE 7.2　Bioterrorism: categories of biological agents of concern to public health

Biological Agent	Disease
CATEGORY A[a]	
Variola major	Smallpox
Bacillus anthracis	Anthrax
Yersinia pestis	Plague
Clostridium botulinum	Botulism
Francisella tularensis	Tularemia
Filoviruses and Arenaviruses (e.g., ebola virus, Lassa virus)	Viral hemorrhagic fevers
CATEGORY B	
Coxiella burnetii	Q fever
Brucella spp.	Brucellosis
Burkholderia mallei	Glanders
Alphaviruses	Encephalitis
Rickettsia prowazekii	Typhus fever
Biotoxins (e.g., ricin, Staphylococcal enterotoxin B)	Toxic syndromes
Chlamydia psittaci	
Foodborne agents (e.g., *Salmonella*)	Psittacosis
	Typhoid and paratyphoid fevers, gastroenteritis
Waterborne agents (e.g., *vibrio cholerae*, caliciviruses, *Cryptosporidium*)	Wide range of illnesses (see Chapter 1 of this book)
CATEGORY C	
Emerging agents (e.g., Nipah virus, hantavirus)	

Source: Adapted from Rotz et al. (2002). Emerg. Infect. Dis. 8: 225–229.
[a]From a public health threat viewpoint: A > B > C.

Cryptosporidium parvum, Giardia lamblia, hepatitis A and E viruses, and caliciviruses.

Category C. This includes emerging pathogens such as hantaviruses, tick-borne hemorrhagic fever viruses, or tick-borne encephalitis viruses.

7.3.3 Major BW Agents

Table 7.1 gives a list of the main BW microbial agents, their estimated infectious dose, and their potential for weaponization.

7.3.3.1 Bacillus anthracis. *Bacillus anthracis* is a gram-positive, spore-forming and rod-shaped bacterium responsible for anthrax, a serious disease affecting both animals and humans. The spores are extremely resistant to environmental stresses and survive for long periods in soils (e.g., survival for up to about 50 years in Gruinard Island). This makes the resistant spores highly desirable for use as biological warfare agents. Although anthrax is primarily a disease of animals, it can also infect humans who contract the disease through contact with infected animals or through occupational exposure to infected animal products. Although person-to-person transmission is rare, *B. anthracis* can be transmitted via inhalation, gastrointestinal, and cutaneous routes. The inhalation route is fatal in most cases with a mortality rate exceeding 80% (Baillie and Read, 2001; Dixon et al., 1999; Franz et al., 1997) although the 2001 bioterrorism-associated outbreak of inhalational anthrax resulted in 40% mortality among patients treated with antibiotics (Jernigan et al., 2001). The mortality rate of cutaneous anthrax is lower and was found to be up to 25%. In 2009–2010, 273 cases of cutaneous anthrax were detected in Bangladesh. Most of the patients had a history of contact with diseased animals (Chakraborty et al., 2012).

The infectious dose for *B. anthracis* is approximately 6000 spores via the inhalation route. If we assume that the infective dose via inhalation and ingestion routes are equivalent, the corresponding spore concentration in drinking water should be 171 spores/L following consumption of 5 L/day over 7 days (Burrows and Renner, 1999). Anthrax is characterized by flu-like symptoms including headache, fever, myalgia, cough, and mild chest discomfort. This is followed by chest wall edema and hemorrhagic meningitis. Death occurs if treatment is delayed beyond 48 hours. Treatment consists of administering the antibiotic ciprofloxacin to patients.

The virulence of *B. anthracis* is associated with the presence of two plasmids, pXO1 and pXO2. pXO1 codes for the production of the two anthrax toxins (lethal toxin and edema toxin) while pXO2 codes for capsule biosynthesis (Dauphin et al., 2008; Okinaka et al., 1999).

The methodology for detecting airborne *B. anthracis* consists of membrane filtration followed by plating the concentrate on blood agar supplemented with polymyxin or on *B. cereus* agar, by detecting components of spore glycoproteins, monoclonal antibodies against surface spore antigens, or via real-time PCR (Dauphin et al., 2008; Fox et al., 2003; Herzog et al., 2009; Makino and Cheun, 2003; Morel et al., 2012).

There are now commercial kits for detecting *B. anthracis* spores in clinical and environmental samples (King et al., 2003). A recently proposed method incorporates an immunoassay, a spore germination test, and PCR analysis and has a detection limit of 10^3 *B. anthracis* spores (Hang et al., 2008). A culture-based PCR method, carried out in a most probable number format in 96-well microplates, is able to detect low numbers of live spores (as low as 1 to 10 spores) of *B. anthracis* or its surrogate *B. atrophaeus*. No significant difference was found between this method and the traditional culture technique. Some advantages of this technique are rapidity (approximately 17 hours) and amenability to automation (Kane et al., 2009).

7.3.3.2 Yersinia pestis. *Yersinia pestis*, a category A gram-negative bacterial pathogen, is the etiological agent of the plague. There are three clinical forms of plague: bubonic, septicemic, and pneumonic. Since *Y. pestis* is amenable to aerosolization, the disease may result in respiratory failure and shock. Death occurs if the patient is not treated within 24 to 36 hours. In addition to its person-to person transmission, it may cause massive casualties when spread via the airborne route. It is transmitted to humans from infected rodents via flea bites and causes bubonic plague. The infectious dose for humans via inhalation is around 100 to 1000 organisms. If ingested, the infectious dose is around 70 organisms (Cooper et al., 1986). Since patient death can occur within the first 24 to 36 hours, there is a need for rapid methodology for detection and identification of *Y. pestis*. An engineered *Y. pestis*-specific bioluminescent reporter phage was proposed for the rapid identification (within hours) of this pathogen and its susceptibility to antibiotics (Schofield et al., 2012).

7.3.3.3 Brucella. *Brucella* is the etiologic agent for brucellosis which causes fever, sweating, and pain, but has a low fatality rate. *Brucella suis* has been weaponized as an aerosolized BW agent. Although *Brucella* (*B. suis*) is mostly pathogenic to livestock, some species are pathogenic to humans. Although less resistant to environmental stresses than *B. anthacis*, *Brucella* is a desirable biological warfare agent due to its low infectious dose (10–100 cells). The mortality rate of brucellosis is around 5% (Eitzen et al., 1998). Since brucellosis can be acquired through consumption of contaminated milk, drinking water may also be considered as a transmission route of this pathogen.

There are 9–10 known *Brucella* species which can be rapidly differentiated, using a multiplex PCR assay (Kang et al., 2011).

7.3.3.4 Francisella tularensis. *Francisella tularensis* is a gram-negative non-spore forming bacterial pathogen which causes tularemia (fever, weight loss, pneumonia). *Francisella tularensis*, subsp. *tularensis* is more virulent than *F. tularensis*, subsp. *holarctica* and has a fatality rate of 5–15%, if left untreated (Dennis et al., 2001). This pathogen attacks mostly animals, and humans may become infected following contact with diseased animals, ingestion of contaminated drinking water or meat, and via inhalation (Ellis et al., 2002). This pathogen has a low infectious dose (10–50 cells) (Eitzen et al., 1998; Franz et al., 1997). The CDC has designated this pathogen as a category A agent. Patients can be treated with doxycycline to fight the infection.

Francisella is a fastidious bacterium that can be detected on commercially available selective growth media containing antibiotics, and the bacterial colonies are confirmed via a slide agglutination assay. Acid treatment prior to plating reduces the competition from other bacteria and improves the recovery of *Francisella* (Humrighouse et al., 2011). Due to its slow growth in culture media, *Francisella* can be rapidly detected via swabing, precessing of the swabs followed by use of real-time PCR (Walker et al., 2010).

7.3.3.5 Burkholderia mallei.
This gram-negative bacterium is the causative agent of glanders, a disease affecting mostly horses, donkeys, and mules, but may also infect humans via the airborne route. It was used during the world wars to infect animals and sometimes humans. An important symptom is the inflammation of mucus membranes of the nose. The septicemic form of glanders, if left untreated, has a 100% mortality rate. The minimum infectious dose is not known.

7.3.3.6 Clostridium botulinum.
Botulism is caused by a neurotoxin produced from the anaerobic, spore-forming bacterium *C. botulinum*. This neurotoxin has a toxic dose of 0.001 μg/kg body weight, and is 15,000 times more toxic than the nerve agent VX and 100,000 times more toxic than sarin (Arnon et al., 2001). It causes nerve and muscle paralysis and may culminate in respiratory failure. The route of exposure to this toxin could be via aerosolization or water contamination although some suggest that food contamination could be the likeliest exposure route (Duerden, 2006). Treatment consists of rapid administration of botulism antitoxin.

7.3.3.7 Smallpox Virus.
Although it was considered eradicated in the 1970s, the smallpox virus is still a serious threat to public health if deliberately released in the environment (Henderson et al., 1999). This virus seems to have a low infectious dose and can be contracted through aerosolization and through person-to-person contact. The fatality rate of this virus is around 30%. It may possibly be weaponized but its water threat is unknown at the present time.

7.3.3.8 Hemorrhagic Fever Viruses.
These viruses (e.g. Marburg, Ebola, Lassa) could also be used in biological warfare. They have a low infectious dose and may cause a high mortality rate among the victims. However, their transmission via the waterborne route is presently unknown.

7.3.4 Biotoxins

Table 7.3 shows the major bacterial, fungal, and algal biotoxins, their potential for weaponization and threats to drinking water supplies.

7.3.4.1 Clostridium botulinum Toxins.
Clostridium botulinum, and more rarely *Clostridium butyricum* and *Clostridium baratii*, are anaerobic spore-forming

TABLE 7.3 Water threat from biotoxins

Biotoxin	Amenability to Weaponization	NOAEL or LD_{50}	Water Threat
Aflatoxins	Yes	LD_{50} = 10–100 mg/person	Yes
Anatoxin A	Unknown	LD_{50} = 200 µg/kg (mice)	Probable
Botulinum toxins	Yes	0.003 µg/kg (mice) 0.006 µg/kg (humans)	Yes
Microcystins	Possible	ID_{50} = 1–10 mg/person NOAEL = 10 µg/L WHO standard = 1 µg/L (lifetime exposure)	Yes
Ricin	Yes	LD_{50} = 20 mg/kg (mice) NOAEL = 2 µg/L	Yes
Saxitoxin	Possible	ID_{50} = 0.3–1 mg/person NOAEL = 0.1 µg/L	Yes
T-2 mycotoxin	Probable		Yes
Tetrodotoxin	Possible	LD_{50} =30 µg/kg NOAEL = 0.1 30 µg/L	Yes

Source: Adapted from Burrows and Renner (1999). Env. Hlth. Persp. 107: 975
LD_{50}, lethal dose affecting 50% of the population; NOAEL, no observed adverse effect level.

bacteria that produce the most potent and deadliest neurotoxins known to humans. The LC_{50} of one of the botulism toxin is less than 0.01 µg/kg body weight. Exposure to this toxin generally occurs through ingestion of contaminated canned food or through the respiratory route and can result in muscle paralysis and death by respiratory failure. An antitoxin is available and is effective against type E toxin. There are seven types (designated by letters A to G) of botulinum neurotoxins and the main ones associated with human botulism are A, B, E, and F (Peck and Stringer, 2005). The traditional method of assessing the toxicity of these neurotoxins is the mouse bioassay. However, multiplex PCR methods were developed for the detection of these toxins with more than 99% accuracy (Lindström et al., 2001; de Medici et al., 2009).

7.3.4.2 Mycotoxins.   Mycotoxins are secondary metabolites produced by fungi such as *Penicillium*, *Fusarium*, or *Aspergillus*. Exposure to these toxins occurs via inhalation, ingestion, or skin contact. They are extremely stable under environmental conditions. *Penicillium* species produce the carcinogens islandotoxin and patulin and the hepatoxicant luteoskyrin. *Fusarium moniliforme* is one of the most common fungi colonizing corn. It produces fusarin, a potent mutagen, and fumonisin, a cancer promoter (Chu and Li, 1994). *Fusarium* also produces the T2 toxin which affects mostly livestock at micrograms or nanograms per gram of food. Signs of exposure of animals to T2 toxin range from feed refusal to cardiovascular shock (Albertson and Oehme, 1994). A dipstick enzyme immunoassay is available for detecting nanograms of T2 toxin (De Saeger and van Peteghem, 1996). This toxin is probably amenable to weaponization and would pose a water threat. However, its use as a bioweapon is limited because it would be unpractical to produce mass quantities of this toxin (Paterson, 2006).

Aflatoxins are produced by *Aspergillus flavus* or *A. parasiticus* and are potent carcinogens, mutagens, and teratogens. Humans and animals come into contact with these chemicals through consumption of contaminated corn, peanuts, and other agricultural crops. Humid local storage conditions are responsible for the high aflatoxin concentrations in food crops. In the United States, the action level for total aflatoxin is 20 µg/kg of food. Aflatoxins have been weaponized and pose a threat to the safety of drinking water (Zilinskas, 1997).

7.3.4.3 Ricin Toxin.
This toxin is produced by the castor plant *R. communis*. It is toxic (LD_{50} = 20 mg/kg for mice) via the oral and respiratory routes. Ricin production is relatively inexpensive but its weaponization for mass destruction is not well known. No vaccine is available against ricin.

7.3.4.4 Saxitoxin.
Dinoflagellates produce an alkaloid, saxitoxin, which acts as a neurotoxin causing paralytic shellfish poisoning. The main exposure routes are ingestion or exposure to aerosols. Ingestion of this biotoxin may lead to paralysis while breathing it may lead to respiratory failure and death. The LD_{50} for this toxin is 0.3–1 mg/person (Burrows and Renner, 1999).

7.3.4.5 Microcystins.
These toxins are produced by cyanobateria (e.g., *Microcystis aeruginosa*) and are highly toxic to humans and animals via the ingestion or respiratory routes (see Chapter 5 for more details). Microcystin–LR would be a prime candidate for weaponization.

7.3.4.6 Tetrodotoxin.
Tetrodotoxin is produced by certain species of puffer fish (fugu fish in Japan). This potent (LD_{50} = 30 µg/kg) neurotoxin accumulates in the fish ovaries and liver and has a 60% mortality rate.

7.4 DELIBERATE CONTAMINATION OF WATER SUPPLIES WITH BW AGENTS OR BIOTOXINS

7.4.1 Introduction

Historically, during World War II, the Japanese contaminated Chinese water supplies with *B. anthacis*, *Shigella* spp., *Salmonella* spp., *Vibrio cholerae*, and *Y. pestis*. Since 2001, several sporadic cases of bioterrorism threats to water supplies around the world have been documented (Ping, 2010). Due to our existential dependence on water for survival, drinking water becomes a highly vulnerable and desirable target for terrorists. The possibility exists that terrorists could strike the water supply sources, water treatment plants, and water distribution, and storage in cities around the world. They can also carry out a cyber attack of the network or bomb water treatment facilities to disrupt the supply of drinking water.

Water security was relatively lax in the past. However, in 1998, Directive 63 signed by President Bill Clinton, recognized drinking water as a critical infrastructure (Linville and Thompson, 2006). The 9/11 attack on US soil led to drastic changes in water security. The U.S. EPA in charge of the security of water and wastewater systems has developed a series of programs to improve the security of utilities.

7.4.2 Actions Taken to Disrupt Water Supply

Different terrorist actions could be taken to adversely disrupt the system (Denileon, 2001):

- *Physical destruction of the system.* This leads to the disruption of water supply to customers due to destruction of equipment (pumps, chlorinators, power source, computers for data acquisition) and structures. The release of chlorine gas could affect the plant personnel as well as the adjacent neighborhoods.
- *Deliberate chemical contamination of the water supply.* The water distribution system is particularly vulnerable to attack because of its accessibility (through distribution reservoirs and fire hydrants).
- *Cyber attack.* This is to disrupt the operation of the plant. Cyber terrorists are trying to get remote access to the monitoring of the operation of our infrastructure, including the drinking water production. Vulnerability to cyber attack has been increased by the transition from proprietary supervisory control of the plant to the SCADA (Supervisory Control and Data Acquisition) system that gathers and analyze real-time data from the plant.
- *Bioterrorist attack.* Drinking water treatment plants have been identified as possible targets of bioterrorism. Although most of biological warfare agents are intended to be delivered via the more efficient aerosol route, the waterborne route would be of no less concern. Deliberate water contamination may occur at the source water, on site at the water treatment plant, or in the distribution networks.

7.4.3 Some Safeguards to the Deliberate Contamination of Drinking Water

Some of the safeguards are (Khan et al., 2001):

- Effect of dilution on the introduced BW agents.
- Physical, chemical, and biological factors contribute to the inactivation of any introduced microorganisms at the source.
- Water treatment plants offer multiple barriers that help in the physical removal (e.g., coagulation/flocculation, sand filtration, activated carbon) and inactivation of pathogens and parasites (e.g., disinfection, water softening) (see Chapters 2 and 3). The US Army mobile water treatment units use reverse osmosis (RO) in addition to the above treatment processes. RO is assumed to remove protozoan

cysts, bacteria, viruses, and biotoxins. However, RO treatment may sometimes fail, resulting in lower removal of pathogens. Some work was undertaken to determine the fate of *B. anthracis* following water disinfection. A wide range of physical and chemical methods have been used to inactivate spores of *B. anthracis* and other bacilli (*B. cereus*, *B. globigii*, *B. subtilis*) in the environment which include water, air, and contaminated surfaces (Spotts et al., 2003). The chemicals used include hypochlorite, hydrogen peroxide, peracetic acid, formaldehyde, glutaraldehyde, sodium hydroxide, ethylene oxide, chlorine dioxide, or ozone. Chemical disinfection achieves from 1- to 6-log inactivation of the spores. As regard the *Ct* values of free chlorine and monochloramine for bacterial bioterrorism agents, *B. anthracis* displays the highest *Ct* values (up to $Ct = 15{,}164$ mg.min/L for monochloramine and 271 mg.min/L for free chlorine for a 3-log inactivation at 5°C (Table 7.4 Rose et al., 2005, 2007).

Bacillus spores can also be inactivated by UV (Nicholson and Galeano, 2003) and gamma irradiation. However, *B. anthracis* spores are quite resistant to UV radiation. A 3-log inactivation of these spores is achieved at a UV fluence (dose) of more than 120 mJ/cm^2 as compared to 3.2–7.9 mJ/cm^2 for the nonspore forming bacterial agents

TABLE 7.4   Inactivation of biowarfare agents with free chlorine and monochloramine (3-log reduction)

BW Agent	Temperature (°C)	Free Chlorine, *Ct* (mg min/L)	Monochloramine *Ct* (mg min/L)
Burkholderia mallei M-9	5	0.2	194.1
	25	0.2	64.6
Burkholderia pseudomallei	5	0.7	156.1
	25	0.6	45.9
Brucella melitensis ATTC 23456	5	0.5	579.5
	25	0.2	116.6
Brucella suis	5	0.4	156.8
	25	0.2	56.1
Francisella tularensis NY98	5	10.3	116.0
	25	3.9	37.1
Francisella tularensis LVS	5	2.4	97.9
	25	1.0	30.4
Yersinia pestis A1122	5	0.7	115.6
	25	0.6	33.1
Yersinia pestis Harbin	5	0.04	91.4
	25	0.04	25.0
Bacillus anthracis Sterne	5	271	15,164
	25	86	1847
Bacillus anthracis Ames	5	339	6813
	25	102	1204

Source: Adapted from Rose et al. (2005). Appl. Environ. Microbiol. 71: 566–568; Rose et al. (2007). Appl. Environ. Microbiol. 73: 3437–3439.

TABLE 7.5   UV fluence required for 3-log inactivation of biowarfare microorganisms

Microorganism	UV Fluence (mJ/cm^2) Required for 3-Log Inactivation
Bacillus anthracis Ames	<120
Bacillus anthracis Sterne	>120
Brucella suis MO562	5.6
Brucella suis KS528	7.9
Brucella melitensis ATTC 23456	7.6
Burkholderia mallei M-9	3.8
Burkholderia mallei M-13	4.1
Burkholderia pseudomallei ATTC 11688	5.5
Burkholderia pseudomallei CA650	4.3
Francisella tularensis LVS	4.8
Francisella tularensis NY98	6.3
Yersinia pestis A1122	3.7
Yersinia pestis Harbin	3.2
Other pathogens or parasites (shown for comparison)	
Escherichia coli	6.7
Cryptosporidium	12
Giardia	11
Adenovirus type 2[a]	119

Source: Adapted from Rose and O'Connell (2009). Appl. Environ. Microbiol. 75: 2987–2990.
[a]Gerba et al. (2002).

(e.g., *F. tularensis, Y. pestis, B. mallei, Brucella suis*) or approximately 10 mJ/cm^2 for *Cryptosporidium* or *Giardia* (Table 7.5 ; Rose and O'Connell, 2009).

Unfortunately, there is a lack of information concerning the removal of biological warfare agents in conventional water treatment plants as well as in hand-held devices based on filtration through activated carbon or reverse osmosis (Burrows and Renner, 1999). In the home environment, point-of-use (POU) devices should also be investigated for their ability to remove BT agents and biotoxins (see Chapters 2 and 8 for more information on these devices).

Despite these safeguards, some pathogens or parasites (e.g., *Cryptosporidium*) can break through the multiple barrier system. An example of the ability of drinking water to cause mass casualties in urban populations is the *Cryptosporidium* outbreak in Milwaukee, USA, in 1993. This outbreak affected 403,000 consumers, of whom 4400 were hospitalized and 54 died of the disease (Kaminski, 1994; MacKenzie et al., 1994). A retrospective analysis of this outbreak showed that the total cost of that outbreak was $96.2 million in medical costs and productivity losses (Corso et al., 2003). Since *Cryptosporidium* oocysts are not effectively inactivated by disinfectants, their removal requires proper sand filtration. Many waterborne outbreaks involving several pathogens or parasites have been documented worldwide. Similarly to municipal tap water, bottled water can also be subject to accidental or intentional contamination with microbial pathogens (see Chapter 9)

There is also little information concerning the fate of biotoxins in drinking water and their removal by water treatment processes. The information we have concerns some algal biotoxins (e.g., microcystins) that cause problems in water treatment plants and reservoirs when using surface waters as the source water (see Chapter 5 for more details).

Most of the risk assessments performed by government and other agencies concerned mostly the military (Rotz et al., 2002). However, civilian populations include a wide range of age and health status groups and this makes the assessment more complex. In addition to the airborne route, civilian populations may also be exposed via the foodborne and waterborne routes.

An example of intentional contamination of the food supply has been documented in an Oregon restaurant where the salad bar was deliberately contaminated with *Salmonella* sp. (Torok et al., 1997).

7.5 EARLY WARNING SYSTEMS FOR ASSESSING THE CONTAMINATION OF SOURCE WATERS OR WATER DISTRIBUTION SYSTEMS

7.5.1 Introduction

Source waters are prone to natural, accidental, or intentional (e.g., bioterrorist activity) contamination. The U.S. EPA has established a contamination warning system that includes online water quality monitoring, and potentially inline sensors (i.e., inside water distribution pipes), sampling and analysis of various contaminants along the distribution system to establish a baseline, enhanced security monitoring to detect and respond to possible contamination, surveillance of consumer complaints, and public health surveillance (Perelman and Ostfeld, 2013; www.epa.gov/watersecurity). This contamination warning system has been simulated in Cincinnati, Ohio, and other large US cities (Fencil and Hartman, 2009).

7.5.2 Early Warning Systems

Early warning systems (EWSs) that monitor the quality of source waters or water distribution systems are a necessity in order to protect the consumers. An ideal EWS should be reliable (few false-positives and negatives), rapid, inexpensive, and requiring low maintenance. EWS must be based on rapid detection technologies to allow implementation of effective response by the local or national public health authorities and emergency management officials (Foran, 2000; Lim et al., 2005; Mikol et al., 2007).

In drinking water treatment plants with intakes from rivers, the most important contaminants are hydrocarbons, chemicals released as a result of industrial and transportation accidents, pesticides from soil runoffs, and microbial pathogens and parasites from untreated or inadequately treated effluents from wastewater treatment plants. The main components of an EWS include pollutant detectors, characterization for confirmation, communication to the proper authorities and the public, and prompt response to the pollution event (Gullick et al., 2003).

If deliberate microbial or chemical contamination is suspected, the U.S. EPA recommends that the utilities conduct field-screening tests which include radioactivity measurement, pH, chlorine, cyanide, and other tests if deemed necessary. Several other analytical tests are available for the preliminary assessment of water safety. A monitoring program may comprise several analytical procedures and approaches (Gullick et al., 2003; States et al., 2004):

- *Physical analyses*: Online probes measure pH, conductivity, dissolved oxygen, or turbidity. The latter has been correlated with the presence of oocysts of hardy parasites such as *Cryptosporidium*.
- *Chemical analyses*: Various online analytical probes are available for the measurement of anions (e.g., chloride, nitrate, nitrite), cations, certain metals (e.g., copper, lead, cadmium), chlorophyll, and organic compounds (e.g., analyses for total organic carbon, surfactants, hydrocarbons, pesticides, or various organics by using gas or liquid chromatography). Portable GC/MS devices can be used for onsite detection of chemical threats.
- *Automated biochemical tests*: Government and commercial laboratories can adapt automated biochemical tests to biothreat agents. These assays are based on substrate utilization patterns (e.g., VITEK or BIOLOG) or fatty acids profiles (e.g., MIDI).
- *Microbial biosensors*: Some examples are microbial biosensors to detect metals or specific organic compounds. There are now sensitive and rapid quantitative real-time PCR tests to detect biological warfare agents or their surrogates in environmental samples (Saikaly et al., 2007).
- *Rapid immunoassays:* BW agents and chemicals (i.e., antigens) may trigger an immune response that can be detected using an antibody. Immunoassays are generally less sensitive than molecular methods. Most of the kits are based on the enzyme-linked immunosorbent assays (ELISA) and immunochromatographic assays or array immunoassays (Emanuel et al., 2000; Fisher et al., 2009; Lim et al., 2005; Peruski and Peruski, 2003; Rowe-Taitt et al., 2000).

    Some commercial test strips allow the detection of *B. anthracis, Y. pestis, F. tularensis, C. botulinum* toxin, ricin, and staphylococcal enterotoxin B (see www.tetracore.com). However, these test strips have a relatively low sensitivity (detection limit is around 10^5 bacteria/mL) and sometimes low specificity.
- *Rapid enzymatic tests:* Enzymatic tests are quite useful in detecting the presence of certain toxic chemicals in environmental samples. An example is a test based on the specific inhibition of acetyl cholinesterase by organophosphates and carbamates.
- *Molecular-based methods*: A wide range of methods (e.g., microchip array technology, PCR) are now available for the detection of human pathogens and parasites (Brettar and Hofle, 2008; Elad et al., 2008b; Ivnitski et al., 2003). PCR-based detection methods are more sensitive than immunoassays. There are now sensitive, specific and rapid quantitative real-time quantitative PCR tests to detect biological warfare agents (Iqbal et al., 2000; Lim et al., 2005) or their surrogates in environmental samples (Saikaly et al., 2007).

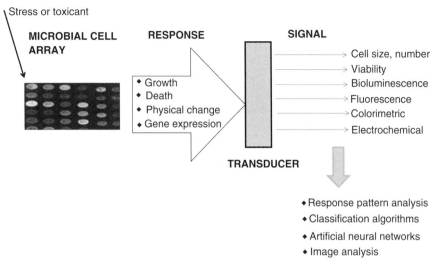

Figure 7.1 Generalized concept of cell array biochip technology. Adapted from Elad et al. (2008a). Microb. Biotechnol. 1: 137–148.

Array platforms are made of silicon, glass, or various polymers and are based on oligonucleotides, enzymes and other proteins, antibodies, or whole live eukaryotic or prokaryotic cells. The array is exposed to various stresses or toxicants and the signal is detected via fluorescence, bioluminescence, colorimetry, electrochemistry, or viability assays. The response pattern or signature is analyzed using statistical methods (Figure 7.1; Elad et al., 2008a).

Whole-cell bacterial arrays consist of batteries of genetically engineered bioluminescent bacteria where the *luxCDABE* operon are fused to stress- or toxic chemical-specific gene promoters. Other popular bioreporters are the *lacZ* (β-galactosidase) and *gfp* (green fluorescent proteins) genes. They can be used to detect categories of toxicants (e.g., heavy metals such as mercury, phenol, polycyclic aromatic hydrocarbons) or stresses (e.g., heat stress, oxidative stress) (Belkin, 2003; Biran et al., 2003; Elad et al., 2008b; Gu et al., 2004; Popovtzer et al., 2005).

Routine use of whole-cell arrays aquatic toxicity testing will necessitate further studies on their activity following long-term storage and shipping (Melamed et al., 2012).

- *Online biomonitors for chemical toxicity, biotoxin concentration, and BW agents*: Use of online biomonitors based on the activity of bacteria, algae, cyanobacteria, mussels, *Daphnia, Ceriodaphnia*, or fish. These EWSs can be helpful in the rapid detection of accidental or deliberate spills from point and nonpoint sources. They measure physiological responses of test organisms, such as ventilatory and swimming movement patterns in fish, swimming in daphnids, or valve closure in clams (Mikol et al., 2007; Storey et al., 2011). For example, the *Daphnia* Toximeter, based on changes of the swimming behavior (e.g.,

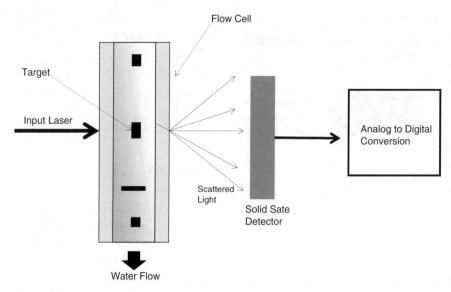

Figure 7.2   Basic MALS system. *Source:* Adams, J.A. and D. McCarty (2007). Intern. J. High Speed Electronics & Systems 17 (4): 643–659.

swimming speed) of *Daphnia magna*, is used to continuously monitor the presence of toxic chemicals and odor compounds in source waters. Results of the bioassay can be obtained in minutes instead of 24–48 hours for the traditional *Daphnia* acute toxicity test (De Hoogh et al., 2006; Lechelt et al., 2000; Watson et al., 2007). An algae online analyzer fluorimeter was proposed as an EWS for the presence of toxic cyanobacteria in source water. This analyzer measures in real-time cyanobacterial chlorophyll *a* which is positively correlated with microcystin concentration (1 μg microcystin is equivalent to 4.8 μg cyanobacterial chlorophyll *a* (Izydorczyk et al., 2009).

Other approaches have been proposed for the continuous, real-time online monitoring of pathogens in the source water and during treatment and distribution. The BioSentry™ Water Monitoring System is an EWS that utilizes a laser beam that is focused on microorganisms in the water. The scattered light is captured by a photodetector and forms a pattern ("fingerprint") that is unique for a given pathogen (Figure 7.2; Adams and McCarty, 2007). This system appear to perform well as regard parameters such as response time, ease of use, automation, sensitivity, and capital and operating cost.

7.6 PROTECTION OF DRINKING WATER SUPPLIES

The Biological and Toxin Weapons Convention (BTWC) of 1972 (http://www.fas.org/nuke/control/bwc/text/bwc.htm) deals with biological warfare agents (covers bacteria, protozoa, and viruses) and biotoxins (e.g., botulinum toxin) produced by microorganisms. It forbids their development, production, and stockpiling. Limited information is available on the safe levels of these agents in drinking water, and no current regulation or guideline covers specifically their presence in drinking water supplies (Pontius, 2002). However, the 9/11/2001 terrorist attack on American soil has led to a closer partnership between the U.S. EPA and the drinking water community to safeguard the safety of the drinking water supplies. They were joined in this effort by the American Water Works Association (AWWA) which helps in the dissemination of information concerning the security of water utilities. In 2002, the Safe Drinking Water Act was amended as the US Congress passed the Public Health Security and Prevention Preparedness Act (also called the Bioterrorism Act) (PL 107-188) which mandates security requirements (vulnerability assessment and emergency response plans) for thousands of water utilities in the United States (Pontius, 2003; Whitler, 2007). Vulnerability assessment includes water collection, pretreatment and treatment, distribution pipes, storage facilities, safety issues related to the use and storage of chemicals used in the plant, and computerized data acquisition systems. Following the vulnerability assessment, the utilities must submit emergency response plans to the U.S. EPA to respond to a potential terrorist attack. The U.S. EPA also proposed a response protocol toolbox that describes the procedures to address contamination threats to drinking water safety. These plans may be different from those addressing natural disasters such as an earthquake or a tornado.

Information Sharing and Analysis Centers (ISACs) were established to promote the exchange of information between water utilities and security agencies. Water-ISAC is a clearinghouse for the flow of information concerning biological, chemical, radiological, and cyber threats to water and wastewater treatment plants. Some Water-ISAC resources include databases on contaminants and emergency contacts, and new search engines. An example of a useful database is the United Kingdom Water Industry Research (UKWIR) concerning toxicity and microbiology data (Sullivan, 2004).

The deliberate contamination of our drinking water should be prevented by protecting water reservoirs and water distribution systems. The government should stockpile, if available, vaccines against microbial pathogens and antidotes against chemical agents.

7.7   DISINFECTION OF BW-CONTAMINATED DRINKING WATER

Most of the research on decontamination of BW-contaminated drinking water supplies deals with the inactivation of *B. anthracis* spores which are notoriously known for their high resistance to disinfectants and environmental adverse conditions. Spore resistance to chemical agents is mostly due to the protective protenaceous spore coat and to the inner membrane which exhibits a low permeability to hydrophilic molecules (Setlow, 2000).

The disinfectants under consideration for their sporicidal action are sodium hypochlorite, chlorine dioxide, hydrogen peroxide, Dichlor (sodium dichloro-*S*-triazinetrione dihydrate), oxone (potassium peroxymonosulfate as the active ingredient), Decon (consisting of hydrogen peroxide, surfactants, and fatty alcohols), L-Gel

TABLE 7.6   Summary of current decontamination
methods for biological warfare agents and biotoxins

Liquids
Alcohol, ethyl
Alcohol, isopropyl
Alcohol, isopropyl + 5% propylene oxide
Glutaraldehyde
Hydrogen peroxide
Phenol
Soap and water
Sodium hypochlorite
VirkonS™ (Antex Corporation)
Sodium hydroxide
Gases and vapors
Ethylene oxide
Paraformaldehyde
Steam
Hydrogen peroxide, vapor phase
Physical agents (energy sources)
Cobalt 60 Incineration
Heat
Ultraviolet light
X-rays

Source: Adapted from Raber and McGuire (2002). J. Hazard. Mater.
B93: 339–352.

(consisting of Oxone and a fumed silica gelling agent), and VirkonS (a triple salt of potassium peroxymonosulfate, potassium hydrogen sulfate, and potassium sulfate) (Raber and McGuire, 2002; Rose et al., 2005, 2007; U.S. EPA, 2007a; Young and Setlow, 2004). We mentioned previously that the Ct values for 3-log inactivation by free chlorine or monochloramine were orders of magnitude higher for *B. anthracis* than for the other BW agents (Rose et al., 2005, 2007). Hydrogen peroxide (5% concentration), Dichlor, and sodium hypochlorite (2% concentration) were the most effective against *B. atrophaeus* and *B. thuringiensis* (surrogates of *B. anthracis*) (Raber and Burklund, 2010). The mechanism of spore inactivation by Decon and Oxone is the damage to the spore inner membrane (Young and Setlow, 2004).

Table 7.6 shows summary of decontamination methods for biological agents and biotoxins (Raber and McGuire, 2002).

WEB RESOURCES

http://water.epa.gov/infrastructure/watersecurity/index.cfm
(U.S. EPA water security home)

http://www.bt.cdc.gov/bioterrorism
(Center for Disease Control and Prevention page on bioterrorism)

www.tetracore.com
(test strips for the detection of bioterrorism agents and biotoxins)

http://www.waterisac.org/
(webpage for WaterISAC concerning the flow of information between utilities and security agencies)

http://www.fas.org/nuke/control/bwc/text/bwc.htm
(Biological and Toxin Weapons Convention)

www.epa.gov/nhsrc/pubs/600r09076.pdf
(Online Water Quality Monitoring in Water Distribution Systems: Sensor Technology)

FURTHER READING

Atlas, R.M. 2002. Bioterrorism: from threat to reality. *Ann. Rev. Microbiol.* 56:167–185.

Burrows, I.D., and S.E. Renner. 1999. Biological warfare agents as threats to potable water. *Environ. Health Persp.* 107:975–984.

Christopher, G.W, T.J. Cieslak, J.A. Pavlin, and E.M. Eitzen, Jr. 1997. Biological warfare: a historical perspective. *J. Am. Med. Assoc.* 278(5):412–417.

Elad, T., J.H. Lee, S. Belkin, and M.B. Gu. 2008. Microbial whole-cell arrays. *Microb. Biotechnol.* 1:137–148.

Hawley, R.L., and E.M. Eitzen, Jr. 2001. Biological weapons – A primer for microbiologists. *Ann. Rev. Microbiol.* 55:235–253.

Khan, A. L., D.L. Swerdlow and D.D. Juranek. 2001. Precautions against biological and chemical terrorism directed at food and water supplies. *Pub. Health Rep.* 116:3–14.

Lim, D.V., J.M. Simpson, E.A. Kearns, and M.F. Kramer. 2005. Current and developing technologies for monitoring agents of bioterrorism and biowarfare. *Clin. Microbiol. Rev.* 18(4):583–607.

Rose, L.J., and H. O'Connell. 2009. UV light inactivation of bacterial biothreat agents. *Appl. Environ. Microbiol.* 75:2987–2990.

Rotz, L.D., A.S. Khan, S.R. Lillibridge, S.M. Ostroff, and J. M. Hughes. 2002. Public health assessment of potential biological terrorism agents. *Emerg. Infect. Dis.* 8(2):225–229.

Sheehan, T. 2002. Bioterrorism. In *Encyclopedia of Environmental Microbiology*, G. Bitton (editor-in-chief). Wiley-Interscience, New York, pp. 771–782.

8

WATER TREATMENT TECHNOLOGIES FOR DEVELOPING COUNTRIES

8.1 INTRODUCTION: WATER FOR A THIRSTY PLANET

According to the United Nations Educational Scientific and Cultural Organization (UNESCO), 97.5% of the planet water resources is saltwater while the remaining 2.5% is freshwater. About two-thirds of the freshwater resources is in glaciers and the permanent ice covers in the Arctic and Antarctic. The remaining one-third is made of underground water, some of which is accessible, and surface waters (rivers, streams, and lakes). Thus, less than 1% of the Planet water resources can be used by humans (Rooy, 2003; UNESCO, 2009). Moreover, 80% of the water available for human use is utilized by agriculture, livestock, and energy production, further exacerbating the water shortage in the world (Shannon et al., 2007). Unfortunately, the world water supply is also being greatly affected by climate change which mainly results in changes in precipitation patterns, increased drought, and decreased snow cover.

For a thirsty planet, water is a precious essential resource that is needed for domestic, livestock, industrial, and irrigation uses. Developing countries use much less water than developed ones. As regard domestic use, per capita water consumption is twice higher in developed countries than in developing ones. Figure 8.1 (Rosegrant et al., 2002) shows the per capita water consumption by world region and the projections for 2025. Access to safe water and sanitation is one of the 10 public health achievements between 2001 and 2010. The other achievements are reduction in child mortality, vaccine-preventable diseases (e.g., polio, measles, diphtheria, tetanus, hepatitis B), prevention and control of HIV, malaria prevention and control, tuberculosis control, tropical diseases control, tobacco control, global road safety, and improved preparedness and response to global health threats (e.g., influenza A [H1N1] pandemic) (CDC, 2011).

Microbiology of Drinking Water Production and Distribution, First Edition. Gabriel Bitton.
© 2014 John Wiley & Sons, Inc. Published 2014 by John Wiley & Sons, Inc.

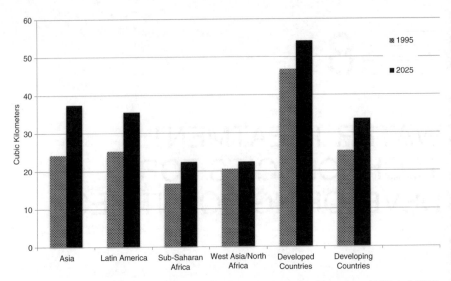

Figure 8.1 Per capita worldwide domestic water consumption by region, 1995 and 2025. Adapted from Rosegrant et al. (2002). *Global Water Outlook to 2025: Averting an Impending Crisis.* Food Policy Report. Inter. Food Policy Res. Inst. (IFPRI) and Intern. Water Management Inst. (IWMI), 28pp.

8.2 SOME STATISTICS OF WATERBORNE DISEASES IN DEVELOPING COUNTRIES

An estimated one billion of the world's population does not have access to safe clean water, and approximately 2.6 billion people lack improved sanitation (WHO/UNICEF, 2000, 2004, 2010; WHO, 2003). Another sign of poor sanitation in developing countries is the finding that 1.1 billion people still defecate in the open (WHO/UNICEF, 2010). Consequently, more than two million people die each year from waterborne diseases. Ninety percent of diarrheal cases are related to unsafe water and poor sanitation and hygiene. Several epidemiological studies have been carried out in developing countries in Africa, Asia, and South America. There are approximately 4 billion cases of diarrheal disease per year, resulting in up to 2.5 million deaths/year from endemic diarrhea (Kosek et al., 2003). Endemic diarrhea causes 1.7 billion episodes/year and accounts for approximately 17% of all deaths among children younger than 5 years (Black et al., 2010; WHO, 2005). Diarrhea and intestinal worm infections account for 117 million disability adjusted-life year (DALY). A study in Yaounde, Cameroon, revealed a diarrhea prevalence of 14.4% linked to contaminated drinking water (Yongsi, 2010).

The United Nations-sponsored Millennium Development Goal (MDG) (http://www.un.org/millenniumgoals) aims at increasing the access to safe drinking water and sanitation in developing countries. MDG aims at decreasing by 50% the number of people without access to safe drinking water and sanitation by 2015

(WHO/UNICEF, 2010). Except for certain regions such as Sub-Saharan Africa, WHO/UNICEF (2012) data show that the MDG drinking water target has been met in 2010 as 2 billion people gained access to safe drinking water sources between 1990 and 2010. Thus, about 11% of the world population is still using unsafe drinking water sources and about 2.5 billion people do not have access to improved sanitation.

The accomplishment of these goals necessitates the development of low cost, efficient, sustainable, and socially acceptable drinking water treatment technologies.

8.3   SOME HWT METHODS OR TECHNOLOGIES IN USE IN DEVELOPING COUNTRIES

Household water treatment (HWT) comprises technologies, devices, or methods to treat water in homes or at point-of-use (POU) in schools, hospitals, and other facilities (WHO, 2011b). Several technologies and interventions have been proposed to improve drinking water quality in developing countries. These point-of-use or community-scale water treatment technologies must be simple, low cost, low maintenance and should require locally available materials. Water, sanitation, and hygiene interventions or their combination have a similar degree of impact on the reduction of the risk of diarrheal diseases in developing countries. These interventions are summarized in Figure 8.2 (Fewtrell et al., 2005). An important intervention is the use of POU devices which allow individuals with no access to safe water to improve the quality of their drinking water.

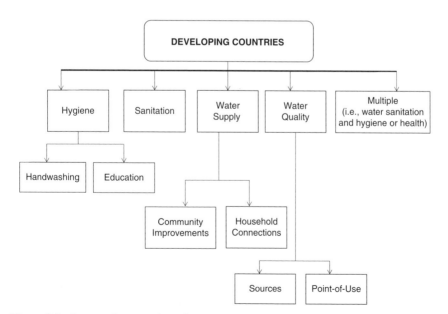

Figure 8.2   Interventions to reduce diarrheal diseases in developing countries. Adapted from Fewtrell et al. (2005). Lancet Infect. Dis. 5:42–52.

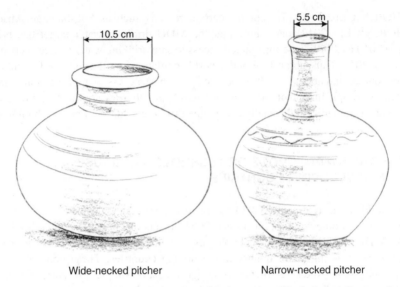

Wide-necked pitcher   Narrow-necked pitcher

Figure 8.3   Traditional wide-necked and modified narrow-necked pitcher. *Source*: Jensen et al. (2002). Trop. Med. Internat. Health 7: 604–609.

Since community water treatment plants are practically inexistent in some of those areas, it is imperative to develop efficient and sustainable technology to safely treat and store drinking water. Additionally, measures must be taken to avoid the recontamination of the treated water at home. One simple and efficient measure is the use of narrow-necked jars instead of the wide-necked ones which allow water recontamination with dirty hands (Figure 8.3; Jensen et al., 2002).

It appears that treating water at home is more effective than improving the quality of the source water. Hence, the use of the household methods has been recommended to reduce diarrheal diseases in the developing world. Household water treatment methods are mainly based on boiling, filtration, solar disinfection, chlorination, or flocculation in combination with chlorination. Some of the POU household water treatment technologies that meet the effectiveness and sustainable use criteria are summarized below (Powers et al., 1994; Sobsey, 2002; Sobsey et al., 2008).

8.3.1   Boiling Water

Heat-based technologies include boiling and pasteurization (temperature higher than 63°C for 30 minutes). Boiling water is perhaps the ultimate means of disinfecting drinking water. Boiling offers the advantage of killing most pathogens and parasites. Results from 67 national surveys around the world revealed that boiling is practiced in 21% (598 million consumers) of the households surveyed (Rosa and Clasen, 2010). A Chinese survey in 6948 sites covering 65,839 households showed that 85% of rural households boiled their water while 5% treated water by other means such as filtration (Zhang et al., 2009c cited by Yang et al., 2012). Studies have been undertaken on the effectiveness of this practice in developing countries. In a rural community in Guatemala, boiling water has led to 86% mean reduction of total thermotolerant coliforms (TTC). Almost three-fourth of the boiled water samples met the WHO guideline of 0 TTC/100 mL (Rosa et al., 2010). Similarly, in Cambodia, collection of 369 matched and boiled water samples showed a 98.5% inactivation of *Escherichia coli* (Brown and Sobsey, 2012). The incomplete inactivation may have been due to incomplete boiling or to postcontamination.

However, some disadvantages of this practice are the inability to reduce turbidity, fuel cost, and the lack of protection from postcontamination (Clasen et al., 2007; Psutka et al., 2011). Storing the boiled water in a covered container reduced the chance of postcontamination (Brown and Sobsey, 2012).

The monthly average cost of boiling water, including fuel and labor costs, was estimated at $1.50 for people using liquid fuel gas and $3.21 for those using wood (Clasen et al., 2008). Furthermore, in developing countries, this practice of burning biomass (e.g., wood, cattle dung) may increase indoor pollution and environmental degradation such as deforestation and the production of greenhouse gases at the global level. Boiling the daily supply of drinking water (approximately 10 L) requires three times the fuel needed to cook the daily meals (Gadgil, 2008). The practice of boiling water and cooking indoors leads to an estimated annual 1.6 million excess deaths from smoke inhalation (WHO, 2008).

Since cost is an issue regarding the practice of heating with wood and fossil fuels, solar heating (solar cooking) using opaque containers is used in areas where sun is plentiful. Solar heating can raise the temperature to more than 60°C, a level at which water is pasteurized, leading to pathogen inactivation.

A wide range of solar cookers are available for pasteurization of food or drinking water. One of the first solar cooker used around the world is the "box cooker." It is an insulated box made of various materials such as cardboard, wood, metal, or plastic and includes a reflector that boosts the solar power. Several other types of cookers are described elsewhere (Pejack, 2011; http://www.solarcookers.org/index.html).

8.3.2 Biosand Filter

Granular media filters are beds packed with sand, diatomaceous earth or other materials to purify water. One popular granular media filter is Biosand which is a household device that was developed by David Manz of the University of Calgary, Canada. It is based on intermittent slow sand filtration described in Chapter 2, with an idle time of 18–22 hours, and is being considered for people with no access to safe water in developing countries. Figure 8.4 shows a diagram of a Biosand filter (http://www.waterforcambodia.org/projects/biosand-filters). The sand filter can be housed in a concrete or plastic container. The plastic housing is of light weight and can be easily transported to remote locations. Some advantages of Biosand are low cost, ease of use and maintenance, sturdy design, relatively high flow rates (3–60 L/h), and ability to treat turbid waters (Tiwari et al., 2009). As of 2009, more than 200,000 Biosand units were installed in 70 countries around the world (http://www.cawst.org/en/resources/biosand-filter).

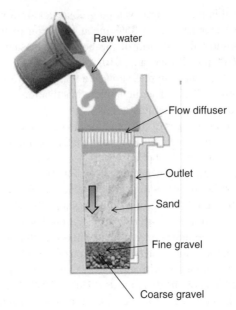

Figure 8.4   Biosand filter. Adapted from http://www.waterforcambodia.org/projects/biosand-filters. (For a color version, see the color plate section.)

The reduction of *E. coli* by Biosand averaged 94% under laboratory conditions but fluctuated between 0% and 99.7% under field conditions in the Dominican Republic. A study of 55 Biosand filters between 1999 and 2010 in Haiti showed an average 92% removal of *E. coli*. Several filters were still in use after 12 years of operation (Sisson et al., 2013). In a trial in Cambodia, the geometric mean reduction of *E. coli* in a plastic Biosand filter was 99.3%. A positive association was observed between water quality parameters (e.g., *E. coli* levels) and the incidence of diarrheal disease. Biosand use in Kenya, Dominican Republic, and Cambodia led to a significant reduction in diarrhea (Stauber et al., 2009, 2012; Tiwari et al., 2009).

The microbial community in the Biosand filter plays a role in microbial removal as biologically mature filters perform better than fresh filters (Bitton, 2011; Stauber et al., 2006). Sand size, residence time in the filter, and media aging (i.e., filter ripening due to the buildup of a biologically active layer, the *schmutzdecke*) appear to play an important role in bacterial and virus removal by Biosand, and an aging period of two or more weeks is necessary for bacterial and virus attenuation by the filter. It is thought that microbial activity leading to the production of proteolytic enzymes may be responsible for virus decay (Ahammed and Davra, 2011; Elliott et al., 2011; Jenkins et al., 2011). More than 99% of *Giardia* cysts and *Cryptosporidium* oocysts were removed following treatment by intermittent slow sand filtration (Palmateer et al., 1999).

Sand impregnation with silver helps in bacterial removal although some of the metal was detected in the filtrate (Bielefeldt et al., 2009). Similarly, Biosand modification by addition of iron oxide-coated sand appears to improve fecal coliforms and *E. coli* removal (Ahammed and Davra, 2011). Biosand amended with iron oxide removed more than 4 logs of rotavirus and MS2 phage (Bradley et al., 2011).

8.3.3 Ceramic Water Filters

POU filtration technologies include cloth and fiber filters (e.g., Indian sari used to remove copepod-associated *Vibrio cholerae*), membrane filters, and ceramic water filters. The latter were first introduced in England in the nineteenth century and several alternative designs (ceramic pot, candle, ceramic disk) are now marketed. This POU device consists of a ceramic filter made of clay, a combustible material (flour, rice husks, or sawdust), and fired in a kiln at 900°C. It is impregnated with colloidal or nanosize silver to help in the inactivation of microbial pathogens (Lantagne, 2001; Lv et al., 2009). Nanosize silver is added to the filter via painting, impregnation of the filter after firing, or mixed with the clay and combustible material before firing. The latter method resulted in the lowest release of silver in the effluent (Ren and Smith, 2013). A diagram of the ceramic filter is shown in Figure 8.5 (van Halem et al., 2009). Ceramic filters removed from 97.8% to 100% of *E. coli* (Oyanedel-Craver and Smith, 2008) and 75.1% of thermotolerant coliforms in three remote communities in Colombia (Clasen et al., 2005). The overall percentage of control and intervention households within the WHO risk categories is shown in Table 8.1 (Clasen et al., 2005). Ceramic filters are capable of removing turbidity and color as well as bacteria and protozoan cysts and oocysts. When challenged with polystyrene fluorescent microspheres of different sizes, the removal of virus-sized particles (0.02–0.1 μm) was highly variable and ranged between 63% and 99.6% while more than 99.6% of (oo)cysts-sized spheres were removed (Bielefeldt et al., 2010). This explains why less

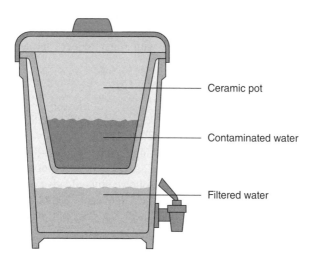

Figure 8.5 Ceramic filtration. *Source*: Van Hallem, et al. (2009). Physics and Chem. of the Earth 34 (1-2): 36–42. (For a color version, see the color plate section.)

TABLE 8.1 Overall percentage of samples from control and intervention households using ceramic filters, according to WHO risk categories

	Percentage of Samples by WHO Risk Category			
	0 TTC/100 mL	1-10 TTC/100 mL	11-100 TTC/100 mL	101-1000 TTC/100 mL
Control group	0.86%	7.30%	37.34%	54.51%
Intervention group	47.66%	24.22%	17.58%	10.55%

Source: Adapted from Clasen, T., G. Garcia Parra, S. Boisson, and S. Collin (2005). Amer. J. Trop. Med. Hyg. 73 (4): 790–795.
TTC, thermotolerant coliforms.

than 2-log removal was observed with MS2 phage (Brown, 2007). However, when ceramic filters based on diatomateous earth were modified by addition of positively charged materials such as magnesium oxyhydroxide, phage (MS2, PhiX174) removal increased by 4 logs (Michen et al., 2013).

A 6-month study in three communities in Colombia showed that the incidence of diarrhea was 60% lower in households using ceramic water filters than in controls (Clasen et al., 2005). A 5-month survey of 60 households in a rural community in Bolivia showed about 50% reduction (Clasen et al., 2006). Another 6-month intervention trial in South Africa and Zimbabwe using ceramic filtration showed an improvement of water quality (zero *E. coli* in drinking water of 57% of the households) and a reduction in bloody and nonbloody diarrhea in children younger than 5 years (du Preez et al., 2008). A recent regression analysis of the data on household water treatment methods revealed that ceramic filters provided more long-term protection to consumers than Biosand, chlorination with safe storage, combined coagulant–chlorine disinfection or SODIS (see Section 8.3.4) (Hunter, 2009). A potential problem associated with ceramic filters is the presence of trace arsenic levels in the filters (arsenic is associated with the clay used in the filter fabrication). However, thorough rinsing of the filters with water reduces the arsenic levels in treated water (Archer et al., 2011).

8.3.4 Solar Disinfection

In remote areas with no access to safe drinking water, solar radiation has been considered for disinfecting drinking water (Odeymi, 1990). The daily UV radiation that reaches the Earth contains UV-A (320–400 nm) and UV-B (280–320 nm) radiations which cause damage to cells and cell components via absorption by intracellular chromophores (e.g., L-tryptophan, flavins) followed by oxidative damage due to the generation of reactive oxygen species (ROS) (Malato et al., 2009). The UV-C portion of UV radiation is absorbed by the atmosphere and does not play a role in water disinfection.

The Solar Water Disinfection (SODIS) process, sponsored by EAWAG (Swiss Federal Institute for Environmental Science and Technology), is a POU treatment system that uses solar radiation to inactivate waterborne pathogens. SODIS has been adopted by more than 2 million users in several developing countries in Latin America, Asia, and Africa and will probably be helpful in reducing waterborne diseases in those countries (Clasen, 2009).

The SODIS process is relatively simple and consists of partially filling transparent plastic bottles with relatively low turbidity water (<30 NTU), shaking the bottles to saturate water with oxygen, and exposing them to solar radiation for 6 hours if the sky is bright or partially cloudy, or for two consecutive days if the sky is 100% cloudy. The combination of sunlight and oxygenation is called photooxidation. The exposure time is also dependent on ambient temperature and sunlight intensity. The plastic bottles are made of polyethylene terephthalate (PET), polycarbonate, polystyrene, Tritan (a bisphenol A (BPA)-free copolyester), or polypropylene copolymer. The latter was shown to perform best as regard the inactivation of *E. coli, Enterococcus* spp., or MS2 phage (Fisher et al., 2012).

Pathogen inactivation is due to the synergistic effects of UV-A (wavelength 315–400 nm) and increased water temperature which should reach 50–60°C, although radiation is probably the predominant factor affecting inactivation, particularly at temperatures lower than 40°C. Furthermore, inactivation can also be due to the highly reactive species such as superoxides, ($O_2{}^-$), hydrogen peroxides (H_2O_2), and hydroxyl radicals (OH·). Using the SODIS system, bacterial inactivation experiments were carried out in Bolivia and Spain under natural sunlight (maximum global irradiance of approximately 1050 W/m^2) and temperature varying between 28°C and 38°C (Figure 8.6; Boyle et al., 2008). Total inactivation was obtained after an exposure time of 20 minutes for *Campylobacter jejuni*, 90 minutes for enteropathogenic *E. coli*, and 150 minutes for *Yersinia enterocolitica*. *Bacillus subtilis* endospores were not inactivated after more than 6-hour exposure to sunlight. SODIS was also quite effective in the inactivation of *Shigella dysenteriae* type 1 even under overcast conditions

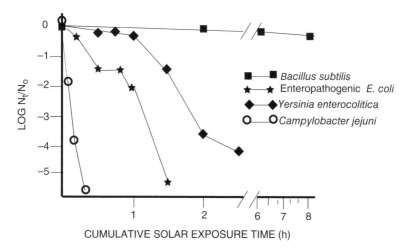

Figure 8.6 Inactivation kinetics of bacterial populations exposed to real sunlight conditions. Adapted from Boyle et al. (2008). Appl. Environ. Microbiol. 74: 2997–3001.

(Kehoe et al., 2004). *Bacillus subtilis* endospores were, however, much more resistant than the vegetative bacteria. It took a 2-day exposure time (equivalent to 16-hour exposure to intense sunlight) to cause 96% inactivation of the spores. Comparatively, less is known about the impact of SODIS on viral inactivation. A 1-day exposure to sunlight was necessary for 90% inactivation of human adenovirus 2 stored in a PET bottle which allows exposure of the water to mostly UV-A radiation (Bosshard et al., 2013).

SODIS efficacy can be reduced by low ambient air temperature, high turbidity (turbidity should be lowered by sedimentation or filtration and should be less than 30 NTU), high color, and topography (lower inactivation at higher elevations) (Fontan-Sainz et al., 2012; Oates et al., 2003; www.sodis.ch; Wegelin et al., 1994; Wegelin and Sommer, 1998). A mathematical model, based on satellite-derived data, was developed to simulate monthly mean, minimum, and maximum 5-hour averaged peak solar radiation intensities in Haiti. Based on the threshold of 3–5 hour of solar radiation above 500 W/m^2, the model suggested that SODIS can be used all year round in Haiti (Oates et al., 2003).

A recent improvement of SODIS is the SOCO-DIS process which consists of concentrating solar radiation with a solar collector and reducing the pH of the water to 3 with lemon juice (0.25% final concentration) and/or vinegar (0.17%). This system improves significantly the disinfection efficiency when compared to SODIS (Amin and Han, 2011). As compared to SODIS, the SOCO-DIS system improved the disinfection efficiency of roof-harvested rainwater by 20–30% (Amin and Han, 2009). Additives such as lime juice or riboflavin and sodium percarbonate in combination with citric acid or trace amounts of copper plus ascorbate enhance SODIS efficiency of inactivation of bacterial and fungal pathogens, protozoan parasites, and phage (Fisher et al., 2012; Harding and Schwab, 2012; Heaselgrave and Kilvington, 2010). Since most of the SODIS designs are based on batch exposure, a continuous flow reactor, using UV-B radiation instead of UV-A, was designed and includes a compound parabolic solar collector which helps amplify the UV-B fluence rate (Mbonimpa et al., 2012).

Several studies have addressed the role of POU devices on the reduction of disease incidence in populations exposed to contaminated drinking water. Use of the SODIS process helped reduce diarrheal morbidity among children (younger than 5 years) in southern India. A 6-month study showed that the use of solar radiation to disinfect drinking water led to a 40% reduction of the risk of diarrhea among children (Table 8.2; Rose et al., 2006). SODIS use led to a reduction of diarrheal disease among Maasai children in Kenya (Conroy et al., 1996), a significant reduction of cholera among children younger than 6 years (Conroy et al., 2001), and a reduction of dysentery and non-dysentery diarrhea among <5-year-old children in Kenya (du Preeze et al., 2011). However, a survey of SODIS use in 40 households in Kathmandu Valley in Nepal showed that the degree of acceptability of SODIS was quite low (9%) (Rainey and Harding, 2005). This points to the need of educating women (the main water gatherers) and sometimes children on public health issues in general, and on the necessity to adequately treat water to reduce diarrheal diseases in their communities.

TABLE 8.2   Effect of SODIS process on reduction of diarrheal morbidity in study and control children during a 6-month follow-up period

	Study Group ($n = 100$)	Control Group ($n = 100$)
Diarrheal incidence	53	75
Mean number of episodes	1.7	2.9
Mean duration of episodes (days)	2.1 ± 2.9	3.6 ± 3.4
Episodes requiring intravenous rehydration 13.9%	34.1%	

Source: Adapted from Rose, A., S. Roy, V. Abraham, G. Holmgren, K. George, V. Balraj, S. Abram, J. Muliyil, A. Joseph, and G. Kang. (2006). Arch. Dis. Child. 91: 139–141.

8.3.5   Water Purification Tablets

Iodine tablets have been traditionally used by the military to purify canteen water. However, these tablets do not remove turbidity and give a medicinal taste and odor to the treated water. Long-term use of iodine as a disinfectant is not recommended due to adverse effects, particularly on the thyroid gland. Aquatabs (Medentech, Wexford, Ireland) contain sodium dichloroisocyanurate (NaDCC) used for low turbidity waters. They dissolve in the water to release hypochlorous acid (HOCl) which is the active ingredient (Clasen and Edmondson, 2006):

$$NaCl_2(NCO)_3 + 2H_2O \leftrightarrow 2HOCl + NaH_2(NCO)_3 \qquad (8.1)$$

NaDCC was approved for emergency disinfection of drinking water by WHO and the USEPA. It is distributed as tablets for treating various volumes of water. Its effectiveness was demonstrated in a 4-month study involving 100 households in Dhaka, Bangladesh. The TTC counts were significantly lower in households using NaDCC than in the control group. About 62% of the samples in the NaDCC group met the WHO guidelines of zero TTC (Clasen et al., 2007). NaDCC is stable, easy to use, and is relatively safe.

8.3.6   Coagulation–Flocculation–Sedimentation

Coagulation–flocculation is a process designed to remove suspended solids, including microorganisms. It is generally followed by sedimentation, filtration, or decanting to remove the flocs. Chlor-Floc is an inexpensive commercial tablet or powder that mainly contains a coagulant–flocculant (e.g., Fe or Al salts) to remove turbidity, suspended organic matter and microorganisms, and chlorine (sodium dichloroiso-cyanurate) to disinfect the water. One tablet is generally sufficient for 10 L of water. Turbidity and color are well removed, leading to a more efficient disinfection and making the water more aesthetically pleasing. Filtration through a cloth is necessary to remove the flocculated sediment. The coagulation–flocculation step can be negatively affected at cold temperatures ($\sim 5°C$) (Marois-Fiset et al., 2013). Table 8.3 shows the inactivation of bacteria (*E. coli*, *Pseudomonas aeruginosa*, *Klebsiella terrigena*), viruses (poliovirus 1, rotavirus SA-11), and *Giardia muris* cysts by iodine and

TABLE 8.3 Inactivation of bacteria, viruses, and protozoan cysts with 1 tablet/L of Chlor-Floc or iodine

Microbial Agent	Time (min)	Temperature (°C)	MRLR	LR (mean)
Chlor-Floc Tablets				
Bacteria (*Escherichia coli*, *Klebsiella terrigena*, *Pseudomonas aeruginosa*)	5	5, 15, 25	6	>6
Rotavirus	5	5, 15, 25	4	>4
Poliovirus	20	5	4	2.5
Giardia muris	10-40	5, 15, 25	3	>3.0
Iodine Tablets				
Bacteria (*E. coli*, *K. terrigena*, *P. aeruginosa*)	5	5, 15, 25	6	>6
Rotavirus	30	5	4	4.9
Poliovirus	30–90	5	4	1.2
Giardia muris	30–90	5	3	≤2.5

Source: Adapted from Powers et al. (1994). Appl. Environ. Microbiol. 60: 2316–2323.
MRLR, Minimum required log_{10} reduction; LR, Log_{10} reduction.

Chlor-Floc tablets (Powers et al., 1994). Both tablets exceed the minimum required log reduction (MRLR) of bacteria and rotaviruses but fail to do so for poliovirus 1. For *G. muris*, only the Chlor-Floc tablets exceeded the MRLR for protozoan parasites. The Chlor-Floc tablets were recommended by the US military as a safe alternative to iodine tablets (Powers et al., 1994). A survey showed that the combined use of coagulation–flocculation and chlorine to treat water in rural Guatemala helped reduce the incidence of diarrhea (Chiller et al., 2006; Reller et al., 2003). A comparative study showed that the flocculant-disinfectant tablets and *N*-halamine beads (see Section 8.3.7) were superior to Aquatab tablets as regard the inactivation of several categories of indicator microorganisms (Figure 8.7; McLennan et al., 2009).

8.3.7 Chlorination with Safe Storage

The Safe Water System (SWS) is a program that was instituted by CDC and the Pan American Health Organization (PAHO) to reduce the incidence of diarrheal diseases and improve drinking water quality in rural areas of developing countries. This household program consists of treating water with a dilute chlorine solution (Na or Ca hypochlorite), safe storage of the treated water in a container with a narrow lip (Figure 8.3) and a spigot to prevent recontamination, and behavioral change programs to promote compliance among the consumers (http://www.cdc.gov/safewater/; Quick et al., 2002; Ram et al., 2007). Although chlorine has the advantage of leaving a residual that protects consumers against postcontamination of the water, it is not efficient in virus and parasite cyst and oocysts inactivation, and can cause taste and odor problems for consumers (see Chapter 3). The importance of storage on water safety is illustrated in a study in Mali, which showed that water storage overnight resulted in an inadequate residual chlorine level of <0.2 mg/L and occurrence of

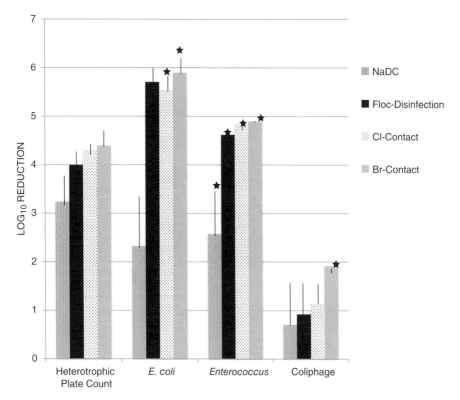

Figure 8.7 Average log reduction (LR) of HPC, *E. coli*, *Enterococcus* and coliphage after 30 min for sewage-contaminated well water with the use of four POU disinfection systems (* indicates that effluent was below the limit of detection for all samples. Limit of detection was substituted to calculate LR and actual reductions may be greater than shown). Adapted from McLennan et al. (2009). Appl. Environ. Microbiol. 75: 7283–7286.

coliforms and *E. Coli* in 48% and 8% of stored water samples, respectively (Baker et al., 2013).

For water treatment plants in the United States, the recommended MCL (maximum contaminant level) for free chlorine is 4 mg/L (U.S. EPA, 2006). A free chlorine residual of 0.2 to 0.5 mg/L at the end of the distribution system is generally sought by conventional water treatment plant operators. For POU chlorination, it was recommended to adopt a free chlorine residual of <2 mg/L after 1-hour contact time and >0.2 mg/L after 24-hour contact time. A higher chlorine concentration is necessary at turbidities above 10 NTUs (Lantagne, 2008). SWS program was successful in reducing the incidence of diarrheal diseases in several developing countries in Africa, Asia, and Latin America. Following a survey of the literature on the impact of chlorination on consumer health in developing countries, it was concluded that the median length of the trials was 30 weeks and that the impact of chlorination was diminished in longer trials. This may be partially due to consumer noncompliance in longer trials (Arnold and Colford, 2007). Compliance, defined as the proportion

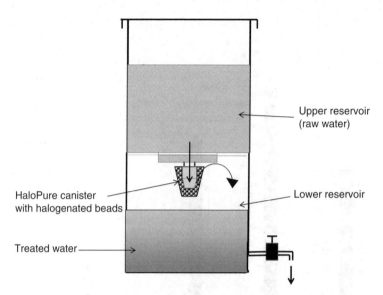

Figure 8.8   AquaSure® system: Arrows represent the flow of water in the system. Adapted from Coulliette et al. (2010). Amer. J. Trop. Med. Hygiene 82: 279–288. (For a color version, see the color plate section.)

of water treated by a given community, is a crucial factor in the effectiveness of household water treatment systems (Enger et al., 2013). A conceptual model linking quantitative microbial risk assessment (QMRA; see Chapter 10) to epidemiological studies showed that in an intervention study on the efficacy of Life Straw Family device (see Section 8.3.9), compliance was an important factor in the device efficacy (Enger et al., 2012).

   Disinfection can also be provided via the use of canisters of N-halamine polymeric beads that contain chlorine or bromine (Chen et al., 2004). The Halopure® canisters of contact disinfectant beads are housed in the Halosure™ water purifier which is described in Figure 8.8. The contaminated water is poured through the canister and is collected in a storage tank. Microorganisms are inactivated by the chlorine or bromine released from the beads. The Halopure beads were evaluated for their capacity to inactivate bacterial pathogens and phage MS2 (Coulliette et al., 2010, 2013). For *Salmonella typhimurium* and *V. cholerae*, the mean $\log_{10}$ CFU reduction was 5.44 and 6.07, respectively, following treatment with chlorine beads. Phage MS2 inactivation by chlorine and bromine beads was 2.98 and 5.02, respectively (Table 8.4). The Halopure® bead technology is consumer-friendly and avoids the need to add chlorine and estimate of the disinfection period by untrained consumers.

   The total number of liters treated by the various HWTs discussed above was estimated at 15.5 billion liters for the year 2007. Na hypochlorite helped in the disinfection of 50% of the volume of water while the flocculant-disinfectant system accounted for about 1.5% of the volume (Figure 8.9; Clasen, 2009). The number of users for the year 2007 was estimated at 18.8 million users. Most of the users (56.6%)

**TABLE 8.4   Inactivation of bacterial pathogens and phage MS2 by Halopure®
chlorine and bromine contact disinfectant units**

Microorganism	Mean Log_{10} CFU Reduction ($\pm$SE)
Salmonella typhimurium (chlorine)	5.44 ($\pm$0.98)
Vibrio cholerae (chlorine)	6.07 ($\pm$0.09)
Phage MS2 (chlorine)	2.98 ($\pm$0.26)
Phage MS2 (bromine)	5.02 ($\pm$0.19)

Source: Adapted from Coulliette et al. (2010). Amer. J. Trop. Med. Hyg. 82: 279–288; Coulliette et al. (2013). Intern. J. Hyg. Environ. Health 216: 355–361.

used Na hypochlorite as the treatment of choice. Unfortunately, the number of users is still a small percentage of the millions worldwide who lack access to improved water supplies.

Using a scoring system to assess the effectiveness of HWT treatments, it was found that ceramic filters and biosand have the highest potential for sustained use by consumers (Sobsey et al., 2008). However, this scoring system was criticized as being subject to bias (Lantagne et al., 2009).

A critical review of the literature revealed that the average reduction in diarrheal diseases is around 30–40% but may vary between no effect and 85%. There is evidence that estimates of diarrheal disease reductions may be biased (responder and observer bias) (Schmidt and Cairncross, 2008).

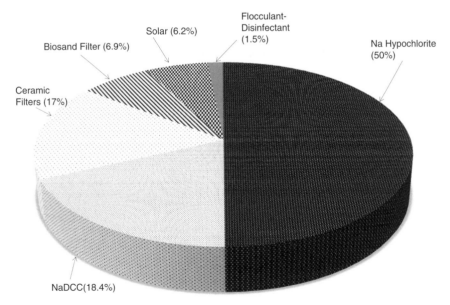

Figure 8.9   Percentage of water treated by various HWT technologies (the total number of liters treated in 2007 was 15.5 billion liters). *Source*: Clasen, T. (2009). *Scaling Up Household Water Treatment Among Low-Income Populations*. WHO Report, WHO/HSE/WSH/09.02 Geneva, Switzerland.

8.3.8 The AQUAPOT Project

The AQUAPOT Project, conceived at the Polytechnic University of Valencia, Spain, focuses on the construction of decentralized, low cost water treatment plants based essentially on ultrafiltration followed by chlorination to prevent recontamination of treated water (Arnal et al., 2009). Two such projects were carried out successfully in a rural area in Ecuador (Arnal et al., 2007) and in Mozambique, an African country with a child mortality rate of 152/1000 births (as compared to 10.2/1000 births in developed countries), and where access to safe drinking water and sanitation is 43% and 32%, respectively (Arnal et al., 2010).

Water treated by the devices or processes discussed above can be recontaminated during transport and storage at home. Therefore, the containers must be thoroughly cleaned and the consumers must have adequate hygiene education to reduce the chances of postcontamination of the water (Rufener et al., 2010).

8.3.9 LifeStraw[R] Family

This POU water treatment device was conceived in Switzerland (http://www .vestergaard-frandsen.com/lifestraw/lifestraw-family) for the purpose of meeting the drinking water needs of vulnerable populations in developing countries. It is essentially a gravity-fed ultrafiltration device that does not require electricity nor batteries. This water purifier is described as follows (Figure 8.10): Contaminated water is

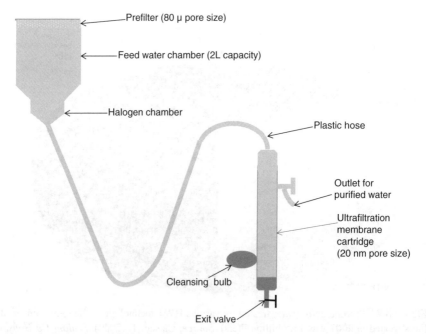

Figure 8.10 Life Straw[R] Family device. Adapted from http://www.vestergaard-frandsen .com/lifestraw/lifestraw-family/functioning.

poured into a bucket that contains a 80 µm prefilter that removes the coarse particles from the water and is easily washable. The water flows by gravity through a 1-meter plastic hose to reach a membrane cartridge that houses an ultrafiltration hollow fiber membrane with a 20 nm porosity. The treated water is collected from the blue tap while the dirt particles are disposed of via the exit valve. The device has a cleaning bulb for backwashing of the cartridge and a small halogen chamber to prevent hose and membrane fouling. It has a maximum design capacity of 18,000 L and allows a flow rate of 12–15 L/h.

LifeStraw was tested for the removal of bacteria, viruses, and protozoan parasites. In laboratory testing, it achieved a log removal of 6.9 for *E. coli*, 4.7 for MS2 bacterial phage and 3.6 for *Cryptosporidium* oocysts (Clasen et al., 2009).

8.3.10 UV Treatment Units

Ultraviolet radiation at the germicidal wavelength of 254 nm is quite efficient in the inactivation of pathogens and parasites in water and is an efficient disinfectant in full-scale water treatment plants and household units (this disinfectant is discussed in more details in Chapter 3). UV Waterworks unit, developed at the Lawrence Berkeley National Laboratory (Berkeley, CA) by Ashok Gadgil, is another means for treating water in developing countries. This is a rapid, low cost, and low maintenance UV unit for disinfecting water. It uses a UV dose of 120 mJ/cm^2, a flow rate of 15 L/min, and brings the cost of drinking water to up to 5 cents/1000 L (Gadgil, 2008).

Bacterial, viral, and protozoan reduction/inactivation by the various household water treatment technologies discussed above are summarized in Table 8.5 (WHO, 2011c).

Household treatment technologies can be improved by adopting the multi-barrier approach practiced in full-scale water treatment plants. For example, SODIS treatment can be followed by chlorination and safe storage (Sobsey, 2002).

8.4 PERSONAL PORTABLE WATER TREATMENT SYSTEMS FOR TRAVELERS AND HIKERS

Travelers to areas in the world with little or no sanitation as well as hikers and campers in wilderness areas, when consuming untreated water, often run the risk of contracting waterborne diseases caused by bacterial and viral pathogens, and protozoan parasites. It is estimated that more than 60% of travelers to tropical and subtropical regions may experience diarrhea. Between 2000 and 2005, more than 6000 travelers to countries around the world were observed for gastrointestinal infections. Gastrointestinal infection rates were highest in South Asia, Sub-Saharan Africa, and South America (Figure 8.11; Greenwood et al., 2008). Nepal and India showed the highest rate of gastrointestinal infections. Travelers to those regions may thus resort to bottled water (which may not always be safe, depending on the area) or, if possible,

TABLE 8.5 Bacterial, viral, and protozoan parasite reduction/inactivation by household water treatment technologies

Treatment Process	Enteric Pathogen	Baseline Removal (LRV)	Maximum Removal (LRV)
Chemical disinfection			
Free chlorine disinfection	Bacteria	3	6
	Viruses	3	6
	Protozoa (non-*Cryptosporidium*)	3	5
Filtration			
Porous ceramic and carbon block filtration	Bacteria	2	6
	Viruses	1	4
	Protozoa	4	6
Membrane filtration (Microfiltration, Ultrafiltration, nanofiltration, reverse osmosis)	Bacteria	2(MF); 3(UF, NF or RO)	4(MF); 6(UF, NF or RO)
	Viruses	0(MF); 3(UF, NF or RO)	4(MF); 6(UF, NF or RO)
	Protozoa	2(MF); 3(UF, NF or RO)	6(MF); 6(UF, NF or RO)
Fiber and fabric filtration (e.g., sari cloth)	Bacteria	1	2
	Viruses	0	0
	Protozoa	0	1
Granular media filtration			
Rapid granular, diatomaceous earth, granular and powdered activated carbon	Bacteria	1	4+
	Viruses	1	4+
	Protozoa	1	4+
Intermittent slow sand filtration	Bacteria	1	3
	Viruses	0.5	2
	Protozoa	2	4
Solar disinfection	Bacteria	3	5+
	Viruses	2	4+
	Protozoa	2	4+
UV inactivation	Bacteria	3	5+
	Viruses	2	5+
	Protozoa	3	5+
Heating (e.g., boiling)	Bacteria	6	5+
	Viruses	6	5+
	Protozoa	6	5+
Sedimentation	Bacteria	0	9+
	Viruses	0	9+
	Protozoa	0	9+
			0.5
			0.5
			1

Source: Adapted from WHO (2011c). Guidelines for Drinking-Water Quality. 4th Ed. World Health Organization, Geneva, Switzerland. http://www.who.int/water_sanitation_health/publications/2011/dwq_guidelines/en/.
LRV, log reduction value; MF, microfiltration; UF, ultrafiltration; NF, nanofiltration; RO, reverse osmosis.

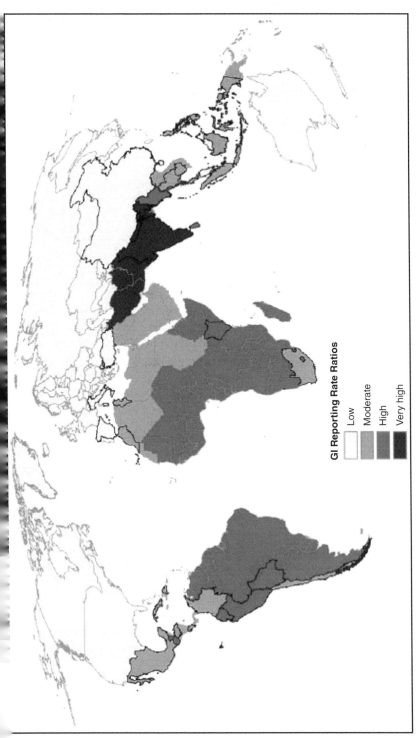

Figure 8.11 A profile map of relative rates of acquisition of gastrointestinal infection by destination. Global distribution of reporting rate ratios for all gastrointestinal infections in ill-returned travelers presenting to 30 GeoSentinel clinics on six continents. Twenty-eight countries with available country-specific data are outlined. *Source*: Greenwood et al. (2008). J. Travel Med. 15 (4): 221–228.

TABLE 8.6 **Inactivation/removal of bacteria by portable water treatment systems for travelers**

Disinfectant or Filter Device	Log_{10} Reduction in Clear Water	Log_{10} Reduction in Turbid Water
Drinkwell chlorine	2.0–2.8	2.1–2.4
Hydroclonazone	0.5–1.2	NT
Aquatabs	1.8–2.8	1.0–2.0
2% iodine in ethanol	1.9–2.8	1.6–2.1
Outdoor M1-E	1.3	0.4–1.7
Traveler	2.2–2.4	NA
Pres2Pure	0.9–2.4	NT
Mini Ceramic	3.2–3.3	1.7–4.9
First Need Deluxe	2.6–3.3	3.9–4.7
WalkAbout	2.4–3.0	2.9–3.8

Source: Adapted from Schlosser et al. (2001). J. Travel Med. 8: 12–18.
NT, not tested; NA, non applicable.

may boil their drinking water and store it in clean containers. There are, however, several commercial disinfectants in solution or in tablets, and filtration systems may also be used to treat water (Table 8.6; Schlosser et al., 2001). Although more expensive, the filtration devices display a better performance than the disinfectants in solution or in tablets as regard the removal/inactivation of active bacteria from clear or turbid waters. Coliforms and/or *E. coli* were detected only in turbid waters when using disinfectants (Schlosser et al., 2001).

Other portable devices are the LifeStraw portable ultrafilter described in Section 8.3.9 and the PUR Scout purification system which consists of a microfilter for cyst and oocyst removal followed by an iodine bed resin which release iodine into water. The average removal/inactivation achieved by this device was $6 \log_{10}$ for *K. terrigena*, $4 \log_{10}$ for poliovirus 1 and rotavirus SA-11, and $3 \log_{10}$ for *Cryptosporidium* oocysts (Naranjo et al., 1997). Nine filtration-based water purification devices were evaluated for bacterial, oocyst, phage, and microcystin removal. The devices achieved 3.6 to $6.9 \log_{10}$ removal for bacteria and *C. parvum* oocysts but were inefficient (except one device based on reverse osmosis) in the removal of phage MS2 and microcystin (Hörman et al., 2004). Simple preventive measures should be taken by travellers in areas of questionable drinking water quality (WHO, 2011c):

- Drink only bottle water and other beverages in sealed tamper-proof containers
- Drink only water that has been treated by an effective POU device
- Do not use ice in drinking water or cocktail
- Avoid salads that have been prepared or washed with a questionable water
- Brush teeth only with safe water

WEB RESOURCES

http://www.who.int/en/
(World Health Organization)

http://www.who.int/water_sanitation_health/publications/2011/dwq_guidelines/en/
(WHO, 2011c. Guidelines for Drinking-Water Quality)

www.who.int/water_sanitation_health
(2000. Global water supply and sanitation assessment 2000).

http://www.solarcookers.org/index.html
(solar cooking; Solar Cookers International publishes a *Solar Cookers Review*)

http://www.waterhealth.com
(Water Health International)

http://www.epa.gov/safewater/contaminants/index.html
(Drinking water contaminants: U.S. EPA)

http://www.ifpri.org/book-6330/ourwork/researcharea/water-policy
(International Food Policy Research Institute (IFPRI)

http://www.lbl.gov/tt/success_stories/articles/WHI_more_no_Mex.html
(UV sterilization unit for developing countries; Lawrence Berkeley National Laboratory, CA)

http://www.who.int/maternal_child_adolescent/en/
(WHO, 2008. Measuring child mortality)

http://www.youtube.com/user/VestergaardFrandsen
(LifeStraw Youtube Clips)

http://www.manzwaterinfo.ca
(Official site of Biosand)

http://www.cawst.org/en/resources/biosand-filter
(Biosand)

http://www.waterforcambodia.org/projects/biosand-filters
(Biosand: How a Biosand filter works?)

http://www.bigberkey.com
(Ceramic cartridges)

http://web.mit.edu/watsan/
(Global water and sanitation projects: Massachusetts Institute of Technology)

http://www.who.int/water_sanitation_health/publications/2011/dwq_guidelines/en/
(WHO Guidelines for drinking water quality)

FURTHER READING

Clasen, T. 2009. *Scaling Up Household Water Treatment Among Low-Income Populations.* WHO Report, WHO/HSE/WSH/09.02, Geneva, Switzerland.

Pejack, E. 2011. Solar pasteurization. In *Drinking Water Treatment, Strategies for Sustainability*, C. Ray and R. Jain (Eds). Springer, New York.

Rosegrant, M.W., X. Cal, and S.A. Cline. 2002. *Global Water Outlook to 2025: Averting an Impending Crisis.* Food Policy Report. Inter. Food Policy Res. Inst. (IFPRI) and Intern. Water Management Inst. (IWMI), 28pp.

Sobsey, M.D. 2002. *Managing Water in the Home: Accelerated Health Gains from Improved Water Supply.* WHO Report, WHO/SDE/WSH/02.07, Geneva, Switzerland.

WHO. 2011. *Guidelines for Drinking-Water Quality*, 4th editon. World Health Organization, Geneva, Switzerland.

WHO. 2011. *Evaluating Household Water Treatment Options: Health-Based Targets and Microbiological Performance Specifications.* World Health Organization, Geneva, Switzerland.

WHO/UNICEF. 2012. *Progress on Drinking Water and Sanitation. 2012 Update.* WHO/UNICEF Joint Monitoring Programme for Water Supply and Sanitation.

BOTTLED WATER MICROBIOLOGY

9.1 INTRODUCTION

Good marketing and advertisement have helped the growth of the bottled water industry around the world, even in developed industrial countries equipped with modern water treatment plants. People consume bottled water for a variety of reasons, ranging from complaints about bad taste of tap water to concerns about the presence of toxic chemicals and pathogens and parasites. Bottled water is perceived by consumers as pure and safe but numerous studies have also revealed the occasional presence of pathogenic microorganisms and toxic chemicals with potential deleterious health effects on humans. In 1999, bottled water sales were approximately 35 billion dollars worldwide, with 5 billion dollars for the United States. A report of the International Bottled Water Association showed that the US market share in 2004 jumped to 11 billion US dollars. In Europe, packaged water sales represent 44% of the total market for nonalcoholic beverages. Mineral and spring water sources represent 97% of the market volume in 2011 (http://efbw.eu). The average per capita consumption of bottled water in the European Union was estimated at 104 L/year in 2011. The highest consumers were Italy (179 L/person per year) and Germany (171 L/person per year) while the lowest was Finland (18 L/person per year) (Figure 9.1; European Federation Bottled Water, 2011, http://efbw.eu/bwf.php?classement=07). US Statistics for 2002 show that the average daily consumption of bottled water was about 0.4 L/day (equivalent to 146 L/year) as compared to 310 L/year for tap water (http://www.fda.gov).

In Europe, the bottled water industry has traditionally projected an image of health. Drinking and bathing in mineral waters has been claimed to cure many human ailments (e.g., migraine headaches, arthritis, kidney infections, excess cholesterol, gout, obesity, and the improvement of metabolism in general). Spas across Europe have enjoyed the reputation for offering relaxation and fitness to their customers.

Microbiology of Drinking Water Production and Distribution, First Edition. Gabriel Bitton.
© 2014 John Wiley & Sons, Inc. Published 2014 by John Wiley & Sons, Inc.

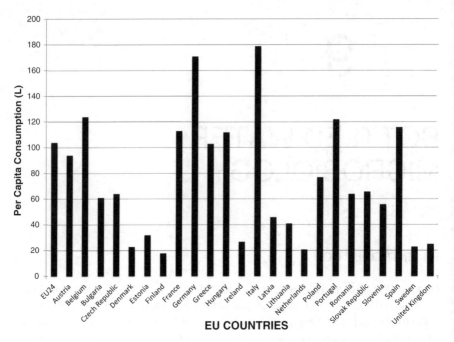

Figure 9.1   Per capita consumption of bottled water in countries of the European Union (2011). Adapted from: http://efbw.eu/bwf.php?classement=07

Today, bottled water is primarily used for drinking but has also been considered for the preparation of baby formula, cleaning contact lens or used in humidifiers (Warburton, 1993). Bottled water is generally perceived by consumers as being of better quality and better taste than tap water. However, the higher quality of bottled water is sometimes a misconception. Several investigators reported that bottled water is of no better quality than municipal drinking water. It is also hundred times more expensive than tap water. Furthermore, the collection, packaging, transport, and disposal of plastic bottles have a negative impact on the environment (Parag and Roberts, 2009). In the United States, some 60 million plastic bottles are disposed of in solid waste landfills. The plastic materials are often quite resistant to biodegradation in the environment. Moreover, a few imported bottled water brands have a sizeable carbon footprint.

For a more complete picture of bottled water safety, one has to mention briefly the occasional presence of chemical contaminants in bottled water. We note that these contaminants may also occur in municipal drinking water. We are focusing on chemicals that may leach out from the bottles into the water. More emphasis has been placed on the microbiological quality of bottled water than hazards from toxic chemicals. The origin of these chemicals is the source water, introduction during bottling, or migration from the bottle to the water.

Some of the chemical contaminants of concern are the following:

- *Excessive amounts of trace elements*: As regard trace elements (e.g., As, F, Rn), only 15 out of 56 brands did not exceed the European Economic Community (EEC) drinking water regulations (Misund et al., 1999). Antimony (Sb) is used as a Sb_2O_3 catalyst in the production of polyethylene terephthalate (PET) bottles. Concern was raised over the leaching of Sb into drinking water and beverages, particularly after storage of the bottles at room temperature (Shotyk et al., 2006; Westerhoff et al., 2008).
- *Lead (Pb) in bottled water*: A Canadian survey indicated that some brands exceeded the Pb guideline of 10 µg/L (Dabeka et al., 2002; Pip, 2000). The bottle type influences the Pb levels as more metal is leached from glass than PET bottles (Misund et al., 1999; Shotyk and Krachler, 2007).
- *Perchlorate and bromate*: Perchlorate was detected in 10 out of 21 bottled water samples with concentrations ranging from 0.07 to 0.74 µg/L. Bromate was detected in 12 out of 21 bottled water samples with concentrations ranging from 0.57 to 76 µg/L (Snyder et al., 2005).
- *Endocrine disrupters*: Some additives used in the manufacture of plastic bottles are endocrine disrupters that may migrate into the drinking water. Eighteen brands of bottled (nine in plastic and nine in glass bottles) mineral waters were examined for their estrogenic potential (Wagner and Oehlmann, 2009). The estrogenic potency, expressed as 17-β estradiol equivalent, varied from below detection limit to 75.2 ng/L. Estrogenic activity was more frequent in PET (78% of the brands) than in glass bottles (33% of the brands). Hard plastic bottles used to dispense baby formulas or water to babies may contain bisphenol A (BPA), a known endocrine disrupter in animals and potentially humans. BPA is also found in the inside lining of some food cans. BPA and other endocrine disrupters may leach out into water or foods. Some BPA-free commercial brands, including bottles, drinking cups, nipples, and pacifiers are now available to consumers (http://www.newbornfree.com/). The U.S. EPA has established a reference dose for BPA at 0.05 mg/kg body weight/day.
- *Volatile organic compounds (VOCs)*: VOCs (e.g., styrene) were detected in bottled water (Al-Mudhaf et al., 2009). In ozonated bottled water, the free radicals produced by ozone disinfection interact with the plastic material to form VOCs such as aldehydes and ketones responsible for taste and odors in the bottled water.

9.2 SOURCES AND CATEGORIES OF BOTTLED WATER

Bottled water comprises any one of the following categories (FDA, 2013; Gray, 1994; Rosenberg, 2002):

9.2.1 Mineral Water

It contains dissolved cations and anions (e.g., Ca^{2+}, Mg^{2+}, Na^+, HCO_3^-). The mineral content varies from one source to another and may be sometimes lower than community water supplies. Mineral water is drawn from underground source that contains at least 250 mg/L of total dissolved solids. Minerals and trace elements must come from the source of the underground water and cannot be added later.

Natural mineral water is water that is drawn from a protected source. It becomes enriched with minerals as it flows through underground formations. According to European Union regulations, manufacturers are not allowed to alter the mineral content of this type of water nor disinfect it to alter the microbial flora of this natural resource. The only treatments allowed are filtration to remove particles or iron, manganese, sulfur or arsenic as well as addition or removal of CO_2. Furthermore, it must be monitored to prevent contamination from allochthonous microorganisms, including pathogens and parasites. In France, "natural mineral water" must also have a demonstrated therapeutic value.

9.2.2 Spring Water

Water is derived from an underground formation that flows naturally to the earth surface. In 2011, spring water accounted for 14% of total sales in Europe, as compared to 83% for mineral water. Approximately 25% of the bulk water sold in the United States is spring water. This water may be disinfected with ozone.

9.2.3 Sparkling Water

It is "naturally" carbonated with carbon dioxide. It can also be made bubbly by adding dissolved carbon dioxide. Some brands of bottled water (e.g., Perrier) are made bubbly by harvesting the "natural" carbon dioxide at the source and reinjecting it into the water during the bottling operation.

9.2.4 Distilled or Purified Water

Water from which minerals have been removed by processes such as distillation (i.e., water vaporized and then condensed), deionization (i.e., passage of water through exchange resins), or reverse osmosis (i.e., passage of water through special membranes which retain the minerals). Reverse osmosis units have been installed in some water treatment plants across the world and for treating water at home (see Chapter 2).

9.2.5 Tap Water

It is also bottled with, sometimes, additional treatment. In the United States, 25–40% of bottled water is bottled tap water.

9.3 BOTTLED WATER MICROORGANISMS

9.3.1 Source Water

Indigenous aquatic microorganisms and potential pathogens or parasites may originate from the source water or may be introduced at the bottling plant through contaminated equipment (the microbiology of source water has been covered in Chapter 2). Culture-dependent and molecular-based techniques have helped in the identification of the autochthonous microorganisms in natural mineral waters and spring waters. A study of three natural mineral sources in Spain has shown that *Pseudomonas* spp. was the predominant bacteria in the source waters. Other predominant bacteria were the nontuberculous mycobacteria and *Aeromonas hydrophila*, some of which are human opportunistic pathogens. Other autochthonous bacteria were *Acidovorax facilis*, *Delftia* sp., *Sphingomonas* sp., *Herminiimonas glaciei*, *Ochrobactrum intermedium*, and *Stenotrophomonas maltophilia* (Casanovas-Massana and Blanch, 2012).

9.3.2 Water Treatment before Bottling

Depending on the source, water may be treated via activated carbon, sand filtration, microfiltration, distillation, deionization, softening, mineral adjustment, ultraviolet radiation, carbonation, ozonation, or reverse osmosis (see Chapters 2 and 3 for further details on treatments and disinfection).

9.3.3 Microorganisms Found in Bottled Water

The diverse bacterial community in bottled waters has been investigated around the world. The identified bacteria are typical aquatic or soil bacteria, some opportunistic pathogens (e.g., *Mycobacterium*) and, occasionally, primary pathogens (*Yersinia*, *Vibrio cholerae*, *Staphylococcus*) (Blake et al., 1977; Hunter, 1993; Warburton, 1993). Mean bacterial counts in three bottled water brands varied from 9.6×10^2 CFU/mL (Evian water from France) to 2×10^4 CFU/mL (Mount Franklin water from Australia) (Jayasekara et al., 1999). Similarly, bacterial count in bottled mineral water from France (Vittel and Hepar) and Belgium (Orée du Bois and Spa) varied from 2.8×10^2 CFU/mL for Vittel to 3.4×10^4 CFU/mL for Hepar (Dewettinck et al., 2001). Microbiological analysis (total culturable counts, microbial indicators, and *Pseudomonas aeruginosa*) of 1082 bottles representing 17 brands was carried out in the United Kingdom. It was shown that natural mineral waters have a lower colony counts than other types of waters and that glass bottles generally harbored less bacteria than plastic bottles. No *Escherichia coli* or fecal streptococci was found in any of the samples. Total coliforms, sulfite-reducing clostridia or *P. aeruginosa* were detected in only 18 samples (Fewtrell et al., 1997). An examination of 35 brands of bottled water in Texas showed no major contamination with bacteria. No bacteria were detected in carbonated or distilled water brands (Saleh et al., 2008). In Greece, 15.6% of 150 samples of bottled water contained environmental mycobacteria. The mycobacteria were identified as *Mycobacterium chelonae*, *M. phlei*, *M. gordonae*, and *M. flavescens*. Some of these mycobacteria (*M. Chelonae, M. flavescens*) are opportunistic pathogens that may be harmful to immunologically deficient consumers (Papapetropoulou et al., 1997). A recent study using plate counts and 16S RNA analysis showed that most of the 238 isolates from three brands of bottled mineral waters belonged to the proteobacteria. The bacteria displayed resistance to antibiotics and some isolates (e.g., *Variovorax, Bosea, Ralstonia, Curvibacter, Afipia*, and *Pedobacter*) were multiple-antibiotic resistant (Falcone-Dias et al., 2012). Twenty-two samples of noncarbonated bottled water in England harbored *Stenotrophomonas maltophilia* and *Pseudomonas* species. *Stenotrophomonas maltophilia* is an antibiotic-resistant nosocomial pathogen that causes septicemia and sepsis in patients. The source of this pathogen in bottled water was not known (Wilkinson and Kerr, 1998).

Bacteria present in bottled water may attach to the bottle inner surfaces (side, bottom, and caps) and form biofilms (see Chapter 4 for more details on biofilm formation), as they use the small amounts of organic compounds in the water to reach levels as high as 10^4–10^5 bacteria/mL. A comparison of three bottled mineral waters showed that the number of planktonic cells varied between 1.8×10^4 and 1.2×10^5 CFU/ mL, while biofilm cells varied between 11 and 632 CFU/cm^2. Thus, the biofilm cells account for 0.03–1.79% of the total bacteria in the bottled water (Figure 9.2; Jones et al., 1999). Bacterial microcolonies are sometimes observed on the inner surface of the bottles instead of a confluent biofilm (Figure 9.3; Jayasekara et al., 1999). The factors controlling the adsorption of microorganisms to the bottle surface include surface electrostatic interaction followed by anchoring with polysaccharides, hydrophobicity, and roughness (see Chapter 4). Some have concluded that surface roughness is the most important physicochemical characteristic controlling the

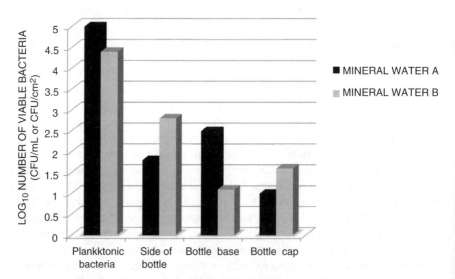

Figure 9.2 The distribution of the viable microflora between the planktonic and biofilm cells for 2 mineral bottled waters. Adapted from Jones et al. (1999). J. Appl. Microbiol. 86: 917–927.

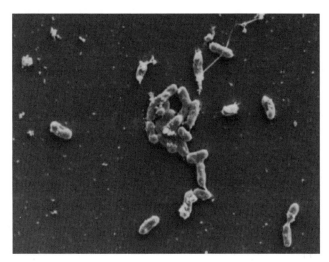

Figure 9.3 Scanning electron micrograph of a bacterial microcolony attached to the surface of a bottle of mineral water. *Source*: Jayasekara et al. (1999). Food Microbiol. 16: 115–128.

adhesion process. Microbial cell surface hydrophobicity and charge also play a role in adhesion to inner bottle surface.

Laboratory experiments have shown that viruses (e.g., HAV, rotaviruses, norovirus) also adsorb well to terephthalate (PET) and glass bottle walls. Following 20 days of storage, the percent virus adsorption was 85%, 90%, and 80% for norovirus, HAV, and rotavirus, respectively (Butot et al., 2007).

More than 50 years ago, Buttiaux and Boudier (1960) first reported the growth of autochthonous bacteria in bottled water. Due to the nutritional stress in bottled water, some bacteria may enter into the viable but not culturable (VBNC) state. Several factors control bacterial growth in bottled water (Bischofberger et al., 1990; Rosenberg, 2002). Storage at room temperature allows higher bacterial growth than when stored under refrigeration (i.e., do not store bottled water in the garage). Moreover, plastic containers allow higher bacterial growth than glass bottles. The lower bacterial survival and growth in carbonated than in noncarbonated water is due to the lower pH of carbonated water (Hunter, 1993). Bottled water may also harbor microorganisms (e.g., *Pseudomonas* sp., *Acantamoeba*, caliciviruses) which can be harmful to consumers. For example, 2 of 13 brands of mineral waters from Brazil harbored *Cryptosporidium* at concentrations of 0.2–0.5 oocysts/L (Franco and Neto, 2002). Bottled water may also harbor green algae (e.g., *Scenedesmus*), diatoms (e.g., *Cyclotella*), and cyanobacteria (e.g., *Anabaena*) (El-Zanfaly, 1990).

Mineral waters are derived from subsurface environments which have very low concentrations of bioavailable organic matter. The assimilable organic carbon (AOC) of five Norwegian brands of bottled water varied between 13.5 and 115 µg/L (Otterholt and Charnock, 2011). The oligotrophic and starved bacteria living under these conditions develop various survival strategies. Bacterial survival in bottled water improves at higher AOC values (Otterholt and Charnock, 2011). As discussed previously, one important strategy is the VBNC state. Several pathogenic bacteria have been shown to enter into the VBNC state and retain infectivity (Koch, 2002; Roszak and Colwell, 1987).

The bacterial composition of mineral waters cannot be given solely by culture techniques. Molecular culture-independent techniques allow a better picture of the bacterial flora. For example, PCR in combination with DGGE (denaturing gradient gel electrophoresis) analysis showed that each bottled mineral water had its own specific bacterial community (i.e., specific molecular fingerprint) which was stable over a period of several months (Dewettinck et al., 2001). The use of fluorescence *in situ* hybridization (FISH) technique for bottled noncarbonated mineral waters in Germany showed that beta-proteobacteria (*Hydrogenophaga*, *Aquabacterium*, and *Polaromonas*) represented 80–98% of all bacteria (Figure 9.4; Loy et al., 2005). Examination of five Norwegian brands of bottled water showed the presence of proteobacteria such as *Sphingomonas* and *Methylobacterium* species (Otterholt and Charnock, 2011).

Following bottling of mineral water, the heterotrophic bacteria experience the so called "bottle effect" or "volumetric bottle effect" as their number increases in bottles with high surface-to-volume ratio. This phenomenon was demonstrated using

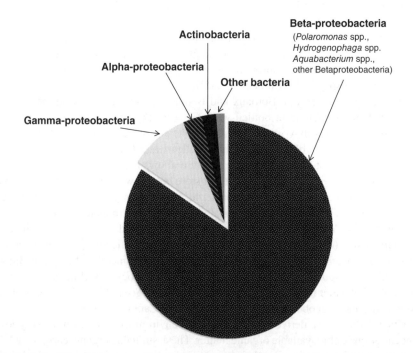

Figure 9.4 Microbial community composition of a natural mineral water approximately 3 weeks after bottling as determined by FISH and microscopic counting. Adapted from Loy, et al. (2005). Appl. Environ. Microbiol. 71: 3624–3632.

both culture methods and molecular techniques (e.g., FISH) (Leclerc and Moreau, 2002). However, more recently, using three counting techniques, no evidence of a "volumetric bottle effect" was found (Hammes et al., 2010). The increase in bacterial numbers is probably due to a concentration of carbon sources on the inner surface of the bottles.

As regard virus presence in bottled water, a 1-year monitoring of mineral waters in Switzerland showed the presence of Norwalk-like virus sequences in 33% of 159 samples analyzed (Beuret et al., 2002). However, more recently, a method using an elution step to take into account the viruses adsorbed to the bottle walls followed by real-time RT-PCR assay did not detect any virus in 294 bottled water samples from 25 countries (Butot et al., 2007). Polyethylene terephtalate (PET) is a widely used polymer by the bottled water industry. Since PET is hydrophobic, it is conceivable that this material may allow the adsorption of microorganisms with a hydrophobic surface. Overall, 10–30% of *Mycobacterium avium*, an opportunistic pathogen with a hydrophobic surface layer made of glycopeptidolipids and mycolic acids, adsorbed to the PET surface, as compared to 1–2% for *Campylobacter jejuni*, a bacterial pathogen causing gastroenteritis (Tatchou-Nyamsi-König et al., 2008). This points out to the need of analyzing both the bulk water as well as the inner bottle surface.

9.4 REGULATIONS CONCERNING BOTTLED WATER

Regulations concerning bottled water appear to be generally much stricter in Europe than in the United States. From a regulatory viewpoint, bottled water sold in interstate commerce is considered in the United States as a food item and, as such, is regulated by the federal Food and Drug Administration (FDA) under the Food, Drug and Cosmetic Act. FDA sets regulations that specifically target bottled water (FDA, 1997). Tap water produced by municipal water treatment plants is regulated by the US Environmental Protection Agency (EPA) under the Safe Drinking Water Act (SDWA). Similarly, Canada regulates bottled water as food under the Canadian Food and Drug Act.

In Europe, the production of natural mineral water is tightly regulated by the Ministry of Health or an equivalent agency. For example, in France, natural mineral water must originate from groundwater free from pollution sources. It is subjected to chemical and microbiological analyses by health authorities. No chemical treatment including disinfection (except for CO_2 addition) is allowed. The water is generally bottled at the source which must be free of pathogens and parasites. If contamination occurs, the operation must be shut down until the source of contamination is identified and corrective measures taken by the responsible authorities. In the United Kingdom, the standards applied to natural mineral water and other types of bottled water are shown in Table 9.1 (Fewtrell et al., 1997). Some companies add fluoride to bottled water which should be labeled as such.

At the international level, the Codex Alimentarius Commission of the World Health Organization (WHO) and the U.N. Food and Agriculture Organization (FAO), has developed standards for mineral waters and bottled/packaged waters

TABLE 9.1   Microbiological standards applied to bottled waters in the United Kingdom

Microorganism	Maximum Concentration
Total coliforms	0/100 mL
Fecal coliforms	0/100 mL
Fecal streptococci	0/100 mL
Sporulated sulfite-reducing clostridia (MPN)	≤1/20 mL
Total viable counts at 20–22°C, 72 hours	100/mL
Total viable counts at 37°C, 24 hours	20/mL

Source: Adapted from Fewtrell et al. (1997). Water Sci. Technol. 35 (11–12): 47–53.

(www.codexalimentarius.net). Packaged ice is under the same regulations as bottled water.

In conclusion, municipal drinking is 500 to more than 1000 times less expensive than bottled water. Tap water contains fluorides which protect consumers from tooth decay and is often monitored by local, State, and federal agencies. Information on the local drinking water quality is available online from the US EPA site: http://www.epa.gov/safewater/dwinfo/index.html. However, bottled water is a convenient resource that can be used while traveling (do not leave bottled water inside a car, especially in warm climates) and can be shipped to communities in emergency situations (earthquakes, floods, hurricanes, and tornadoes).

WEB RESOURCES

www.niehs.nih.gov
(National Institute of Environmental Health)

http://www.fda.gov/ForConsumers/ConsumerUpdates/ucm203620.htm
(Bottled water basics)

http://www.worldwater.org
(General information about water, including bottled water)

www.codexalimentarius.net
(International food standards: FAO & WHO)

www.globaldrinks.com
(online beverage data & forecasts)

http://www.efbw.eu
(European Federation of Bottled Waters)

FURTHER READING

Bischofberger, T., S.K. Cha, R. Schmitt, B. Konig, W. Schmidt-Lorenz. 1990. The bacterial flora of non-carbonated, natural mineral water from the springs to reservoir and glass and plastic bottles. *Int. J. Food Microbiol.* 11:51–71.

Casanovas-Massana, A. and A.R. Blanch. 2012. Diversity of the heterotrophic microbial populations for distinguishing natural mineral waters. *Int. J. Food Microbiol.* 153:38–44.

European Federation of Bottled Water (EFBW). 2011. http://efbw.eu/bwf.php?classement=07.

Loy, A., W. Beisker, and H. Meier. 2005. Diversity of bacteria growing in natural mineral water after bottling. *Appl. Environ. Microbiol.* 71:3624–3832.

Rosenberg, F.A. 2002. Bottled water, microbiology. In *Encyclopedia of Environmental Microbiology*, G. Bitton (editor-in-chief). Wiley-Interscience, New York, pp. 795–802.

Wilkinson, F.H. and K.G. Kerr. 1998. Bottled water as a source of multi-resistant *Stenotrophomonas* and *Pseudomonas* species for neutropenic patients. *Eur. J. Cancer Care* 7:12–14.

10

INTRODUCTION TO MICROBIAL RISK ASSESSMENT FOR DRINKING WATER

Risk assessment is the qualitative or quantitative estimation of adverse health risks following exposure to physical, chemical, or microbial hazards (Haas et al., 1999). Quantitative risk assessment was initiated by the US National Academy of Science (NAS) and the Environmental Protection Agency to address the risks due to hazardous chemicals (NAS, 1983). Risk assessment pertaining to drinking water is undertaken to predict the burden of waterborne diseases on a given community, set tolerable limits of waterborne disease, help improve drinking water treatment (i.e., set specific levels of treatment to remove/inactivate pathogens or target microorganisms), identify means to reduce the risks to the consumers or determine measures to protect the watershed.

10.1 HEALTH-BASED TARGETS FOR DRINKING WATER

Various hazards (microbial, chemical, radiological) may be associated with drinking water. The burden of disease associated with these hazards can be quantified (duration of the disease, probability, severity) and compared using a common "metric" called the disability-adjusted life year (DALY) adopted by WHO (Havelaar and Melse, 2003; WHO, 2011c). As compared to other metrics (e.g., annual mortality rates), the DALY metric takes into account both lethal and nonlethal (i.e., morbidity) endpoints (Havelaar and Melse, 2003). The tolerable burden of disease was set at 10^{-6} DALY/person per year. However, due to potential multiple exposure routes the burden of disease from waterborne exposure has been set at 10^{-4} DALY/person per year.

Microbiology of Drinking Water Production and Distribution, First Edition. Gabriel Bitton.

Health-based targets aim at improving drinking water quality and protecting human health. The targets should take into account other exposure sources such as air, food, or person-to-person contact.

There are four categories of health-based targets:

- Health outcome targets expressed as DALY/person per year. Quantitative microbial risk assessment serves as a basis for health-based targets (see Section 10.2).
- Water quality targets addressing chemical hazards.
- Performance targets expressed, for example, as log_{10} reduction of pathogens and parasites in a given water treatment plant.
- Specified technology targets.

Performance targets help in selecting control measures to obtain safe drinking water (see Chapter 2). They are expressed as log_{10} reductions of pathogens or surrogates following a given water treatment or a combination of treatments. It is quite difficult to derive performance targets for all waterborne pathogens. Therefore, reference pathogens are generally used to represent bacterial (*Campylobacter jejuni*), viral (rotavirus), and protozoan (*Cryptosporidium*) pathogens. Alternative reference microorganisms have been recommended for *C. jejuni* (e.g., *E. coli*, *Enterococcus*), rotavirus (e.g., echovirus 12, MS2), and *Cryptosporidium* (e.g., bacterial spores). For example, a study of 66 waterworks across China showed that the estimated health burden due to the presence of *Cryptosporidium* in drinking water was 8.31×10^{-6} DALYs for conventionally treated water, 0.74×10^{-6} DALYs for ozonated water, and 0.47×10^{-6} DALYs for microfiltered water (Figure 10.1; Xiao et al., 2012). Thus, the health burden of conventionally treated water is higher than the WHO safe burden of 10^{-6} DALYs/person per year.

The World Health Organization (WHO, 2011b) recommends a tiered approach for setting the performance targets, expressed as log_{10} reduction of the pathogens. The top-tier standard is highly protective and is equivalent to a level of risk of 10^{-6} DALY/person per year. The second tier standard is protective and is equivalent to a risk level of 10^{-4} DALY/person per year. The bottom tier (interim standard) is applied to technologies with a demonstrated health improvement and which are protective for two classes of pathogens. Table 10.1 (WHO, 2011b) shows the log_{10} reduction required to achieve the highly protective, protective, and interim criteria for home water treatment (HWT) technologies.

10.2 QUANTITATIVE MICROBIAL RISK ASSESSMENT (QMRA)

Risk assessment can be accomplished via epidemiological studies or quantitative microbial risk assessment (QMRA). It is based on data from epidemiological studies and literature results from experimental studies on animals or humans. The epidemiological approach is, however, more time-consuming and more expensive than QMRA.

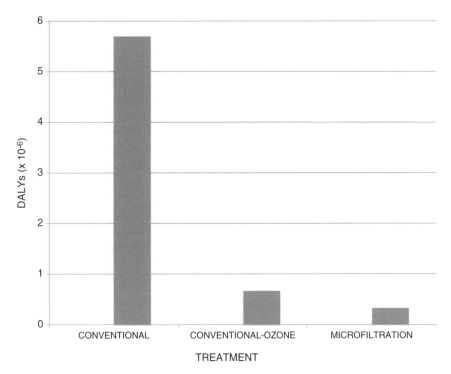

Figure 10.1 Estimated health burden (in DALYs) due to *Cryptosporidium* in drinking water by exposure route (direct drinking) and type of water treatment. Adapted from Xiao et al. (2012). Water Res. 46: 4272–4280.

TABLE 10.1 Performance requirements for HWT technologies and associated $\log_{10}$ reduction criteria for "interim," "protective," and "highly protective"

Reference Pathogen Used in Dose-Response Model	Assumed Number of Microorganisms/ Liter	Pathogen Class	$\log_{10}$ Reduction Required		
			Interim	Protective	Highly Protective
			Requires correct, consistent and continuous use to meet performance levels		
Campylobacter jejuni	1	Bacteria	Achieves protective target for two classes of pathogens and results in health gains	≥ 2	≥ 4
Rotavirus	1	Viruses		≥ 3	≥ 5
Cryptosporidium	0.1	Protozoa		≥ 2	≥ 4

Source: Adapted from WHO (2011b). Evaluating Household Water Treatment Options: Health-based targets and microbiological performance specifications. World Health Organization, Geneva, Switzerland.

The acceptable risk proposed by U.S. EPA is 10^{-4} (1 infection per 10,000 exposed consumers/year) (U.S. EPA, 1989).

QMRA specifically addresses the adverse health risks (infection, illness, death) associated with microbial agents. It helps estimate the risk of contracting an infection or disease, based on the occurrence of a given pathogen in drinking water and the volume of drinking water consumed in liters per day (Haas et al., 1999; Post et al., 2011).

Risk assessment consists of five steps (Haas and Eisenberg, 2001; Haas et al., 1999; Hunter et al., 2003; Hrudey et al., 2006; Post et al., 2011; U.S. EPA, 2012; WHO, 2011c).

10.2.1   Hazard Identification and Characterization

The goal is to provide a qualitative evaluation of the microbial hazard. It identifies the microbial agent and its adverse health effects (host–pathogen interaction, virulence, pathogenicity, infection, morbidity, mortality) on a given population. Disease outbreaks and other epidemiological data as well as water quality data provide useful information at this stage.

10.2.2   Exposure Assessment

The goal of exposure assessment is to determine the size of the exposed population, route, frequency, and the dose of exposure to the pathogen over a given period (e.g., over a day or a year). Fate and transport studies and models, along with adequate methods to detect the pathogens in the environment, are also needed at this stage. To calculate the dose one needs to know about the volume of ingested drinking water (generally estimated at 2 L/person per day) and the pathogen concentration in the water. However, since the pathogen concentration in treated drinking water is generally very low, especially for viruses and protozoan parasites, one can indirectly estimate it by determining the pathogen level in nontreated source water and correcting it for log removal following passage through a water treatment plant. The $\log_{10}$ removal is obtained via experiments with indicator microorganisms.

10.2.3   Dose-Response Assessment

It is a mathematical expression that describes the relationship between the number of pathogens people are exposed to and the probability of adverse health effects. The dose-response is derived from studies on exposure of animals or, sometimes, healthy human volunteers to different doses of the pathogen or from disease outbreak data (Teunis et al., 2008). Unfortunately, it does not always address certain segments of the population such as the immunocompromised, the young, the elderly, and pregnant women. Furthermore, dose-response assessment was carried out only for a few pathogens. Factors influencing the dose-response are the exposure route, exposure medium, pathogen strain host, endpoint, and the data source (Messner et al., 2001;

Teunis et al., 2002; U.S. EPA, 2012). The data are modeled (e.g., exponential model) to extrapolate the probability of infection at low doses.

Among a wide range of proposed dose-response models, the two most commonly used are the exponential model and the beta-Poisson model:

The **exponential model** is given by the following equation:

$$P_i(D) = 1 - e^{-rD} \tag{10.1}$$

where

P_i (D) is the probability of infection in a population following exposure to dose D;

D is the ingested dose of pathogens. It is given by the pathogen concentration in water × daily water intake (generally assumed to be 2 L/day);

r is the pathogen "infectivity constant" or the fraction of the surviving pathogens that causes infection. This constant expresses the interaction between the pathogen and the host.

The exponential model assumes a random distribution (Poisson distribution) of the pathogens in water, infection caused at least by one single surviving pathogen in the host (no minimum infectious dose), and a constant probability of infection per ingested pathogen. However, the model does not take into account the variability of individuals within the population.

The **beta-Poisson model** is as follows (Haas et al., 1999):

$$P_i(D) = 1 - (1 + (D/\beta)^{-\alpha} \tag{10.2}$$

where

$P_i(D)$ is the probability of infection following exposure to dose D;

D is the ingested dose of pathogens. It is given by the pathogen concentration in water × daily water intake (generally assumed to be 2 L/day);

β and α are model constants.

This model has similar assumptions as the exponential model, except that the probability of infection per ingested pathogen may vary within the exposed population due to variation in human response or pathogen competence (Haas et al., 1999).

These models were used to estimate the risk of infection following ingestion (1 to 10^6 cysts) by human volunteers of drinking water containing *Giardia*. The average r value for *Giardia* was 0.01982. It was found that the yearly risks of infection were 4.8×10^{-3} for polluted source water and 1.3×10^{-4} for pristine source water in water treatment plants achieving 10^{-3} treatment reduction (Rose et al., 1991a). The models were also used to evaluate the potential health risks caused by low levels of viruses in drinking water. Based on the acceptable risk of 10^{-4} proposed by the U.S. EPA, it was concluded that poliovirus and rotavirus levels should not exceed 0.1 and 0.3 PFU/100 L, respectively (Rose and Gerba, 1991).

Some reported values for r, α, and β for various bacterial, viral, and protozoan pathogens were compiled (McBride et al., 2002; U.S. EPA, 2012).

10.2.4 Risk Characterization

This step takes into account the data collected in the preceding steps to estimate the risk of infection or illness. The potential for adverse health effects can be expressed as an individual or population-level risk estimate. Risk characterization is mostly quantitative but can be qualitative if certain data are missing (U.S. EPA, 2012).

10.2.5 Risk Management and Communication

Risk management is a decision-making process based on social, economical, ethical, political, technical, and regulatory factors.

Table 10.2 (WHO, 2011c) illustrates the calculation of infection and illness risks and the disease burden of three reference pathogens (*Campylobacter*, rotavirus, and *Cryptosporidium*).

Compared to risk assessment for chemicals, the infectious disease process is more complex, population-based and dynamic. In addition to the risk from direct exposure to pathogens, one must also take into account the risk of secondary infection from infected individuals or asymptomatic carriers within the population. Another factor to take into account is the immunity to the pathogen under study (Haas and Eisenberg, 2001).

A computational tool for quantitative microbial risk assessment, called QMRAspot, was developed for rapid and automatic undertaking of QMRA for drinking water produced from surface water as a source. QMRAspot uses four index pathogens (*Campylobacter*, enterovirus, *Cryptosporidium*, *Giardia*) (Schijven et al., 2011).

10.3 SOME EXAMPLES OF USE OF RISK ASSESSMENT TO ASSESS THE RISK OF INFECTION OR DISEASE FROM EXPOSURE TO MICROBIAL PATHOGENS

QMRA dose-response models help determine the treatments required to reduce the pathogens/parasites levels in source waters to achieve an acceptable risk such as the U.S. EPA 10^{-4} risk or the WHO 10^{-6} DALYs/person per year (see Rose and Gerba, 1991). Following are some examples of risk assessment of exposure to bacterial and viral pathogens and protozoan parasites in drinking water.

10.3.1 *Legionella*

In Chapter 1, we showed that *Legionella* is transmitted via the airborne route. Infections may occur following exposure to bioaerosols generated by cooling towers,

TABLE 10.2   Linking tolerable disease burden and raw water quality for reference pathogens: example calculation

River Water (Human and Livestock Pollution)	Units	*Cryptosporidium*	*Campylobacter*	Rotavirus
Raw water quality (C_R)	Organisms/L	10	100	10
Treatment needed to reach tolerable risk (PT)	Log_{10} reduction	5.89	5.98	5.96
Drinking-water quality (CD)	Organisms/L	1.3×10^{-5}	1.05×10^{-4}	1.1×10^{-5}
Consumption of unheated drinking water (V)	L/day	1	1	1
Exposure by drinking Water (E)	Organisms/day	1.3×10^{-5}	1.05×10^{-4}	1.1×10^{-5}
Dose-response (r)[a]	Probability of infection/ organism	2.0×10^{-1}	1.9×10^{-2}	5.9×10^{-2}
Risk of infection ($P_{inf,d}$)	Per day	2.6×10^{-6}	2.0×10^{-6}	6.5×10^{-6}
Risk of infection ($P_{inf,y}$)	Per year	9.5×10^{-4}	7.3×10^{-4}	2.4×10^{-3}
Risk of diarrheal illness given infection ($P_{ill,inf}$)	Probability of illness per infection	0.7	0.3	0.5
Risk of diarrheal illness (P_{ill})	Per year	6.7×10^{-4}	2.2×10^{-4}	1.1×10^{-3}
Disease burden (db)	DALY per case	1.5×10^{-3}	4.6×10^{-3}	1.2×10^{-2}
Susceptible fraction (f_s)	Percentage of population	100	100	6
Health outcome target (HT)	DALY per year	1×10^{-6}	1×10^{-6}	1×10^{-6}

Formulas:
$$C_D = \frac{C_R}{10^{PT}}$$
$$P_{inf,d} = E \times r$$
$$HT = \frac{P_{ill} \times db \times f_s}{100}$$
$$E = C_D \times V$$
$$P_{ill} = P_{inf,y} \times P_{ill,inf}$$

DALY, Disability-adjusted life year.
[a]Dose-response for *Campylobacter* and Rotavirus from Haas et al. (1999).
Source: WHO (2011a). Guidelines for Drinking-Water Quality. 4th Ed. World Health Organization, Geneva, Switzerland. http://www.who.int/water_sanitation_health/publications/2011/dwq_guidelines/en/.

shower heads, spas, humidifiers, or air conditioning. The risk of infection from *Legionella* aerosols generated from spas was estimated at 8.9×10^{-4}, based on a *Legionella* air concentration of $5–180$ CFU/m³ generating a retained dose of $1.3–34.5$ CFU (Armstrong and Haas, 2008). In in-premise plumbing, exposure to *Legionella* during showering may be via planktonic, biofilm-associated, protozoan host-associated (pathogen inside trophozoite or cyst forms) cells or *Legionella* cells inside vesicles generated by the protozoan host (Buse et al., 2012). An exposure model for biofilm-associated *Legionella* was proposed to predict the water and air

TABLE 10.3   Risks associated with coxsackievirus in drinking water

Type of Risk[a]	5×10^{-3} MPN CPU/L Surface Water		0.13 PFU/L Groundwater	
	Per day	Per year	Per day	Per year
Risk of infection	7.75×10^{-5}	2.79×10^{-2}	2.01×10^{-3}	5.21×10^{-1}
Risk of illness	5.81×10^{-5}	2.09×10^{-2}	1.51×10^{-3}	3.91×10^{-1}
Risk of death	3.43×10^{-7}	1.23×10^{-4}	8.91×10^{-6}	2.30×10^{-3}

Source: Adapted from Mena et al. (2003).
PFU, plaque forming unit; CPU, cytopathic unit.
[a]General population: 2 L drinking water/day per person.

pathogen concentrations that could lead to infection during showering (Figure 1.2; Schoen and Ashbolt, 2011). The best estimated predictions were 3.5×10^7 CFU/L for water and 3.5×10^2 CFU/m³ for air.

10.3.2 Viruses

Ingestion of even low levels of some pathogens, particularly viruses, may pose some risk of infection, clinical illness, or even mortality to susceptible populations (Gerba and Haas, 1986) (see Chapter 1). An example of risk associated with coxsackievirus in drinking water is shown in Table 10.3   (Mena et al., 2003). The exponential risk assessment model described in Section 10.2 was used to assess the risk associated with the presence of coxsackieviruses in drinking water. The model for the general population assumes a consumption of 2 L of drinking water per day and a virus concentration of 0.13 PFU/L (highest concentration) or 10^{-3} MPN CPU (most probable number of cytopathic units/L) (lowest concentration). The daily risk of infection from drinking treated surface water was 10^{-5} whereas the annual risk was 10^{-2}. A higher risk was associated with drinking nondisinfected groundwater (Mena et al., 2003).

The estimated risk for norovirus infection was $10^{-2.1}$ infection/person-year, assuming an ID_{50} of 10. It is higher than the acceptable risk (10^{-4} infection/person-year) proposed by the U.S. EPA (Masago et al., 2006).

Using an acceptable risk of 10^{-4} and a 5-log removal/inactivation of different emerging pathogens in water treatment plants, Table 10.4   gives their estimated threshold concentrations (numbers/L) in source water (Haas et al., 1999; Regli et al., 1991).

10.3.3 Protozoan Parasites

Following a survey of infectious *Cryptosporidium* in finished water from 82 filter water plants in the United States, it was concluded that the daily risk of infection with *Cryptosporidium* was 1.5×10^{-5}. This represents an annual risk of 52 infections/10,000 people, a risk much higher than the EPA recommended 1/10,000 annual risk (Aboytes et al., 2004). In a conventional water treatment plant with a chlorination step, the annual risk of infection with *Cryptosporidium* oocysts (8.4×10^{-4}) was eight

TABLE 10.4   Estimated thresholds for finished and raw water concentrations of pathogens to meet the U.S. EPA average risk of infection of 10^{-4}

Organism	Average Finished Water Concentration (#/L)	Average Log Treatment Removal	Estimated Average Raw Water Concentration Threshold (#/L)
Salmonella	0.0013	5	130
Poliovirus 1	0.0000151	5	1.51
Poliovirus 3	0.000000265	5	0.0265
Echovirus 12	0.0000675	5	6.75
Rotavirus	0.000000222	5	0.0222
Entamoeba coli	0.000000265	5	0.0625
Cryptosporidium	0.003	3	0.075^a
Giardia	0.00000675	5	0.675

Source: Adapted from Crockett (2007). Water Environ. Res. 79: 221–232; Regli et al. (1991). Amer. Water Works Assoc. J. 83(11): 76–84; Haas et al. (1999). Quantitative Microbial Risk Assessment. J. Wiley and Sons: New York.
[a] Value established by the U.S. EPA to reflect analytical recovery from environmental samples.

times higher than the risk caused by *Giardia* cysts (1.1×10^{-4}). Thus, the *Giardia* risk is close to the U.S. EPA guideline. A cost-benefit analysis was used to estimate to daily risk of illness from exposure to *Cryptosporidium*. The risk was estimated at 9×10^{-4} and corresponds to a finished water concentration of 0.046 oocysts/L (Ryan et al., 2013). The fate of the immunodeficient population was addressed in a study of 66 water treatment plants across China. The annual diarrhea morbidity due to *Cryptosporidium* was 2701 cases/100,000 population for immunodeficient patients as compared to 148 cases/100,000 for the immunocompetent population (Xiao et al., 2012). The disease burden, expressed as DALYs, was estimated at 8.31×10^{-6} DALY/person per year for conventionally treated water. This burden is more than eight times higher than the WHO guideline of 10^{-6} DALYs/person per year (the DALY concept is explained in more details in Section 10.1). The risk due to *Cryptosporidium* decreases following disinfection with ozone or microfiltration (Barbeau et al., 2000; Xiao et al., 2012; see Figure 10.1). However, a more recent study of 14 surface water treatment plants showed no infectious oocysts in any of the 370 samples. The calculated risk was lower than the EPA recommended risk of 1 in 10,000 people (Rochelle et al., 2012).

Giardia is recognized as one of the most important etiological agents in waterborne disease outbreaks. An exponential risk assessment model was considered for the estimation of the risk of infection following exposure of human volunteers to *Giardia* cysts (Rose et al., 1991). The model revealed that the yearly risk for polluted source waters was 4.8×10^{-3} while the yearly risk for pristine source waters was 1.3×10^{-4}.

Risk incidents due to pathogens and parasites in water treatment plant are associated with inefficient particle removal, disinfection failures or malfunctions, cross connections in distribution systems, microbial pollution of reservoirs, and pathogen concentration in source water. Using *C. jejuni*, rotavirus, and *Cryptosporidium* as model pathogens, a risk assessment study was carried out in a Swedish drinking water system. It was found that the microbial yearly infections were mostly due to normal operation of the plant and not to the failures or malfunctions mentioned above (Westrell et al., 2003).

WEB RESOURCES

http://books.nap.edu/openbook.php?record_id=11728&page=R1
(National Research Council, 2006. *Drinking Water Distribution System: Assessing and Reducing Risks*)

http://www.camra.msu.edu/qmra.html
(Center for advancing microbial risk assessment)

http://www.ucd.ie/microbialrisknetwork/
(Irish microbial risk assessment network)

http://www.epa.gov/athens/research/qmra.html
(QMRA from EPA)

FURTHER READING

Haas, C.N., J.B. Rose, and C.P. Gerba. 1999. *Quantitative Microbial Risk Assessment*. John Wiley & Sons, New York, 449 p.

Haas, C.N., and J.N.S. Eisenberg. 2001. Risk assessment. In *Water Quality: Guidelines, Standards and Health*, L. Fewtrell and J. Bartram (Eds). WHO by IWA Publishing (ISBN: 924154533X), pp. 161–193, http://www.who.int/water_sanitation_health/dwq/whoiwa/en/

Havelaar, A.H., and J.M. Melse. 2003. *Quantifying Public Health Risks in the WHO Guidelines for Drinking-Water Quality: A Burden of Disease Approach*. National Institute for Public Health and the Environment, Bilthoven (RIVM Report 734301022/2003).

National Research Council. 2006. *Drinking Water Distribution Systems: Assessing and Reducing Risks*. NRC Washington, DC, 400 p. http://www.nap.edu/catalog/11728.html

U.S. EPA. 2012. *Microbial Risk Assessment Guideline: Pathogenic Microorganisms with Focus on Food and Water*. EPA/100/J-12/001 and USDA/FSIS/2012-001.

WHO. 2011. *Guidelines for Drinking-Water Quality*, 4th edition. World Health Organization, Geneva, Switzerland.

WHO. 2011. *Evaluating Household Water Treatment Options: Health-Based Targets and Microbiological Performance Specifications*. World Health Organization, Geneva.

REFERENCES

Abbaszadegan, M., M.S. Huber, C.P. Gerba, and I.L. Pepper. 1997. Detection of viable Giardia cysts by amplification of heat shock-induced mRNA. *Appl. Environ. Microbiol.* 63:324–328.

Abbaszadegan, M., B.E. Mayer, H. Ryu, and N. Nwachuku. 2007. Efficacy of removal of CCL viruses under enhanced coagulation conditions. *Environ. Sci. Technol.* 41:971–977.

Abbaszadegan, M., P. Montero, N. Nwachuku, A. Alum, and H. Ryu. 2008. Removal of adenovirus, calicivirus, and bacteriophages by conventional drinking water treatment. *J. Environ. Sci. Health A Tox. Hazard. Subst. Environ. Eng.* 43:171–177.

Aboagye, G. and M.T. Rowe. 2011. Occurrence of *Mycobacterium avium* subsp. *paratuberculosis* in raw water and water treatment operations for the production of potable water. *Water Res.* 45:3271–3278.

Aboytes, R., G.D. Giovanni, F.A. Abrams, C. Rheinecker, W. Mcelroy, N. Shaw, and M.W. Lechevallier. 2004. Detection of infectious *Cryptosporidium* in filtered drinking water. *J. Amer. Water Wks. Assoc.* 96(9):88–98.

Adal, K.A., C.R. Sterling, and R.L. Guerrant. 1995. *Cryptosporidium* and related species. In *Infections of the Gastrointestinal Tract*, M.J. Blaser, P.D. Smith, J.I. Ravdin, H.B. Greenberg, and R.L. Guerrant (Eds). Raven Press, New York, pp. 1107–1128.

Adams, J.A. and D. McCarty. 2007. Real-time, on-line monitoring of drinking water for waterborne pathogen contamination warning. *Int. J. High Speed Electron. Syst.* 17(4):643–659.

Afzal, A., T. Oppenlander, J.R. Bolton, and M.G. El-Din. 2010. Anatoxin-a degradation by advanced oxidation processes: vacuum-UV at 172 nm, photolysis using medium pressure UV and UV/H_2O_2. *Water Res.* 44:278–286.

Aguilar, A., B. Jimenez, J.E. Becerril, and L.P. Castro. 2008. Use of nanofiltration for potable water from an aquifer recharged with wastewater. *Water Sci. Technol.* 57(6):927–933.

Ahammed, M.M. and K. Davra. 2011. Performance evaluation of biosand filter modified with iron oxide-coated sand for household treatment of drinking water. *Desalination* 276:287–293.

Ahmed, W., A. Vieritz, A. Goonetilleke, and T. Gardner. 2010a. Health risk from the use of roof-harvested rainwater in Southeast Queensland, Australia, as potable or nonpotable water, determined using quantitative microbial risk assessment. *Appl. Environ. Microbiol.* 76:7382–7391.

Microbiology of Drinking Water Production and Distribution, First Edition. Gabriel Bitton.
© 2014 John Wiley & Sons, Inc. Published 2014 by John Wiley & Sons, Inc.

Ahmed, W., A. Goonetilleke, and T. Gardner. 2010b. Implications of faecal indicator bacteria for the microbiological assessment of roof-harvested rainwater quality in southeast Queensland, Australia. *Can J. Microbiol.* 56:471–479.

Aieta, E.M. and J.D. Berg. 1986. A review of chlorine dioxide in drinking water treatment. *J. Am. Water Works Assoc.* 78:62–72.

Al-Adhami, B.H., R.A.B. Nichols, J.R. Kusel, J. O'Grady, and H.V. Smith. 2007. Detection of UV-induced thymine dimers in individual *Cryptosporidium parvum* and *Cryptosporidium hominis* oocysts by immunofluorescence microscopy. *Appl. Environ. Microbiol.* 73:947–955.

Alagappan, A., P.L. Bergquist, and B.C. Ferrari. 2009. Development of a two-color fluorescence *in situ* hybridization technique for species-level identification of human-infectious *Cryptosporidium* spp. *Appl. Environ. Microbiol.* 75:5996–5998.

Al-Ani, M.Y., D.W. Hendricks, G.S. Logsdon, and C.P. Hibler. 1986. Removing *Giardia* cysts from low turbidity waters by rapid rate filtration. *J. Am. Water Works Assoc.* 78:66-73.

Albertson, J.S. and F.W. Oehme. 1994. Animal and plant toxins. In *Basic Environmnetal Toxicology*, L.G. Cockerham and B.S. Shane (Eds). CRC Press, Boca Raton, FL, pp. 263–283.

Albinana-Gimenez, N., P. Clemente-Casares, S. Bofill-Mas, F. Ribas, and R. Girones. 2006. Distribution of human polyomaviruses, adenoviruses, and hepatitis E virus in the environment and in a drinking-water treatment plant. *Environ. Sci. Technol.* 40:7416–7422.

Alkan, U., A. Teksoy, A. Atesli, and H.S. Baskaya. 2007. Influence of humic substances on the ultraviolet disinfection of surface waters. *Water Environ. J.* 21:61–68.

Allard, A., B. Albinsson, and G. Wadell. 1992. Detection of adenoviruses in stools from healthy persons and patients with diarrhea by two-step polymerase chain reaction. *J. Med. Virol.* 37:149–157.

Allegra, S., F. Grattard, F. Girardot, S. Riffard, B. Pozzetto, and P. Berthelot. 2011. Longitudinal evaluation of the efficacy of heat treatment procedures against *Legionella* spp. in hospital water systems by using a flow cytometric assay. *Appl. Environ. Microbiol.* 77:1268–1275.

Allen, S.J. 1996. Types of adsorbent materials. In *Use of Adsorbents for the Removal of Pollutants from Wastewater*, G. McKay (Ed.). CRC Press, Boca Raton, FL, pp. 59–97.

Al-Mudhaf, H.F., F.A. Alsharifi, and A.-S. Abu-Shady. 2009. A survey of organic contaminants in household and bottled drinking waters in Kuwait. *Sci. Total Environ.* 407:1658–1668.

Alrousan, D.M.A., P.S.M. Dunlop, T.A. McMurray, and J.A. Byrne. 2009. Photocatalytic inactivation of *E. coli* in surface water using immobilised nanoparticle TiO_2 films. *Water Res.* 43:47–54.

Altekruse, S.F., M.L. Cohen, and D.L. Swerdlow. 1997. Emerging foodborne diseases. *Emerg. Infect. Dis.* 3:285–293.

Altmar, R.L., F.H. Neill, J.L. Romalde, F. LeGuyader, C.M. Woodley, T.G. Metcalf, and M.K. Estes. 1995. Detection of Norwalk virus and hepatitis A virus in shellfish tissues with the PCR. *Appl. Environ. Microbiol.* 61:3014–3018.

Alvarez, M.E. and R.T. O'Brien. 1982. Mechanism of inactivation of poliovirus by chlorine dioxide and iodine. *Appl. Environ. Microbiol.* 44:1064–1071.

American Water Works Association. 2000. Committee report: disinfection at large and medium-sized systems. *J. Am. Water Works Assoc.* 92:32–33.

American Water Works Association. 2006. *Fundamentals and Control of Nitrification in Chloraminated Drinking Water Distribution Systems*, 1st editon. AWWA, Denver, CO.

Amin, M.T. and M.Y. Han. 2009. Roof-harvested rainwater for potable purposes: application of solar collector disinfection (SOCO-DIS). *Water Res.* 42:5225–5235.

Amin, M.T. and M.Y. Han. 2011. Improvement of solar based rainwater disinfection by using lemon and vinegar as catalysts. *Desalination* 276:416–424.

Anaissie, E.J., R.T. Kuchar, J.H. Rex, A. Francesconi, M. Kasai, F.M. Müller, M. Lozano-Chiu, R.C. Summerbell, M.C. Dignani, S.J. Chanock, and T.J. Walsh. 2001. Fusariosis associated with pathogenic *Fusarium* species colonization of a hospital water system: a new paradigm for the epidemiology of mold infections. *Clin. Infect. Dis.* 33:1871–1878.

Anaissie, E.J., S.R. Penzak, and M.G. Dignani. 2002. The hospital water supply as a source of nosocomial infections. *Arch. Intern. Med.* 162:1483–1492.

Anderson, W.B., R.M. Slawson, and C.I. Mayfield. 2002. A review of drinking-water-associated endotoxin, including potential routes of human exposure. *Can. J. Microbiol.* 48:567–587.

Anderson, W.B., P.M. Huck, D.G. Dixon, and C.I. Mayfield. 2003. Endotoxin inactivation in water by using medium-pressure UV lamps. *Appl. Environ. Microbiol.* 6:3002–3004.

Andersson, A., P. Laurent, A. Kihn, M. Prévost, and P. Servais. 2001. Impact of temperature on nitrification in biological activated carbon (BAC) filters used for drinking water treatment. *Water Res.* 35:2923–2934.

Anonymous. 1903. The purification of water supplies by slow sand filtration. *J. Am. Med. Assoc.* 41:850–853.

Antopol, S.C. and P.D. Ellner. 1979. Susceptibility of *Legionella pneumophila* to ultraviolet radiation. *Appl. Environ. Microbiol.* 38:347–348.

Anwar, H. and J.W. Costerton. 1992. Effective use of antibiotics in the treatment of biofilm-associated infections. *ASM News* 58:665–668.

Appenzeller, B.M.R., C. Yanez, F. Jorand, and J.C. Block. 2005. Advantage provided by iron for *Escherichia coli* growth and cultivability in drinking water. *Appl. Environ. Microbiol.* 71:5621–5623.

Archer, A.R., A.C. Elmore, E. Bell, and C. Rozycki. 2011. Field investigation of arsenic in ceramic pot filter-treated drinking water. *Water Sci. Technol.* 63(10):2193–2198.

Arkhangelsky, E. and V. Gitis. 2008. Effect of transmembrane pressure on rejection of viruses by ultrafiltration membranes. *Sep. Purif. Technol.* 62:619–628.

Armstrong, T.W. and C.N. Haas. 2007. Quantitative microbial risk assessment model for Legionnaires' disease: assessment of human exposures for selected spa outbreaks. *J. Occup. Environ. Hyg.* 4:634–646.

Armstrong, T.W. and C.N. Haas. 2008. Legionnaires' disease: evaluation of a quantitative microbial risk assessment model. *J. Water Health* 6(2):149–166.

Arnal, J.M., M. Sancho, B. Garcıa Fayos, J. Lora, and G. Verdú. 2007. Aquapot: UF real applications for water potabilization in developing countries. Problems, location and solutions adopted. *Desalination* 204:316–321.

Arnal, J., B. Garcia-Fayos, G. Verdu, and J. Lora. 2009. Ultrafiltration as an alternative membrane technology to obtain safe drinking water from surface water: 10 years of experience on the scope of the AQUAPOT project. *Desalination* 248(1–3):34–41.

Arnal, J.M., B. García-Fayos, M. Sancho, G. Verdú, and J. Lora. 2010. Design and installation of a decentralized drinking water system based on ultrafiltration in Mozambique. *Desalination* 250:613–617.

Arnold, B.F. and J.M. Colford. 2007. Treating water with chlorine at point-of-use to improve water quality and reduce child diarrhea in developing countries: a systematic review and meta-analysis. *Am. J. Trop. Med. Hyg.* 76(2):354–364.

Arnon, S.S., R. Schechter, T.V. Inglesby, D.A. Henderson, and J.G. Bartlett; Working Group on Civilian Biodefense. 2001. Botulinum toxin as a biological weapon: medical and public health management. *J. Am. Med. Assoc.* 285:1059–1070.

Arora, H. and M.W. LeChevallier. 2002. Occurrence of protozoa in spent filter backwash water. In *Encyclopedia of Environmental Microbiology*, G. Bitton (Editor-in-Chief). Wiley-Interscience, New York, pp. 2261–2267.

Arvanitidou, M., K. Kanellou, T.C. Constantinides, and V. Katsouyannopoulos. 1999. The occurrence of fungi in hospital and community potable waters. *Lett. Appl. Microbiol.* 29:81–84.

Ashokkumar, M., T. Vu, F. Grieser, A. Weerawardena, N. Anderson, N. Pilkington, and D.R. Dixon. 2003. Ultrasonic treatment of *Cryptosporidium* oocysts. *Water Sci. Technol.* 47(3):173–177.

Atlas, R.M. 2002. Bioterrorism: from threat to reality. *Annu. Rev. Microbiol.* 56:167–185.

Atmar, R.L. and M.K. Estes. 2006. The epidemiologic and clinical importance of norovirus infection. *Gastroenterol. Clin. North Am.* 35:275–290.

AWWA Disinfection Systems Committee. 2008. Committee Report: Disinfection Survey, Part 2–Alternatives, experiences, and future plans. *J. Am. Water Works Assoc.* 100(11):110–124.

AWWA "Organisms in Water" Committee. 1987. Committee report: microbiological consideration for drinking water regulations revisions. *J. Am. Water Works Assoc.* 79:81–88.

AWWA Subcommittee on Periodical Publications of the Membrane Process Committee. 2008. Microfiltration and ultrafiltration membranes for drinking water. *J. Am. Water Works Assoc.* 100(12):84–97.

Bachmann, R.T. and R.G.J. Edyvean. 2006. Biofouling: an historic and contemporary review of its causes, consequences and control in drinking water distribution systems. *Biofilms* 2:197–227.

Bader, T.F. 1995. *Viral Hepatitis: Practical Evaluation and Treatment*. Hogrefe & Huber Publishers, Seattle, WA, 234 pp.

Badireddy, A.J., E.M. Hotze, S. Chellam, P. Alvarez, and M.R. Wiesner. 2007. Inactivation of bacteriophages via photosensitization of fullerol nanoparticles. *Water Res.* 41:6627–6632.

Bae, B.U., H.S. Shin, and J.J. Choi. 2007. Taste and odour issues in South Korea's drinking water industry. *Water Sci. Technol.* 55(5):203–208.

Bai, J., X. Shi, and T.G. Nagaraja. 2010. A multiplex PCR procedure for the detection of six major virulence genes in *Escherichia coli* O157:H7. *J. Microbiol. Methods* 82:85–89.

Baillie, L. and T.D. Read. 2001. *Bacillus anthracis*, a bug with attitude!. *Curr. Opin. Microbiol.* 4(1):78–81.

Baker, M.N. and M.J. Taras. 1981. *The Quest for Pure Water: The History of the Twentieth Century*, Vols. 1 and 2. American Water Works Association, Denver, CO.

Baker, K.K., S.O. Sow, K.L. Diarra, C.E. O'Reilly, E. Mintz, S. Panchalingam, Y. Wu, W.C. Blackwelder, and M.M. Levine. 2013. Quality of piped and stored water in households with children under five years of age enrolled in the Mali site of the Global Enteric Multi-Center Study (GEMS). *Am. J. Trop. Med. Hyg.* 89:214–222.

Baldursson, S. and P. Karanis. 2011. Waterborne transmission of protozoan parasites: review of worldwide outbreaks—an update 2004–2010. *Water Res.* 45:6603–6614.

Bancroft, K., P. Chrostowski, R.L. Wright, and I.H. Suffet. 1984. Ozonation and oxidation competition values. Relationship to disinfection and microorganisms regrowth. *Water Res.* 18:473-478.

Barbaree, J.M., B.S. Fields, J.C. Feeley, G.W. Gorman, and W.T. Martin. 1986. Isolation of protozoa from water associated with a legionellosis outbreak and demonstration of intracellular multiplication of *Legionella pneumophila*. *Appl. Environ. Microbiol.* 51:422–424.

Barbeau, B., P. Payment, J. Coallier, B. Clement, and M. Prevot. 2000. Evaluating the risk of infection from the presence of *Giardia* and *Cryptosporidium* in drinking water. *Quant. Microbiol.* 2:37–54.

Barbot, V., V. Migeot, N. Quellard, M.H. Rodier, and C. Imbert. 2012. *Hartmannella vermiformis* can promote proliferation of *Candida* spp. in tap-water. *Water Res.* 46:5707–5714.

Barker, J., M.R.W. Brown, P.J. Collier, I. Farrell, and P. Gilbert. 1992. Relationship between *Legionella pneumophila* and *Acanthamoeba polyphaga*: physiological status and susceptibility to chemical inactivation. *Appl. Environ. Microbiol.* 58:2420–2425.

Barnaud, E., S. Rogée, P. Garry, N. Rose, and N. Pavio. 2012. Thermal inactivation of infectious hepatitis E virus in experimentally contaminated food. *Appl. Environ. Microbiol.* 78:5153–5159.

Baron, J. 1997. Repair of wastewater microorganisms after ultraviolet disinfection under seminatural conditions. *Water Environ. Res.* 69:992–996.

Barraud, N., M.V. Storey, Z.P. Moore, J.S. Webb, S.A. Rice, and S. Kjelleberg. 2009. Nitric oxide-mediated dispersal in single- and multi-species biofilms of clinically and industrially relevant microorganisms. *Microb. Biotechnol.* 2:370–378.

Barrette, W.C., Jr., D.M. Hannum, W.D. Wheeler, and J.K. Hurst. 1988. General mechanism for the bacterial toxicity of hypochlorous acid: abolition of ATP production. *Biochemistry* 28:9172–9178.

Bartley, T.D., T.J. Quan, M.T. Collins, and S.M. Morrison. 1982. Membrane filter technique for the isolation of *Yersinia enterocolitica*. *Appl. Environ. Microbiol.* 43:829–834.

Batterman, S., J. Eisenberg, R. Hardin, M.E. Kruk, M.C. Lemos, A.M. Michalak, B. Mukherjee, E. Renne, H. Stein, C. Watkins, and M.L. Wilson. 2009. Sustainable control of water-related infectious diseases: a review and proposal for interdisciplinary health-based systems research. *Environ. Health Perspect.* 117:1023–1032.

Battigelli, D.A., D. Lobe, and M.D. Sobsey. 1993. Inactivation of hepatitis A virus and other enteric viruses in water by ultraviolet. *Water Sci. Technol.* 27:339–342.

Bauer, M., L. Mathieu, M. Deloge-Abarkan, T. Remen, P. Tossa, P. Hartemann, and D. Zmirou-Navier. 2008. *Legionella* bacteria in shower aerosols increase the risk of Pontiac fever among older people in retirement homes. *J. Epidemiol. Community Health* 62(10):913–920.

Bauer, R., H. Dizer, I. Graeber, K.-H. Rosenwinkel, and J.M. Lopez-Pila. 2011. Removal of bacterial fecal indicators, coliphages and enteric adenoviruses from waters with high fecal pollution by slow sand filtration. *Water Res.* 45:439–452.

Bazri, M.M., B. Barbeau, and M. Mohseni. 2012. Impact of UV/H$_2$O$_2$ advanced oxidation treatment on molecular weight distribution of NOM and biostability of water. *Water Res.* 46:5297–5304.

Beech, I.B. and J. Sunner. 2004. Biocorrosion: towards understanding interactions between biofilms and metals. *Curr. Opin. Biotechnol.* 15:181–186.

de Beer, D., R. Srinivasan, and P.S. Stewart. 1994. Direct measurement of chlorine penetration into biofilms during disinfection. *Appl. Environ. Microbiol.* 60:4339–4344.

Behets, J., F. Seghi, P. Declerck, L. Verelst, L. Duvivier, A. Van Damme, and F. Ollevier. 2003. Detection of *Naegleria* spp. and *Naegleria fowleri*: a comparison of flagellation tests, ELISA and PCR. *Water Sci. Technol.* 47(3):117–122.

Belkin, S. 2003. Microbial whole-cell sensing systems of environmental pollutants. *Curr. Opin. Microbiol.* 6:206–212.

Bell, F.A. 1991. Review of effects of silver-impregnated carbon filters on microbial water quality. *J. Am. Water Works Assoc.* 83:74–76.

Bell, K., M.W. LeChevallier, M. Abbaszadegan, G.L. Amy, S. Sinha, M. Benjamin, and E.A. Ibrahim. 2000. *Enhanced and Optimized Coagulation for Particulate and Microbial Removal*. AWWA Research Foundation and American Water Works Association, Denver, CO.

Bellamy, W.D., G.P. Silverman, D.W. Hendricks, and G.S. Logsdon. 1985a. Removing *Giardia* cysts with slow sand filtration. *J. Am. Water Works Assoc.* 77:52–60.

Bellamy, W.D., D.W. Hendricks, and G.S. Logsdon. 1985b. Slow sand filtration: influences of selected process variables. *J. Am. Water Works Assoc.* 77:62–66.

Bellona, C., J.E. Drewes, P. Xu, and G. Amy. 2004. Factors affecting the rejection of organic solutes during NF/RO treatment—a literature review. *Water Res.* 38:2795–2809.

Berg, J.D., P.V. Roberts, and A. Matin. 1986. Effect of chlorine dioxide on selected membrane functions of *Escherichia coli*. *J. Appl. Bacteriol.* 60:213–220.

Berger, S.B. and R.K. Oshiro. 2002. Source water protection: microbiology of source water. In *Encyclopedia of Environmental Microbiology*, G. Bitton (Editor-in-Chief). Wiley-Interscience, New York, pp. 2967–2978.

Berk, S.G., G. Faulkner, E. Garduno, M.C. Joy, M.A. Ortiz-Jimenez, and R.A. Garduno. 2008. Packaging of live *Legionella pneumophila* into pellets expelled by *Tetrahymena* spp. does not require bacterial replication and depends on a dot/Icm-mediated survival mechanism. *Appl. Environ. Microbiol.* 74:2187–2199.

Berney, M., F. Hammes, F. Bosshard, H.-U. Weilenmann, and T. Egli. 2007. Assessment and interpretation of bacterial viability by using the LIVE/DEAD BacLight kit in combination with flow cytometry. *Appl. Environ. Microbiol.* 73:3283–3290.

Berney, M., M. Vital, I. Hülshoff, H.-U. Weilenmann, T. Egli, and F. Hammes. 2008. Rapid, cultivation-independent assessment of microbial viability in drinking water. *Water Res.* 42:4010–4018.

Berry, D., C. Xi, and L. Raskin. 2006. Microbial ecology of drinking water distribution systems. *Curr. Opin. Biotechnol.* 17:297–302.

Berry, D., C. Xi, and L. Raskin. 2009. Effect of growth conditions on inactivation of *Escherichia coli* with monochloramine. *Environ. Sci. Technol.* 43:884–889.

Best, M.G., A. Goetz, and V.L. Yu. 1984. Heat eradication measures for control of nosocomial Legionnaires' disease: implementation, education and cost analysis. *Am. J. Infect. Control* 12:26–30.

Betts, K.S. 2002. New drinking water hazard. *Environ. Sci. Technol.* 36:92A–93A.

Beumer, A., D. King, M. Donohue, J. Mistry, T. Covert, and S. Pfaller. 2010. Detection of *Mycobacterium avium* subsp. *paratuberculosis* in drinking water and biofilms by quantitative PCR. *Appl. Environ. Microbiol.* 76:7367–7370.

Beuret, C., D. Kohler, A. Baumgartner, and T.M. Lüthi. 2002. Norwalk-like virus sequences in mineral waters: one-year monitoring of three brands. *Appl. Environ. Microbiol.* 68:1925–1931.

Beuret, C., A. Baumgartner, and J. Schluep. 2003. Virus-contaminated oysters: a three-month monitoring of oysters imported to Switzerland. *Appl. Environ. Microbiol.* 69:2292–2297.

Beydoun, D., R. Amal, G. Low, and S. McEvoy. 1999. Role of nanoparticles in photocatalysis. *J. Nanopart. Res.* 1:439–458.

Beyenal, H., Z. Lewandowski, C. Yakymyshyn, B. Lemley, and J. Wehri. 2000. Fiber-optic microsensors to measure backscattered light intensity in biofilms. *Appl. Opt.* 39(19):3408–3412.

Bichai, F., P. Payment, and B. Barbeau. 2008. Protection of waterborne pathogens by higher organisms in drinking water. *Can. J. Microbiol.* 54(7):509–524.

Bichai, F., B. Barbeau, and P. Payment. 2009. Protection against UV disinfection of *E. coli* bacteria and *B. subtilis* spores ingested by *C. elegans* nematodes. *Water Res.* 43:3397–3406.

Bichai, F., S. Léveillé, and B. Barbeau. 2011. Comparison of the role of attachment, aggregation and internalisation of microorganisms in UVC and UVA (solar) disinfection. *Water Sci. Technol.* 63(9):1823–1831.

Bielefeldt, A.R., K. Kowalski, and R.S. Summers. 2009. Bacterial treatment effectiveness of point-of-use ceramic water filters. *Water Res.* 43:3559–3565.

Bielefeldt, A.R., K. Kowalski, C. Schilling, S. Schreier, A. Kohler, and R. Scott Summers. 2010. Removal of virus to protozoan sized particles in point-of-use ceramic water filters. *Water Res.* 44:1482–1488.

Biran, I., D.M. Rissina, E.Z. Ron, and D.R. Walt. 2003. Optical imaging fiber-based live bacterial cell array biosensor. *Anal. Biochem.* 315:106–113.

Bischofberger, T., S.K. Cha, R. Schmitt, B. Konig, and W. Schmidt-Lorenz. 1990. The bacterial flora of non-carbonated, natural mineral water from the springs to reservoir and glass and plastic bottles. *Intern. J. Food Microbiol.* 11:51–71.

Bitton, G. 1980. *Introduction to Environmental Virology.* John Wiley & Sons, New York, 326 pp.

Bitton, G. 1999. *Wastewater Microbiology*, 2nd edition. Wiley.

Bitton, G. 2005. *Wastewater Microbiology*, 3rd edition. Wiley-Liss, New York, 746 pp.

Bitton, G. 2011. *Wastewater Microbiology*, 4th edition. Wiley-Blackwell, Hoboken, NJ, 781 pp.

Bitton, G. and V. Freihoffer. 1978. Influence of extracellular polysaccharides on the toxicity of copper and cadmium towards *Klebsiella aerogenes*. *Microb. Ecol.* 4:119–125.

Bitton, G. and C.P. Gerba (Eds). 1984. *Groundwater Pollution Microbiology.* John Wiley & Sons, New York.

Bitton, G. and K.C. Marshall (Eds). 1980. *Adsorption of Microorganisms to Surfaces.* Wiley-Interscience, New York.

Bitton, G., Y. Henis, and N. Lahav. 1972. Effect of several clay minerals and humic acid on the survival of *Klebsiella aerogenes* exposed to ultraviolet irradiation. *Appl. Microbiol.* 23:870–874.

Bitton, G., S.R. Farrah, C. Montague, and E.W. Akin. 1986. Global survey of virus isolations from drinking water. *Environ. Sci. Technol.* 20:216–222.

Biyela, P.T., H. Ryu, A. Brown, A. Alum, M. Abbazadegan, and B. Rittmann. 2012. Distribution systems as reservoirs of *Naegleria fowleri* and other amoebae. *Water Works Assoc. J.* 104(1):49–50.

Black, B.D., G.W. Harrington, and P.C. Singer. 1996. Reducing cancer risks by improving organic carbon removal. *J. Am. Water Works Assoc.* 88:40–52.

Black, R.E., S. Cousens, H.L. Johnson, J.E. Lawn, I. Rudan, D.G. Bassani, P. Jha, H. Campbell, C.F. Walker, R. Cibulskis, T. Eisele, L. Liu, and C. Mathers. 2010. Global, regional, and national causes of child mortality in 2008: a systematic analysis. *Lancet* 375:1969–1987.

Blake, P.A., M.L. Rosenberg, J. Florencia, J. Bandeira Costa, L. do Prado Quintino, and E.J. Gangarosa. 1977. Cholera in Portugal, 1974. II: transmission by bottled mineral water. *Am. J. Epidemiol.* 105:344–348.

Block, J.C., L. Mathieu, P. Servais, D. Fontvielle, and P. Werner. 1992. Indigenous bacterial inocula for measuring the biodegradable dissolved organic carbon (BDOC) in waters. *Water Res.* 26:481–486.

Block, J.C., K. Haudidier, J.L. Paquin, J. Miazaga, and T. Lévi. 1993. Biofilm accumulation in drinking water distribution system. *Biofouling* 6:333–343.

Block, J.C., I. Sibille, D. Gatel, D.J. Reasoner, B. Lykins, and R.M. Clark. 1997. Biodiversity in drinking water distribution systems: a brief review. In *The Microbiological Quality of Water*, D. Sutcliffe (Ed.). Royal Society for Public Health Hygiene, London, UK, pp. 63–71.

Bobadilla Fazzini, R.A., M.E. Skindersoe, P. Bielecki, J. Puchałka, M. Givskov, and V.A.P. Martins dos Santos. 2013. Protoanemonin: a natural quorum sensing inhibitor that selectively activates iron starvation response. *Environ. Microbiol.* 15:111–120.

Boccia, D., A.E. Tozzi, and B. Cotter. 2002. Waterborne outbreak of Norwalk-like virus gastroenteritis at a tourist resort, Italy. *Emerg. Infect. Dis.* 8:563–568.

Bodet, C., T. Sahr, M. Dupuy, C. Buchrieser, and Y. Héchard. 2012. *Legionella pneumophila* transcriptional response to chlorine treatment. *Water Res.* 46:808–816.

Bohrerova, Z. and K.G. Linden. 2006. Ultraviolet and chlormine disinfection of *Mycobacterium* in wastewater: effect of aggregation. *Water Environ. Res.* 78(6):565–571.

Bohrerova, Z. and K.G. Linden. 2007. Standardizing photoreactivation: comparison of DNA photorepair rate in *Escherichia coli* using four different fluorescent lamps. *Water Res.* 41:2832–2838.

Bohrerova, Z., H. Shemer, R. Lantis, C.A. Impellitteri, and K.G. Lindene. 2008. Comparative disinfection efficiency of pulsed and continuous-wave UV irradiation technologies. *Water Res.* 42:2975–2982.

Bollin, G.E., J.F. Plouffe, M.F. Para, and B. Hackman. 1985. Aerosols containing *Legionella pneumophila* generated by shower heads and hot-water faucets. *Appl. Environ. Microbiol.* 50:1128–1131.

Bond, T., E. Roma, K.M. Foxon, M.R. Templeton, and C.A. Buckley. 2013. Ancient water and sanitation systems—applicability for the contemporary urban developing world. *Water Sci. Technol.* 67(5):935–941.

Boone, S.A. and C.P. Gerba. 2007. Significance of fomites in the spread of respiratory and enteric viral disease. *Appl. Environ. Microbiol.* 73:1687–1696.

Bordalo, A.A. and J. Savva-Bordalo. 2007. The quest for safe drinking water: an example from Guinea-Bissau (West Africa). *Water Res.* 41:2978–2986.

Bosshard, F., F. Armand, R. Hamelin, and T. Kohn. 2013. Mechanisms of human adenovirus inactivation by sunlight and UVC light as examined by quantitative PCR and quantitative proteomics. *Appl. Environ. Microbiol.* 79:1325–1332.

Botes, M. and T.E. Cloete. 2010. The potential of nanofibers and nanobiocides in water purification. *Crit. Rev. Microbiol.* 36(1):68–81.

Bourne, D.G., R.L. Blakeley, P. Riddles, and G.J. Jones. 2006. Biodegradation of the cyanobacterial toxin microcystin LR in natural water and biologically active slow sand filters. *Water Res.* 40:1294–1302.

Boxall, A.B., K. Tiede, and Q. Chaudhry. 2007. Engineered nanomaterials in soils and water: how do they behave and could they pose a risk to human health? *Nanomedicine* 2:919–927.

Boyce, T.G., D.L. Swerdlow, and P.M. Griffin. 1995. *Escherichia coli* O157:H7 and the hemolytic-uremic syndrome. *N. Engl. J. Med.* 333:364–368.

Boyle, K.E., S. Heilmann, D. van Ditmarsch, and J.B. Xavier. 2013. Exploiting social evolution in biofilms. *Curr. Opin. Microbiol.* 16:207–212.

Bradley, I., A. Straub, P. Maraccini, S. Markazi, and T.H. Nguyen. 2011. Iron oxide amended biosand filters for virus removal. *Water Res.* 45:4501–4510.

Brady-Estevez, A.S., M.H. Schnoor, S. Kang, and M. Elimelech. 2010. SWNT-MWNT hybrid filter attains high viral removal and bacterial inactivation. *Langmuir* 26:19153–19158.

Braganca, S.M., N.F. Azevedo, L.C. Simoes, C.W. Keevil, and M.J. Vieira. 2007. Use of fluorescent *in situ* hybridisation for the visualisation of *Helicobacter pylori* in real drinking water biofilms. *Water Sci. Technol.* 55(8–9):387–393.

Branda, S.S. and R. Kolter. 2004. Multicellularity and biofilms. In *Microbial Biofilms*, M. Ghannoum and G.A. O'Toole (Eds). ASM Press, Washington, DC, pp. 20–29.

Brayton, P.R., M.L. Tamplin, A. Huk, and R.R. Colwell. 1987. Enumeration of *Vibrio cholerae* 01 in Bangladesh waters by fluorescent-antibody direct viable count. *Appl. Environ. Microbiol.* 53:2862–2865.

Brescia, C.C., S.M. Griffin, M.W. Ware, E.A. Varughese, A.I. Egorov, and E.N. Villegas. 2009. *Cryptosporidium* propidium monoazide-PCR, a molecular biology-based technique for genotyping of viable *Cryptosporidium* oocysts. *Appl. Environ. Microbiol.* 75:6856–6863.

Brettar, I. and M.G. Hofle. 2008. Molecular assessment of bacterial pathogens—a contribution to drinking water safety. *Curr. Opin. Biotechnol.* 19:274–280.

Brightwell, G., E. Mowat, R. Clemens, J. Boerema, D.J. Pulford, and S.L. On. 2007. Development of a multiplex and real time PCR assay for the specific detection of *Arcobacter butzleri* and *Arcobacter cryaerophilus*. *J. Microbiol. Methods* 68:318–325.

Brisou, J.F. 1995. *Biofilms: Methods for Enzymatic Release of Microorganisms*. CRC Press, Inc., New York.

Brooke, E. and M.R. Collins. 2011. Post treatment aeration to reduce THMs. *J. Am. Water Works Assoc.* 103(10):84–96.

Broschat, S.L., D.R. Call, E.A. Kuhn, and F.J. Loge. 2005. Comparison of the reflectance and Crystal Violet assays for measurement of biofilm formation by *Enterococcus*. *Biofilms* 2:177–181.

Brown, L.M. 2000. *Helicobacter pylori*: epidemiology and routes of transmission. *Epidemiol. Rev.* 22:283–297.

Brown, J.M. 2007. *Effectiveness of Ceramic Filtration for Drinking Water Treatment in Cambodia*. Ph.D. Dissertation. University of North Carolina at Chapel Hill.

Brown, R.A. and D.A. Cornwell. 2007. Using spore removal to monitor plant performance for *Cryptosporidium* removal. *J. Am. Water Works Assoc.* 99(3):95–109.

Brown, J. and M.D. Sobsey. 2012. Boiling as household water treatment in Cambodia: a longitudinal study of boiling practice and microbiological effectiveness. *Am. J. Trop. Med. Hyg.* 87:394–398.

Brown, T.S., J.F. Malina, Jr., and B.D. Moore. 1974. Virus removal by diatomaceous-earth filtration—Part 2. *J. Am. Water Works Assoc.* 66(12):735–738.

Brownell, S.A. and K.L. Nelson. 2006. Inactivation of single-celled *Ascaris suum* eggs by low-pressure UV radiation. *Appl. Environ. Microbiol.* 72:2178–2184.

Brunkard, J.M.E., A. Ailes, V.A. Roberts, V. Hill, E.D. Hilborn, G.F. Craun, A. Rajasingham, A. Kahler, L. Garrison, L. Hicks, J. Carpenter, T.J. Wade, M.J. Beach, and J.S. Yoder. 2011. Surveillance for waterborne disease outbreaks associated with drinking water—United States, 2007–2008. *MMWR Surveill. Summ.* 60(12):38–68.

Bryers, J. and W. Characklis. 1981. Early fouling biofilm formation in a turbulent flow system: overall kinetics. *Water Res.* 15:483–491.

Bucheli-Witschel, M., S. Kötzsch, S. Darr, R. Widler, and T. Egli. 2012. A new method to assess the influence of migration from polymeric materials on the biostability of drinking water. *Water Res.* 46:4246–4260.

Bukhari, Z., T.M. Hargy, J.R. Bolton, B. Dussert, and J.L. Clancy. 1999. Medium-pressure UV for oocyst inactivation. *J. Am. Water Works. Assoc.* 91:86–94.

Bukhari, Z., M.M. Marshall, D.G. Korich, C.R. Fricker, H.V. Smith, J. Rosen, and J.L. Clancy. 2000. Comparison of *Cryptosporidium parvum* viability and infectivity assays following ozone treatment of oocysts. *Appl. Environ. Microbiol.* 66:2972–2980.

Burger, J.S., W.O.K. Grabow, and R. Kfir. 1989. Detection of endotoxins in reclaimed and conventionally treated drinking water. *Water Res.* 23:733–738.

Burlage, R.S. 1997. Emerging technologies: bioreporters, biosensors, and microprobes. In *Manual of Environmental Microbiology*, C.J. Hurst, G.R. Knudsen, M.J. McInerney, L.D. Stetzenbach, and M.V. Walter (Eds). ASM Press, Washington, DC, pp. 115–123.

Burrows, I.D. and S.E. Renner. 1999. Biological warfare agents as threats to potable water. *Environ. Health Perspect.* 107:975–984.

Buse, H.Y., M.E. Schoen, and N.J. Ashbolt. 2012. Legionellae in engineered systems and use of quantitative microbial risk assessment to predict exposure. *Water Res.* 46:921–933.

Butot, S., T. Putallaz, C. Croquet, G. Lamothe, R. Meyer, H. Joosten, and G. Sanchez. 2007. Attachment of enteric viruses to bottles. *Appl. Environ. Microbiol.* 73:5104–5110.

Buttiaux, R. and A. Boudier. 1960. Comportement des bacteries autotrophes dans les eaux minerales consevees en recipients clos. *Ann. Inst. Pasteur Lille* 11:43–52.

Cacciò, S.M., M. De Giacomo, F.A. Aulicino, and E. Pozio. 2003. *Giardia* cysts in wastewater treatment plants in Italy. *Appl. Environ. Microbiol.* 69:3393–3398.

Caldwell, K.N., B.B. Adler, G.L. Anderson, P.L. Williams, and L.R. Beuchat. 2003. Ingestion of *Salmonella enterica* serotype Poona by a free-living nematode, *Caenorhabditis elegans*, and protection against inactivation by produce sanitizers. *Appl. Environ. Microbiol.* 69:4103–4110.

Campbell, A.T. and P. Wallis. 2002. The effect of UV irradiation on human-derived *Giardia lamblia* cysts. *Water Res.* 36:963–969.

Campbell, I., S. Tzipori, G. Hutchinson, and K.W. Angus. 1982. Effect of disinfectants on survival of *Cryptosporidium* oocysts. *Vet. Rec.* 111:414–415.

Campbell, A.T., L.J. Robertson, and H.V. Smith. 1992. Viability of *Cryptosporidium parvum* oocysts: correlation of in vitro excystation with inclusion or exclusion of fluorogenic vital dyes. *Appl. Environ. Microbiol.* 58:3488–3493.

Camper, A.K., M.W. Lechevallier, S.C. Broadaway, and G.A. McFeters. 1985. Growth and persistence of pathogens on granular activated carbon filters. *Appl. Environ. Microbiol.* 50:1378–1382.

Camper, A.K., M.W. Lechevallier, S.C. Broadaway, and G.A. McFeters. 1986. Bacteria associated with granular activated carbon particles in drinking water. *Appl. Environ. Microbiol.* 52:434–438.

Camper, A.K., S.C. Broadaway, M.W. LeChevallier, and G.A. McFeters. 1987. Operational variables and the release of colonized granular activated carbon particles in drinking water. *J. Am. Water Works Assoc.* 79:70–74.

Camper, A.K., G.A. McFeters, W.G. Characklis, and W.L. Jones. 1991. Growth kinetics of coliform bacteria under conditions relevant to drinking water distribution systems. *Appl. Environ. Microbiol.* 57:2233–2239.

Camper, A.K., K. Brastrup, A. Sandvig, J. Clement, C. Spencer, and A.J. Capuzzi. 2003. Efffect of distribution systems materials on bacterial regrowth. *J. Am. Water Works Assoc.* 95:107–121.

Campos, L.C., S.R. Smith, and N.J.D. Graham. 2006. Deterministic-based model of slow sand filtration. II: model application. *J. Environ. Eng.* 132:887–894.

Cardenas, V.M., Z.D. Mulla, M. Ortiz, and D.Y. Graham. 2006. Iron deficiency and *Helicobacter pylori* infection in the United States. *Am. J. Epidemiol.* 163:127–134.

Carmichael, W.W. 1989. Freshwwater cyanobacteria (blue-green algae) toxins. In *Natural Toxins: Characterization, Pharmacology and Therapeutics*, C.L. Ownby and G.V. Odell (Eds). Pergamon Press, Oxford, pp. 3–16.

Carraro, E., E.H. Bugliosi, L. Meucci, C. Baiocchi, and G. Gilli. 2000. Biological drinking water treatment processes, with special reference to mutagenicity. *Water Res.* 34:3042–3054.

Carrico, B.A., F.A. Digiano, N.G. Love, P. Vikesland, K. Chandran, M. Fiss, and A. Zaklikowski. 2008. Effectiveness of switching disinfectants for nitrification control. *J. Am. Water Works Assoc.* 100(10):104–115.

Carrington, B.G. 1985. Pasteurization: effects on Ascaris eggs. In *Inactivation of Microorganisms in Sewage Sludge by Stabilization Processes*, D. Strauch, A.H. Havelaar, and P. L'Hermite (Eds.). Elsevier Applied Science, London, UK, pp. 121-125.

Carroll, T., S. King, S.R. Gray, B.A. Bolto, and N.A. Booker. 2000. The fouling of microfiltration membranes by NOM after coagulation treatment. *Water Res.* 34:2861–2868.

Carson, L.A. and N.J. Petersen. 1975. Photoreactivation of *Pseudomonas cepacia* after ultraviolet exposure: a potential source of contamination in ultraviolet-treated waters. *J. Clin. Microbiol.* 1:462–464.

Casanovas-Massana, A. and A.R. Blanch. 2012. Diversity of the heterotrophic microbial populations for distinguishing natural mineral waters. *Int. J. Food Microbiol.* 153:38–44.

Cashdollar, J., N.E. Brinkman, S.M. Griffin, B.R. McMinn, E.R. Rhodes, E.A. Varughese, A.C. Grimm, S.U. Parshionikar, L. Wymer, and G.S. Fout. 2013. Development and evaluation of EPA method 1615 for detection of enterovirus and norovirus in water. *Appl. Environ. Microbiol.* 79:215–223.

Casson, L.W., C.A. Sorber, J.L. Sykora, P.D. Cavaghan, M.A. Shapiro, and W. Jakubowski. 1990. *Giardia* in wastewater—Effect of treatment. *J. Water Pollut. Control Fed.* 62:670–675.

Castro-Hermida, J.A., I. García-Presedo, A. Almeida, M. González-Warleta, J.M. Correia Da Costa, and M. Mezo. 2008. Presence of *Cryptosporidium* spp. and *Giardia duodenalis* through drinking water. *Sci. Total Environ.* 405:45–53.

Cater, S.R., M.I. Stefan, J.R. Bolton, and A. Safarzadeh-Amiri. 2000. UV/H_2O_2 treatment of methyl tert-butyl ether in contaminated waters. *Environ. Sci. Technol.* 34(4):659–662.

Cates, E.L., M. Cho, and J.-H. Kim. 2011. Converting visible light into UVC: microbial inactivation by Pr3þ-activated upconversion materials. *Environ. Sci. Technol.* 45:3680–3686.

Cees, B., J. Zoeteman, and G.J. Piet. 1974. Cause and identification of taste and odour compounds in water. *Sci. Total Environ.* 3:103–115.

Centers for Disease Control and Prevention (CDC). 1991. Outbreaks of diarrheal illness associated with cyanobacteria (blue-green algae)-like bodies: Chicago and Nepal, 1989 and 1990. *MMWR Morb. Mortal. Wkly. Rep.* 40:325–327.

Centers for Disease Control and Prevention (CDC). 1999. Ten great public health achievements—United States 1900–1999. *MMWR Morb. Mortal. Wkly. Rep.* 48(12):241.

Centers for Disease Control and Prevention (CDC). 2001. *Safe Water System Handbook. Safe Water Systems for the Developing World: A Handbook for Implementing Household-Based Water Treatment and Safe Storage Projects*. US Department of Health & Human Services, Centers for Disease Control and Prevention, Atlanta, GA.

Centers for Disease Control and Prevention (CDC). 2010. http://www.cdc.gov/HAI/organisms/acinetobacter.html (accessed May 5, 2014).

Centers for Disease Control and Prevention (CDC). 2013. http://www.cdc.gov/parasites/cyclosporiasis/outbreaks/investigation-2013.html (accessed May 5, 2014).

Cerca, F., G. Trigo, A. Correia, N. Cerca, J. Azeredo, and M. Vilanova. 2011. SYBR green as a fluorescent probe to evaluate the biofilm physiological state of *Staphylococcus epidermidis*, using flow cytometry. *Can. J. Microbiol.* 57:850–856.

Chakraborty, A., S.U. Khan, M.A. Hasnat, S. Parveen, M.S. Islam, A. Mikolon, R.K. Chakraborty, B.-N. Ahmed, K. Ara, N. Haider, S.R. Zaki, A.R. Hoffmaster, M. Rahman, S.P. Luby, and M.J. Hossain. 2012. Anthrax outbreaks in Bangladesh, 2009–2010. *Am. J. Trop. Med. Hyg.* 86:703–710.

Chang, S.D. and P.C. Singer. 1991. The impact of ozonation on particle stability and the removal of TOC and THM precursors. *Am. Water Works. Assoc. J.* 83:71–79.

Chang, Y., C.-W. Li, and M.M. Benjamin. 1997. Iron oxide-coated media for NOM sorption and particulate filtration. *J. Am. Water Works Assoc.* 89:100–113.

Chang, W., D.A. Small, F. Toghrol, and W.E. Bentley. 2005. Microarray analysis of toxicogenomic effects of peracetic acid on *Pseudomonas aeruginosa*. *Environ. Sci. Technol.* 39:5893–5899.

Chang, W., F. Toghrol, and W.E. Bentley. 2006. Toxicogenomic response of *Staphylococcus aureus* to peracetic acid. *Environ. Sci. Toxicol.* 40(16):5124–5131.

Chang, M.W., F. Toghrol, and W.E. Bentley. 2007. Toxicogenomic response to chlorination includes induction of major virulence genes in *Staphylococcus aureus*. *Environ. Sci. Technol.* 41:7570–7575.

Characklis, W.G. (Ed.): 1988. *Bacterial Regrowth in Distribution Systems*, Research Report. AWWA Research Foundation, Denver, CO.

Characklis, W.G. and K.E. Cooksey. 1983. Biofilms and microbial fouling. *Adv. Appl. Microbiol.* 29:93–138.

Characklis, W.G., M.G. Trulear, J.D. Bryers, and N. Zelver. 1982. Dynamic of biofilm processes: methods. *Water Res.* 16:1207–1216.

Charnock, C. and O. Kjønnø. 2000. Assimilable organic carbon and biodegradable dissolved organic carbon in Norwegian raw and drinking water. *Water Res.* 34:2629–2642.

Charoenca, N. and R.S. Fujioka. 1993. Assessment of *Staphylococcus* bacteria in Hawaii recreational waters. *Water Sci. Technol.* 27:283–289.

Charrois, J.W.A. 2010. Private drinking water supplies: challenges for public health. *Can. Med. Assoc. J.* 182(10):1061–1064.

Chauret, C., N. Armstrong, J. Fisher, R. Sharma, S. Springthorpe, and S. Sattar. 1995. Correlating *Cryptosporidium* and *Giardia* with microbial indicators. *J. Am. Water Works Assoc.* 87:76–84.

Chauret, C.P., C.Z. Radziminski, M. Lepuil, R. Creason, and R.C. Andrews. 2001. Chlorine dioxide inactivation of *Cryptosporidium parvum* oocysts and bacterial spore indicators. *Appl. Environ. Microbiol.* 67:2993–3001.

Chen, Y.-S. and J. Vaughn. 1990. Inactivation of human and simian rotaviruses by chlorine dioxide. *Appl. Environ. Microbiol.* 56:1363–1366.

Chen, Y.S., O.J. Sproul, and A. Rubin. 1985. Inactivation of *Naegleria gruberi* cysts by chlorine dioxide. *Water Res.* 19:783–789.

Chen, Y.S., J.M. Vaughn, and R.M. Niles. 1987. Rotavirus RNA and protein alterations resulting from ozone treatment (Abstr. Q-22). In *Annual Meeting of the American Society of Microbiol*, Atlanta, GA.

Chen, G., B.W. Dussert, and I.H. Suffet. 1997. Evaluation of granular activated carbons for removal of methylisoborneol to below odor threshold concentration in drinking water. *Water Res.* 31:1155–1163.

Chen, Y., Y. Chen, S.D. Worley, T.S. Huang, J. Weese, J. Kim, C.-I. Wei, and J.F. Williams. 2004. Biocidal polystyrene beads. III. Comparison of N-halamine and quat functional groups. *J. Appl. Polym. Sci.* 92(1):363–367.

Cheng, Y.W., R.C.Y. Chan, and P.K. Wong. 2007. Disinfection of *Legionella pneumophila* by photocatalytic oxidation. *Water Res.* 41(4):842–852.

Cheng, X., H. Shi, C.D. Adams, T. Timmons, and Y. Ma. 2009. Effects of oxidative and physical treatments on inactivation of *Cylindrospermopsis raciborskii* and removal of cylindrospermopsin. *Water Sci. Technol.* 60:689–698.

Chiang, W.-C., C. Schroll, L.R. Hilbert, P. Møller, and T. Tolker-Nielsen. 2009. Silverpalladium surfaces inhibit biofilm formation. *Appl. Environ. Microbiol.* 75:1674–1678.

Chiller, T.M., C.E. Mendoza, M.B. Lopez, M. Alvarez, R.M. Hoekstra, B.H. Keswick, and S.P. Luby. 2006. Reducing diarrhoea in Guatemalan children: randomized controlled trial of flocculant-disinfectant for drinking-water. *Bull. World Health Organ.* 84: 28–35.

Chisholm, K., A. Cook, C. Bower, and P. Weinstein. 2008. Risk of birth defects in Australian communities with high levels of brominated disinfection by-products. *Environ. Health Perspect.* 116:1267–1273.

Choi, Y. and Y.J. Choi. 2010. The effect of UV disinfection on drinking water quality in distribution systems. *Water Res.* 44:116–122.

Choi, Y.C. and E. Morgenroth. 2003. Monitoring biofilm detachment under dynamic dhanges in shear stress using laser-based particle size analysis and mass fractionation. *Water Sci. Technol.* 47(5):69–76.

Choi, J. and R.L. Valentine. 2002. Formation of N-nitrosodimethylamine (NDMA) from reaction of monochloramine: a new disinfection by-product. *Water Res.* 36:817–824.

Choi, H., M.G. Antoniou, M. Pelaez, A.A. De La Cruz, J.A. Shoemaker, and D.D. Dionysiou. 2007. Mesoporous nitrogen-doped TiO_2 for the photocatalytic destruction of the cyanobacterial toxin microcystin-LR under visible light irradiation. *Environ. Sci. Technol.* 41:7530–7535.

Chorus, I. and J. Bartram (Eds). 1999. *Toxic Cyanobacteria in Water: A Guide to Their Public Health Consequences.* WHO Report. ISBN 0-419-23930-8. http://www.who.int/water_sanitation_health/resourcesquality/toxcyanbegin.pdf (accessed May 5, 2014).

Chow, A.T. 2006. Disinfection byproduct reactivity of aquatic humic substances derived from soils. *Water Res.* 40:1426–1430.

Christensen, S.C.B., E. Nissen, E. Arvin, and H.J. Albrechtsen. 2011. Distribution of *Asellus aquaticus* and microinvertebrates in a non-chlorinated drinking water supply system: effects of pipe material and sedimentation. *Water Res.* 45:3215–3224.

Christensen, S.C.B., E. Nissen, E. Arvin, and H.J. Albrechtsen. 2012. Influence of *Asellus aquaticus* on *Escherichia coli, Klebsiella pneumoniae, Campylobacter jejuni* and naturally occurring heterotrophic bacteria in drinking water. *Water Res.* 46:5279–5286.

Christopher, G.W., T.J. Cieslak, J.A. Pavlin, and E.M. Eitzen, Jr. 1997. Biological warfare: a historical perspective. *J. Am. Med. Assoc.* 278(5):412–417.

Chu, F.S. and G.Y. Li. 1994. Simultaneous occurrence of fumonisin B1 and other mycotoxins in moldy corn collected from the People's Republic of China in regions with high incidence of esophageal cancer. *Appl. Environ. Microbiol.* 60:847–852.

Ciesielski, C.A., M.J. Blaser, and W.L. Wang. 1984. Role of stagnation and obstruction of water flow in isolation of *Legionella pneumophila* from hospital plumbing. *Appl. Environ. Microbiol.* 48:984–987.

Ciglio, S., J. Jiang, C.P. Saint, D.E. Cane, and P.T. Monis. 2008. Isolation and characterization of the gene associated with geosmin production in cyanobacteria. *Environ. Sci. Technol.* 42:8027–8032.

Ciglio, S., W.K.W. Chou, H. Ikeda, S.D.E. Cane, and P.T. Monis. 2011. Biosynthesis of 2-methylisoborneol in cyanobacteria. *Environ. Sci. Technol.* 45:992–998.

Cirillo, J., S. Falkow, L. Tompkins, and L. Bermudez. 1997. Interaction of *Mycobacterium avium* with environmental amoebae enhances virulence. *Infect. Immun.* 65(9):3759–3767.

Clark, R.M., E.J. Read, and J.C. Hoff. 1989. Analysis of inactivation of *Giardia lamblia* by chlorine. *J. Environ. Eng. Div.* 115:80–90.

Clasen, T. 2009. *Scaling Up Household Water Treatment Among Low-Income Populations.* WHO Report. WHO/HSE/WSH/09.02, Geneva, Switzerland.

Clasen, T. and P. Edmondson. 2006. Sodium dichloroisocyanurate (NaDCC) tablets as an alternative to sodium hypochlorite for the routine treatment of drinking water at the household level. *Int. J. Hyg. Environ. Health.* 209:173–181.

Clasen, T., G. Garcia Parra, S. Boisson, and S. Collin. 2005. Household-based ceramic water filters for the prevention of diarrhea: a randomized, controlled trial of a pilot program in Colombia. *Am. J. Trop. Med. Hyg.* 73(4):790–795.

Clasen, T.F., J. Brown, and S.M. Collin. 2006. Preventing diarrhoea with household ceramic water filters: assessment of a pilot project in Bolivia. *Intern. J. Environ. Health Res.* 16:231–239.

Clasen, T., T.F. Saeed, S. Boisson, P. Edmondson, and O. Shipin. 2007. Household water treatment using sodium dichloroisocyanurate (NaDCC) tablets: a randomized, controlled trial to assess microbiological effectiveness in Bangladesh. *Am. J. Trop. Med. Hyg.* 76:187–192.

Clasen, T., C. McLaughlin, N. Nayaar, S. Boisson, R. Gupta, D. Desai, and N. Shah. 2008. Microbiological effectiveness and cost of disinfecting water by boiling in semi-urban India. *Am. J. Trop. Med. Hyg.* 70(3):407–413.

Clasen, T., J. Naranjo, D. Frauchiger, and C.P. Gerba. 2009. Laboratory Assessment of a gravity-fed ultrafiltration water treatment device designed for household use in low-income settings. *Am. J. Trop. Med. Hyg.* 80:819–823,

Cleasby, J.L., D.J. Hilmoe, and C.J. Dimitracopoulos. 1984. Slow sand and direct in-line filtration of a surface water. *J. Am. Water Works Assoc.* 76:44–55.

Cliver, D.O. 1984. Significance of water and environment in the transmission of virus disease. *Monogr. Virol.* 15:30–42.

Coetser, S.E. and T.E. Cloete. 2005. Biofouling and biocorrosion in industrial water systems. *Crit. Rev. Microbiol.* 31:213–232.

Colbourne, J.S., P.J. Dennis, R.M. Trew, C. Berry, and G. Vesey. 1988. *Legionella* and public water supplies. *Proceedings of the International Conference on Water and Wastewater Microbiology*, Vol. 1, Newport Beach, CA.

Collins, M.R., T.T. Eightmy, J. Fenstermacher, and S.K. Spanos. 1992. Removing natural organic matter by conventional slow sand filtration. *J. Am. Water Works Assoc.* 84: 80–90.

Collivignarelli, C., S. Sorlini, and M. Belluati. 2006. Chlorite removal with GAC. *J. Am. Water Works. Assoc.* 98(12):74–79.

Conroy, R.M., M. Elmore-Meegan, T. Joyce, K.G. McGuigan, and J. Barnes. 1996. Solar disinfection of drinking water and diarrhoea in Maasai children: a controlled field trial. *Lancet* 348(9043):1695–1697.

Conroy, R.M., M.E. Meegan, T. Joyce, K. McGuigan, and J. Barnes. 2001. Solar disinfection of drinking water protects against cholera in children under 6 years of age. *Arch. Dis. Child.* 85(4):293–295.

Cook, D., G. Newcombe, and P. Sztajnbok. 2001. The application of powdered activated carbon for MIB and geosmin removal: predicting pac doses in four raw waters. *Water Res.* 35:1325–1333.

Cooper, R.C., A.W. Olivieri, R.E. Danielson, P.G. Badger, R.C. Spear, and S. Selvin. 1986. Evaluation of Military Field-Water Quality. Vol 5, Infectious Organisms of Military Concern Associated with Consumption: Assessment of Health Risks, and Recommendations for Establishing Related Standards. APO 82PP2817, U.S. Army Medical Research and Development Command, Ft. Detrick, MD.

Coral, L.A., A. Zamyadi, B. Barbeau, F.J. Bassetti, F.R. Lapolli, and M. Prèvost. 2013. Oxidation of *Microcystis aeruginosa* and *Anabaena flos-aquae* by ozone: impacts on cell integrity and chlorination by-product formation. *Water Res.* 47:2983–2994.

Cornwell, D.A. and M.J. MacPhee. 2001. Effects of spent filter backwash recycle on *Cryptosporidium* removal. *J. Am. Water Works Assoc.* 93(4):153–162.

Correa, I.E., N. Harb, and M. Molina. 1989. Incidence and prevalence of *Giardia* spp. in Puertorican waters: removal of cysts by conventional sewage treatment plants. In *89th Annual Meeting, American Society Microbiology*, New Orleans, LA, May 14–18, 1989.]

Corsaro, D., G.S. Pages, V. Catalan, J.F. Loret, and G. Greub. 2010. Biodiversity of amoebae and amoeba-associated bacteria in water treatment plants. *Int. J. Hyg. Environ. Health.* 213:158–166.

Corso, P.S., M.H. Kramer, K.A. Blair, D.G. Addiss, J.P. Davis, and A.C. Haddix. 2003. Cost of illness in the 1993 waterborne *Cryptosporidium* outbreak, Milwaukee, Wisconsin. *Emerg. Infect. Dis.* 9:426–431.

Costantini, V., F. Loisy, L. Joens, F.S. Le Guyader, and L.J. Saif. 2006. Human and animal enteric caliciviruses in oysters from different coastal regions of the United States. *Appl. Environ. Microbiol.* 72:1800–1809.

Costerton, J.W. 1980. Some techniques involved in study of adsorption of microorganisms to surfaces. In *Adsorption of Microorganisms to Surfaces*, G. Bitton and K.C. Marshall (Eds). John Wiley & Sons, New York, pp. 403–423.

Costerton, J.W. 1999. Bacterial biofilms: a common cause of persistent infections. *Science* 284:1318–1322.

Costerton, J.W. and G.C. Geesey. 1979. Microbial contamination of surfaces. In *Surface Contamination*, K.L. Mittal (Ed.). Plenum, New York, pp. 211-221.

Costerton, J.W., Z. Lewandowski, D.E. Caldwell, D.R. Korber, and H.M. Lappin-Scott. 1995. Microbial biofilms. *Annu. Rev. Microbiol.* 49:711–745.

Coulliette, A.D., L.A. Peterson, J.A.W. Mosberg, and J.B. Rose. 2010. Evaluation of a new disinfection approach: efficacy of chlorine and bromine halogenated contact disinfection for reduction of viruses and microcystin toxin. *Am. J. Trop. Med. Hyg.* 82:279–288.

Coulliette, A.D., K.S. Enger, M.H. Weir, and J.B. Rose. 2013. Risk reduction assessment of waterborne *Salmonella* and *Vibrio* by a chlorine contact disinfectant point-of-use device. *Int. J. Hyg. Environ. Health* 216:355–361.

Covert, T.C., M.R. Rodgers, A.L. Reyes, and G.N. Stelma, Jr. 1999. Occurrence of nontuberculous mycobacteria in environmental samples. *Appl. Environ. Microbiol.* 65:2492–2496.

Craun, G.F. 1984. Waterborne outbreaks of giardiasis: current status. In *Giadia and Giardiasis*, S.L. Erlandsen and E.A. Meyers (Eds). Plenum, New York, pp. 243–261.

Craun, G.F. 1986. Statistics of waterborne outbreaks in the U.S. (1920-1980). In *Water Diseases in the United States*, G.F. Craun (Ed.). CRC Press, Boca Raton, FL, pp. 73–159.

Craun, G.F. 1988. Surface water supplies and health. *J. Am. Water Works Assoc.* 80:40–52.

Craun, G.F. 2001. Waterborne disease outbreaks caused by distribution system deficiencies. *J. Am. Water Works Assoc.* 93(9):64–75.

Craun, G.F., J.M. Brunkard, J.S. Yoder, V.A. Roberts, J. Carpenter, T. Wade, R.L. Calderon, J.M. Roberts, M.J. Beach, and S.L. Roy. 2010. Causes of outbreaks associated with drinking water the United States from 1971 to 2006. *Clin. Microbiol. Rev.* 23(3):507–528.

Crockett, C.S. 2007. The role of wastewater treatment in protecting water supplies against emerging pathogens. *Water Environ. Res.* 79:221–232.

Cullen, T.R. and R.D. Letterman. 1985. The effect of slow sand filter maintainance on water quality. *J. Am. Water Works Assoc.* 77:48–55.

Current, W.L. 1987. *Cryptosporidium*: its biology and potential for environmental transmission. *Crit. Rev. Env. Control* 17:21.

Current, W.L. 1988. The biology of *Cryptosporidium*. *ASM News* 54:605–611.

Cvitkovitch, D.G. 2004. Genetic exchange in biofilms. In *Microbial Biofilms*, M. Ghannoum and G.A. O'Toole (Eds). ASM Press, Washington, DC, pp. 192–205.

Dabeka, R.W., H.B.S. Conacher, J.F. Lawrence, W.H. Newsome, A. McKenzie, H.P. Wagner, R.K.H. Chadha, and K. Pepper. 2002. Survey of bottled drinking waters sold in Canada for chlorate, bromide, bromate, lead, cadmium and other trace elements. *Food Addit Contam.* 19(8):721–732.

Dadjour, M.F., C. Ogino, S. Matsumura, S. Nakamura, and N. Shimizu. 2006. Disinfection of *Legionella pneumophila* by ultrasonic treatment with TiO_2. *Water Res.* 40(6):1137–1142.

Dankovich, T.A. and D.G. Gray. 2011. Bactericidal paper impregnated with silver nanoparticles for point-of-use water treatment. *Environ. Sci. Technol.* 45:1992–1998.

Dauphin, L.A., B.R. Newton, M.V. Rasmussen, R.F. Meyer, and M.D. Bowen. 2008. Gamma irradiation can be used to inactivate *Bacillus anthracis* spores without compromising the sensitivity of diagnostic assays. *Appl. Environ. Microbiol.* 74:4427–4433.

Day, H.R. and G.T. Felbeck, Jr. 1974. Production and analysis of a humic-like exudate from the aquatic fungus *Aureobasidium pullulans. J. Am. Water Works Assoc.* 66:484–489.

Declerck, P. 2010. Biofilms: the environmental playground of *Legionella pneumophila*. *Environ. Microbiol.* 12(3):557–566.

De Gusseme, B., L. Sintubin, L. Baert, E. Thibo, T. Hennebel, G. Vermeulen, M. Uyttendaele, W. Verstraete, and N. Boon. 2010. Biogenic silver for disinfection of water contaminated with viruses. *Appl. Environ. Microbiol.* 76:1082–1087.

De Hoogh, C.I., A.J. Wagenvoort, F. Jonker, J.A. van Leerdam, and A.C. Hogenboom. 2006. HPLC-DAD and QTOF MS techniques identify cause of *Daphnia* biomonitor alarms in the river Meuse. *Environ. Sci. Technol.* 40:2678–2685.

De Kievit, T.R. 2009. Quorum sensing in *Pseudomonas aeruginosa* biofilms. *Environ. Microbiol.* 11:279–288.

Demazeau, G. and N. Rivalain. 2011. The development of high hydrostatic pressure processes as an alternative to other pathogen reduction methods. *J. Appl. Microbiol.* 110:1359–1369.

De Medici, D., F. Anniballi, G.M. Wyatt, M. Lindström, U. Messelhäußer, C.F. Aldus, E. Delibato, H. Korkeala, M.W. Peck, and L. Fenicia. 2009. Multiplex PCR for detection of botulinum neurotoxin-producing clostridia in clinical, food, and environmental samples. *Appl. Environ. Microbiol.* 75:6457–6461.

Denileon, G.P. 2001. The who, what, why and how of counter terrorism issues. *J. Am. Water Works Assoc.* 93(5):78–85.

Dennis, P.J., D. Green, and B.P. Jones. 1984. A note on the temperature tolerance of *Legionella*. *J. Appl. Bacteriol.* 56:349–350.

Dennis, D.T., T.V. Inglesby, D.A. Henderson, J.G. Bartlett, M.S. Ascher, E. Eitzen, A.D. Fine, A.M. Friedlander, J. Hauer, M. Layton, S.R. Lillibridge, J.E. McDade, M.T. Osterholm, T. O'Toole, G. Parker, T.M. Perl, P.K. Russell, and K. Tonat. 2001. Tularemia as a biological weapon, medical and public health management. *J. Am. Med. Assoc.* 285:2763–2773.

De Saeger, S. and C. van Peteghem. 1996. Dipstick enzyme immunoassay to detect *Fusarium* T-2 toxin in wheat. *Appl. Environ. Microbiol.* 62:1880–1884.

De Serres, G., T.L. Cromeans, B. Levesque, N. Brassard, C. Barthe, M. Dionne, H. Prud'homme, D. Paradis, C.N. Shapiro, O.V. Naiman, and H.S. Margolis. 1999.

Molecular confirmation of hepatitis A virus from well water: epidemiology and public health implications. *J. Infect. Dis.* 179:37–43.

Despins, C., K. Farahbaksh, and C. Leidl. 2009. Assessment of rainwater quality from rainwater harvesting systems in Ontario, Canada. *J. Water Supply Res. Technol.* 58(2):117–134.

Dewettinck, T., W. Hulsbosch, K. Van Hege, E.M. Top, and W. Verstraete. 2001. Molecular fingerprinting of bacterial populations in groundwater and bottled mineral water. *Appl. Microbiol. Biotechnol.* 57:412–418.

Dice, J.C. 1985. Denver's seven decades of experience with chloramination. *J. Am. Water Works Assoc.* 77:34–37.

Didier, E.S., M.E. Stovall, L.C. Green, P.J. Brindley, K. Sestak, and P.J. Didier. 2004. Epidemiology of microsporidiosis: sources and modes of transmission. *Vet. Parasitol.* 126:145–166.

Dittmann, E., D.P. Fewer, and B.A. Neilan. 2013. Cyanobacterial toxins: biosynthetic routes and evolutionary roots. *FEMS Microbiol. Rev.* 37:23–43.

Dixon, C., M. Meselson, T. Guillemin, and P. Hanna. 1999. Anthrax. *N. Engl. J. Med.* 341:815–826.

Doggett, M.S. 2000. Characterization of fungal biofilms within a municipal water distribution system. *Appl. Environ. Microbiol.* 66:1249–1251.

Donlan, R.M. 2002. Biofilms: microbial life on surfaces. *Emerg. Infect. Dis.* 8:880–890.

Dotson, A.D., V.O.S. Keen, D. Metz, and K.G. Linden. 2010. UV/H_2O_2 treatment of drinking water increases post-chlorination DBP formation. *Water Res.* 44:3703–3713.

Dotson, A.O., C.E. Rodriguez, and K.G. Linden. 2012. UV disinfection implementation status in US water treatment plants. *J. Am. Water Works Assoc.* 104(5):77–78.

Douterelo, I., R.L. Sharpe, and J.B. Boxall. 2013. Influence of hydraulic regimes on bacterial community structure and composition in an experimental drinking water distribution system. *Water Res.* 47:503–516.

Driedger, A.M., J.L. Rennecker, and B.J. Mariñas. 2000. Sequential inactivation of *Cryptosporidium parvum* oocysts with ozone and free chlorine. *Water Res.* 34:3591–3597.

Driedger, A., E. Staub, U. Pinkernell, B. Mariñas, W. Köster, and U. von Gunten. 2001. Inactivation of *Bacillus subtilis* spores and formation of bromate during ozonation. *Water Res.* 35:2950–2960.

Drikas, M., C.W.K. Chow, J. House, and M.D. Burch. 2001. Using coagulation, flocculation and settling to remove toxic cyanobacteria. *J. Am. Water Works Assoc.* 93(2):100–111.

Drikas, M., M. Dixon, and J. Morran. 2009. Removal of MIB and geosmin using granular activated carbon with and without MIEX pre-treatment. *Water Res.* 43:5151–5159.

Drikas, M., M. Dixon, and J. Morran. 2011. Long term case study of MIEX pre-treatment in drinking water; understanding NOM removal. *Water Res.* 45:1539–1548.

Dubey, J.P. 2002. *Toxoplasma gondii*. In *Encyclopedia of Environmental Microbiology*, G. Bitton (Editor-in-Chief). Wiley, New York, pp. 3176–3183.

Duchin, J.S., J. Koehler, J.M. Kobayashi, R.M. Rakita, K. Olson, N.B. Hampson, D.N. Gilbert, J.M. Jackson, K.R. Stefonek, M.A. Kohn, J. Rosenberg, D. Vugia, and M. Marchione-Mastroianni. 2000. Legionnaires' disease associated with potting soil—California, Oregon, and Washington, May–June 2000. *MMWR Morb. Mortal. Wkly. Rep.* 49(34):777–778.

Duerden, B.I. 2006. Health protection: microbiology and the public health response to the bioterrorism threat. *Anaerobe* 12:59–62.

Dufour, A., M. Snozzi, W. Koster, J. Bartram, E. Ronchi, and L. Fewtrell (Eds). 2003. *Microbial Safety of Drinking Water: Improving Approaches and Methods*. ISBN 92 4 154630 1(WHO) and 1 84339 036 1(IWA Publishing) © WHO OECD 2003.

Duncan, A. 1988. The ecology of slow sand filters. In *Slow Sand Filtration: Recent Development in Water Treatment Technology*, N.J.D. Graham (Ed.). Ellis Horwood, Chichester, UK, pp. 163–180.

Eaton, J.W., C.F. Kolpin, and H.S. Swofford. 1973. Chlorinated urban water: a cause of dyalysis-induced hemolytic anemia. *Science* 181:463–464.

Eboigbodin, K.E., A. Seth, and C.A. Biggs. 2008. A review of biofilms in domestic plumbing. *J. Am. Water Works Assoc.* 100(10):131–137.

Echigo, S., S. Itoh, T. Natsui, T. Araki, and R. Ando. 2004. Contribution of brominated organic disinfection by-products to the mutagenicity of drinking water. *Water Sci. Technol.* 50(5):321–328.

Edzwald, J.K. (Ed.). 2011. *Water Quality and Treatment_A Handbook on Drinking Water*, 6th edition. AWWA, Denver, CO.

Egli, T. 1995. The ecological and physiological significance of the growth of heterotrophic microorganisms with mixtures of substrates. *Adv. Microb. Ecol.* 14:305–386.

Egli, T. 2010. How to live at very low substrate concentration. *Water Res.* 44:4826–4837.

Eischeid, A.C., J.N. Meyer, and K.G. Linden. 2009. UV disinfection of adenoviruses: molecular indications of DNA damage efficiency. *Appl. Environ. Microbiol.* 75:23–28.

Eitzen, E., J. Pavlin, T. Cieslak, G. Christopher, and R. Culpepper (Eds). 1998. *Medical Management of Biological Casualties Handbook*. USAMRIID, Fort Dietrick, MD.

Elad, T., J.H. Lee, S. Belkin, and M.B. Gu. 2008a. Microbial whole-cell arrays. *Microb. Biotechnol.* 1:137–148.

Elad, T., E. Benovitch, S. Magrisso, and S. Belkin. 2008b. Toxicant identification by a luminescent bacterial bioreporter panel: application of pattern classification algorithms. *Environ. Sci. Technol.* 42:8486–8491.

Elhadi, S.L.N., P.M. Huck, and R.M. Slawson. 2006. Factors affecting the removal of geosmin and MIB in drinking water biofilters. *J. Am. Water Works Assoc.* 98(8):108–119.

Elliott, M.A., F.A. DiGiano, and M.D. Sobsey. 2011. Virus attenuation by microbial mechanisms during the idle time of a household slow sand filter. *Water Res.* 45:4092–4102.

Ellis, J. and W. Korth. 1993. Removal of geosmin and methylisoborneol from drinking water by adsorption on ultrastable zeolite-Y. *Water Res.* 27:535–539.

Ellis, J., P.C. Oyston, M. Green, and R.W. Titball. 2002. Tularemia. *Clin. Microbiol. Rev.* 15:631–646.

El-Zanfaly, H.T. 1990. The concepts of heterotrophic bacteria limit in bottled water as quality criteria. *Riv. Ital. Ig.* 50:440–446.

Emanuel, P.A., J. Dang, J.S. Gebhardt, J. Aldrich, E.A.E. Garber, H. Kulaga, P. Stopa, J.J. Valdes, and A. Dion-Schultz. 2000. Recombinant antibodies: a new reagent for biological agent detection. *Biosens. Bioelectron.* 14:751–759.

Emelko, M.B. and P.M. Huck. 2004. Microspheres as surrogates for *Cryptosporidium* filtration. *J. Am. Water Works. Assoc.* 96(3):94–105.

Emelko, M.B., U. Silins, K.D. Bladon, and M. Stone. 2011. Implications of land disturbance on drinking water treatability in a changing climate: demonstrating the need for "source water supply and protection" strategies. *Water Res.* 41:461–472.

Emerson, S.U. and R.H. Purcell. 2006. Hepatitis E virus. *Rev. Med. Virol.* 13(3):145–154.

Engelbrecht, R.S. 1983. Source, testing and distribution. In *Assessment of Microbiology and Turbidity Standards for Drinking Water*, P.S. Berger and Y. Argaman (Eds). EPA570-9-83001. U.S. Environmental protection Agency, Washington, DC.

Enger, K.S., K.L. Nelson, T. Clasen, J.B. Rose, and J.N.S. Eisenberg. 2012. Linking quantitative microbial risk assessment and epidemiological data: informing safe drinking water trials in developing countries. *Environ. Sci. Technol.* 46:5160–5167.

Enger, K.S., K.L. Nelson, J.B. Rose, and J.N.S. Eisenberg. 2013. The joint effects of efficacy and compliance: a study of household water treatment effectiveness against childhood diarrhea. *Water Res.* 47:1181–1190.

Enriquez, C.E., C.J. Hurst, and C.P. Gerba. 1995. Survival of the enteric adenoviruses 40 and 41 in tap, sea, and wastewater. *Water Res.* 29:2548–2553.

Escobar, I.C. and A.A. Randall. 2001. Case study: ozonation and distribution system biostability. *J. Am. Water Wks. Assoc.* 93(10):77–89.

Escobar, I.C., A.A. Randall, and J.S. Taylor. 2001. Bacterial growth in distribution systems: effect of assimilable organic carbon and biodegradable dissolved organic carbon. *Environ. Sci. Technol.* 35:3442–3447.

European Federation of Bottled Water (EFBW). 2011. http://efbw.eu/bwf.php?classement=07

Fajardo, A. and J.L. Martínez. 2008. Antibiotics as signals that trigger specific bacterial responses. *Curr. Opin. Microbiol.* 11:161–167.

Falcone-Dias, M.F., I. Vaz-Moreira, and C.M. Manaia. 2012. Bottled mineral water as a potential source of antibiotic resistant bacteria. *Water Res.* 46:3612–3622.

Falconer, I.R. 2005. *Cyanobacterial Toxins of Drinking Water Supplies: Cylindrospermopsins and Microcystins.* CRC Press, Boca Raton, FL.

Falconer, I.R. and T.H. Buckley. 1989. Tumour promotion by *Microcystis* sp., a blue-green alga occurring in water supplies. *Med. J. Aust.* 150:351–356.

Falconer, I.R. and A.R. Humpage. 2006. Cyanobacterial (Blue-Green Algal) toxins in water supplies: cylindrospermopsins. *J. Environ. Toxicol.* 21:299–304.

Falkinham, J.O., III. 2002. *Mycobacterium avium* complex. In *Encyclopedia of Environmental Microbiology*, G. Bitton (Editor-in-Chief). Wiley-Interscience, New York, pp. 2112–2120.

Falkinham, J.O., III. 2003. Mycobacterial aerosols and respiratory disease. *Emerg. Infect. Dis.* 9:763–767.

Falkinham, J.O., III, C.D. Norton, and M.W. LeChevallier. 2001. Factors influencing numbers of *Mycobacterium avium*, *Mycobacterium intracellulare*, and other mycobacteria in drinking water distribution systems. *Appl. Environ. Microbiol.* 67:1225–1231.

Falkinham, J.O., M.D. Iseman, P. Haas, and D. Soolingen. 2008. *Mycobacterium avium* in a shower linked to pulmonary disease. *J. Water Health* 6:209–213.

Farahbakhsh, K. and D.W. Smith. 2004. Removal of coliphages in secondary effluent by microfiltration—-mechanisms of removal and impact of operating parameters. *Water Res.* 38:585–592.

Farooq, S. and S. Akhlaque. 1983. Comparative response of mixed cultures of bacteria and virus to ozonation. *Water Res.* 17:809–812.

Feachem, R.G., D.J. Bradley, H. Garelick, and D.D. Mara. 1983. *Sanitation and Disease: Health Aspect of Excreta and Wastewater Management.* John Wiley & Sons, Chichester.

Felfoldi, T., Z. Heeger, M. Vargha, and K. Marialigeti. 2010. Detection of potentially pathogenic bacteria in the drinking water distribution system of a hospital in Hungary. *Clin. Microbiol. Infect.* 16:89–92.

Fenchel, T.M. and B.B. Jorgensen. 1977. Detritus food chain in aquatic ecosystems. *Adv. Microb. Ecol.* 1:1–58.

Fencil, J. and D. Hartman. 2009. Cincinnati's drinking water contamination warning system goes through full-scale exercise. *J. Am. Water Works Assoc.* 101(2):52–56.

Feng, Y., X. Zhao, J. Chen, W. Jin, X. Zhou, N. Li, L. Wang, and L. Xiao. 2011. Occurrence, source, and human infection potential of *Cryptosporidium* and *Giardia* spp. in source and tap water in Shanghai, China. *Appl. Environ. Microbiol.* 77:3609–3616.

Ferguson, D.W., M.J. McGuire, B. Koch, R.L. Wolfe, and E.M. Aieta. 1990. Comparing PEROXONE and ozone for controlling taste and odor compounds, disinfection by-products and microorganisms. *J. Am. Water Works Assoc.* 82:181–191.

Fewtrell, L., D. Kay, M. Wyer, A. Godfree, and G. O'Neill. 1997. Microbiological quality of bottled water. *Water Sci. Technol.* 35(11–12):47–53.

Fewtrell, L., R.B. Kaufmann, D. Kay, W. Enanoria, L. Haller, and J.M. Colford, Jr. 2005. Water, sanitation, and hygiene interventions to reduce diarrhoea in less developed countries: a systematic: review and meta-analysis. *Lancet Infect. Dis.* 5:42–52.

Fields, B.S., E.B. Shotts, Jr., J.C. Feeley, G.W. Gorman, and W.T. Martin. 1984. Proliferation of *Legionella pneumophila* as an intracellular parasite of the ciliated protozoan *Tetrahymena pyriformis*. *Appl. Environ. Microbiol.* 47:467–471.

Fiksdal, L. and T. Leiknes. 2006. The effect of coagulation with MF/UF membrane filtration for the removal of virus in drinking water. *J. Memb. Sci.* 279:364–371.

Fisher, M., Y. Atiya-Nasagi, I. Simon, M. Gordin, A. Mechaly, and S. Yitzhaki. 2009. A combined immunomagnetic separation and lateral flow method for a sensitive on-site detection of *Bacillus anthracis* spores—assessment in water and dairy products. *Lett. Appl. Microbiol.* 48:413–418.

Fisher, M.B., M. Iriarte, and K.L. Nelson. 2012. Solar water disinfection (SODIS) of *Escherichia coli*, *Enterococcus* spp. and MS2 coliphage: effects of additives and alternative container materials. *Water Res.* 46:1745–1754.

Flannery, B., L.B. Gelling, D.J. Vugia, J.M. Weintraub, J.J. Salerno, M.J. Conroy, V.A. Stevens, C.E. Rose, M.R. Moore, B.S. Fields, and R.E. Besser. 2006. Reducing *Legionella* colonization of water systems with monochloramine. *Emerg. Infect. Dis.* 12(4):588–596.

Fleming, K.K., G.W. Harrington, and D.R. Noguera. 2005. Nitrification potential curves: a new strategy for nitrification prevention. *J. Am. Water Works Assoc.* 97(8):90–99.

Flemming, H.-C. and J. Wingender. 2002. Extracellular polymeric substances (EPS): structural, ecological and technical aspects. In *Encyclopedia of Environmental Microbiology*, G. Bitton (Editor-in-Chief). Wiley-Interscience, New York, pp. 1223–1231.

Flemming, H.-C. and J. Wingender. 2010. The biofilm matrix. *Nat. Rev. Microbiol.* 8:623–633.

Flemming, H.-C., U. Szewzyk, and T. Griebe (Eds). 2000. *Biofilms-Investigative Methods & Applications*. Technomic Publishing Co., Lancaster-Basel.

Flewett, T.H. 1982. Clinical features of rotavirus infections. In *Virus Infections of the Gastrointestinal Tract*, D.A. Tyrell and A.Z. Kapikian (Eds). Marcel Dekker, New York, pp. 125–137.

Fliermans, C.B. and R.S. Harvey. 1984. Effectiveness of 1-bromo-3-chloro-5,5-dimethylhydantoin against *Legionella pneumophila* in a cooling tower. *Appl. Environ. Microbiol.* 47:1307–1310.

Focht, D.D. and W. Verstraete. 1977. Biochemical ecology of nitrification and denitrification. *Adv. Microb. Ecol.* 1:135–214.

Fontan-Sainz, M., H. Gomez-Couso, P. Fernandez-Ibanez, and E. Ares-Mazas. 2012. Evaluation of the solar water disinfection process (SODIS) against *Cryptosporidium parvum* using a 25-L static solar reactor fitted with a compound parabolic collector (CPC). *Am. J. Trop. Med. Hyg.* 86:223–228.

Food and Drug Administration (FDA). 1997. The FDA Bottled Water Regulations. Code of Federal Regulations, Parts 129 and 165, Title 21. FDA, Rockville, MD.

Food and Drug Administration (FDA). 2013. http://www.fda.gov/ForConsumers/Consumer Updates/ucm203620.htm (accessed May 5, 2014).

Foran, J.A. 2000. Early warning systems for hazardous biological agents in potable water. *Environ. Health Perspect.* 108:993–996.

Ford, T.E. and R.R. Colwell. 1996. *A Global Decline in Microbiological Safety of Water: A Call for Action.* American Academy of Microbiology, Washington, DC.

Fout, G.S., N.E. Brinkman, J.L. Cashdollar, S.M. Griffin, B.R. McMinn, E.R. Rhodes, E.A. Varughese, M.R. Karim, A.C. Grimm, S.K. Spencer, and M.A. Borchardt. 2010. *Method 1615: Measurement of Enterovirus and Norovirus Occurrence in Water by Culture and RT-qPCR.* Publication no EPA/ 600/R-10/181. US Environmental Protection Agency, Cincinnati, OH.

Fox, K.R., R.J. Miltner, G.S. Logsdon, D.L. Dicks, and L.F. Drolet. 1984. Pilot-plant studies of slow-rate filtration. *J. Am. Water Works Assoc.* 76:62–68.

Fox, A., G.C. Stewart, L.N. Waller, K.F. Fox, W.M. Harley, and R.L. Price. 2003. Carbohydrates and glycoproteins of *Bacillus anthracis* and related bacilli: targets for biodetection. *J. Microbiol. Methods* 54:143–152.

Francis, C.A., K.J. Roberts, J.M. Beman, A.E. Santoro, and B.B. Oakley. 2005. Ubiquity and dversity of ammonia-oxidizing Archaea in water columns and sediments of the ocean. *Proc. Natl. Acad. Sci. U S A* 102(41):14683–14688.

Franco, R.M.B. and R.C. Neto. 2002. Occurrence of cryptosporidial oocysts and *Giardia* cysts in bottled mineral water commercialized in the City of Campinas, State of São Paulo, Brazil. *Mem. Inst. Oswaldo Cruz* 97:205–207.

Franz, D.R., P.B. Jahrling, A.M. Friedlander, D.J. McClain, D.L. Hoover, W.R. Bryne, J.A. Pavlin, G.W. Christopher, and E.M. Eitzen. 1997. Clinical recognition and management of patients exposed to biological warfare agents. *J. Am. Med. Assoc.* 278:399–411.

Frias, J., F. Ribas, and F. Lucena. 1992. A method for the measurement of biodegradable organic carbon in waters. *Water Res.* 26:255–258.

Frost, J.A., I.A. Gillespie, and S.J. O'Brien. 2002. Public health implications of *Campylobacter* outbreaks in England and Wales 1995–9: epidemiological and microbiological investigations. *Epidemiol. Infect.* 128:111–118.

Fulton, G.P. 2000. *Diatomaceous Earth Filtration for Safe Drinking Water.* American Society of Civil Engineers, Reston, VA, 214 pp.

Furness, B.W., M.J. Beach, and J.M. Roberts. 2000. Giardiasis surveillance—United States, 1992–1997. *MMWR CDC Surveill. Summ.* 49(7):1–13.

Gadgil, A. 2008. Safe and affordable drinking water for developing countries. In *Physics of Sustainable Energy, Using Energy Efficiently and Producing It Renewably*, D. Hafemeister, B. Levi, M. Lebine, and P. Schwartz (Eds). American Institute of Physics, pp. 176–191.

Gajardo, R., N. Bouchriti, R.M. Pinto, and A. Bosch. 1995. Genotyping of rotaviruses isolated from sewage. *Appl. Environ. Microbiol.* 61:3460–3462.

Gangadharan, D., K. Harshvardan, G. Gnanasekar, D. Dixit, K.M. Popat, and P. Singh Anand. 2010. Polymeric microspheres containing silver nanoparticles as a bactericidal agent for water disinfection. *Water Res.* 44:5481–5487.

Garcés-Sanchez, G., P.A. Wilderer, J.C. Munch, H. Horn, and M. Lebuhn. 2009. Evaluation of two methods for quantification of *hsp70* mRNA from the waterborne pathogen *Cryptosporidium parvum* by reverse transcription real-time PCR in environmental samples. *Water Res.* 43:2669–2678.

Garcia, A., P. Goñi, J. Cieloszyk, M.T. Fernandez, L. Calvo-Beguería, E. Rubio, M.F. Fillat, M.L. Peleato, and A. Clavel. 2013. Identification of free-living amoebae and amoeba-associated bacteria from reservoirs and water treatment plants by molecular techniques. *Environ. Sci. Technol.* 47:3132–3140.

Garcia-Nunez, M., N. Sopena, S. Ragull, M.L. Pedro-Botet, J. Morera, and M. Sabria. 2008. Persistence of *Legionella* in hospital water supplies and nosocomial Legionnaires' disease. *FEMS Immunol. Med. Microbiol.* 52:202–206.

Geldreich, E.E. 1990. Microbiological quality of source waters for water supply. In *Drinking Water Microbiology*, G.A. McFeters (Ed.). Springer-Verlag, New York, pp. 3–31.

Geldreich, E.E. and D.J. Reasoner. 1990. Home treatment devices and water quality. In *Drinking Water Microbiology*, G.A. McFeters (Ed.). Springer Verlag, New York, pp. 147–167.

Geldreich, E.E., R.H. Taylor, J.C. Blannon, and D.J. Reasoner. 1985. Bacterial colonization of point-of-use water treatment devices. *J. Am. Water Works Assoc.* 77:72–80.

Gelover, S., L.A. Gómez, K. Reyes, and M.T. Leal. 2006. A practical demonstration of water disinfection using TiO_2 films and sunlight. *Water Res.* 40:3274–3280.

Gentry, J., J. Vinje, D. Guadagnoli, and E.K. Lipp. 2009. Norovirus distribution within an estuarine environment. *Appl. Environ. Microbiol.* 75:5474–5480.

Gerba, C.P. 1987. Recovering viruses from sewage, effluents, and water. In *Methods for Recovering Viruses from the Environment*, G. Berg (Ed.). CRC Press, Boca Raton, FL, pp. 1–23.

Gerba, C.P. and C.N. Haas. 1986. Risks associated with enteric viruses in drinking water. In *Progress in Chemical Disinfection*, G.E. Janauer (Ed.). State University of New York, Binghamton, NY, pp. 460–468.

Gerba, C.P. and R.B. Thurman. 1986. Evaluation of the efficacy of Microdyn against waterborne bacteria and viruses. Monograph, University of Arizona, Tucson, AZ.

Gerba, C.P., S.N. Singh, and J.B. Rose. 1985. Waterborne gastroenteritis and viral hepatitis. *CRC Crit. Rev. Environ. Contr.* 15:213–236.

Gerba, C.P., D. Gramos, and N. Nwachuku. 2002. Comparative inactivation of enteroviruses and adenovirus 2 by UV light. *Appl. Environ. Microbiol.* 68:5167–5169.

Gerba, C.P., N. Nwachcuku, and K.R. Riley. 2003. Disinfection resistance of waterborne pathogens on the United States Environmental Protection Agency's Contaminant Candidate List (CCL). *J. Water Supply Res. Technol.* 52:81–94.

Ghannoum, M. and G.A. O'Toole (Eds). 2004. *Microbial Biofilms*. ASM Press, Washington, DC, 426 pp.

Ghosh, M., S. Pathak, and A. Ganguli. 2009. Effective removal of *Cryptosporldium* by a novel bioflocculant. *Water Environ. Res.* 81:160–164.

Gião, M.S., N.F. Azevedo, S.A. Wilks, M.J. Vieira, and C.W. Keevil. 2008. Persistence of *Helicobacter pylori* in heterotrophic drinking-water biofilms. *Appl. Environ. Microbiol.* 74:5898–5904.

Gião, M.S., N.F. Azevedo, S.A Wilks, M.J. Vieira, and C.W. Keevil. 2011. Interaction of *Legionella pneumophila* and *Helicobacter pylori* with bacterial species isolated from drinking water biofilms. *BMC Microbiol.* 11:57–67.

Gibert, O., B. Lefèvre, M. Fernández, X. Bernat, M. Paraira, M. Calderer, and X. Martínez-Lladó. 2013. Characterising biofilm development on granular activated carbon used for drinking water production. *Water Res.* 47:1101–1110.

Gijsbertsen-Abrahamse, A.J., W. Schmidt, I. Chorus, and S.G.J. Heijman. 2006. Removal of cyanotoxins by ultrafiltration and nanofiltration. *J. Memb. Sci.* 276:252–259.

Gomez-Bautista, M., L.M. Ortega-Mora, E. Tabares, V. Lopez-Rodas, and E. Costas. 2000. Detection of infectious *Cryptosporidium parvum* oocysts in mussels (*Mytilus galloprovincialis*) and cockles (*Cerastoderma edule*). *Appl. Environ. Microbiol.* 66:1866–1870.

Gomez-Suarez, C., H.C. van der Mei, and H.J. Busscher. 2002. Adhesion, immobilization and retention of microorganisms on solid substrata. In *Encyclopedia of Environmental Microbiology*, G. Bitton (Editor-in-Chief). Wiley-Interscience, New York, pp. 100–113.

Göttlich, E., W. van der Lubbe, B. Lange, S. Fiedler, I. Melchert, M. Reifenrath, H.C. Flemming, and S. de Hoog. 2002. Fungal flora in groundwater-derived public drinking water. *Int. J. Hyg. Environ. Health* 205:269–279.

Grabow, N.A. and R. Kfir. 1990. Growth of Legionella bacteria in activated carbon filters. Presented at the *International Symposium on Health-Related Microbiology*, Tubingen, W. Germany, April 1–6, 1990.

Grace, R.D., N.E. Dewar, W.G. Barnes, and G.R. Hodges. 1981. Susceptibility of *Legionella pneumophila* to three cooling tower microbicides. *Appl. Environ. Microbiol.* 41:233–236.

Graczyk, T.K., D. Sunderland, L. Tamang, T.M. Shields, F.E. Lucy, and P.N. Breysse. 2007. Quantitative evaluation of the impact of bather density on levels of human-virulent microsporidian spores in recreational water. *Appl. Environ. Microbiol.* 73:4095–4099.

Gratacap-Cavallier, B., O. Genoulaz, K. Brengel-Pesce, H. Soule, P. Innocenti-Francillard, M. Bost, L. Gofti, D. Zmirou, and J.M. Seigneurin. 2000. Detection of human and animal rotavirus sequences in drinking water. *Appl. Environ. Microbiol.* 66:2690–2692.

Gray, N.F. 1994. *Drinking Water Quality: Problems and Solutions*. John Wiley & Sons, Chichester, 315 pp.

Greenwood, Z., J. Black, L. Weld, D. O'Brien, K. Leder, F. Von Sonnenburg, P. Pandey, E. Schwartz, B.A. Connor, G. Brown, D.O. Freedman, and J. Torresi. 2008. Gastrointestinal infection among international travelers globally. *J. Travel Med.* 15(4):221–228.

Gu, M.B., R.J. Mitchell, and B.-C. Kim. 2004. Whole-cell-based biosensors for environmental biomonitoring and application. *Adv. Biochem. Eng. Biotechnol.* 87:269–305.

Guerrant, R.L. and N.M. Thielman. 1995. Types of *Escherichia coli* enteropathogens. In *Infections of the Gastrointestinal Tract*, M.J. Blaser, P.D. Smith, J.I. Ravdin, H.B. Greenberg, and R.L. Guerrant (Eds). Raven Press, New York, pp. 687–690.

Gullick, R.W., W.M. Grayman, R.A. Deininger, and R.M. Males. 2003. Design of early warning systems for source waters. *J. Am. Water Works Assoc. J.* 95(11):58–72.

Gullick, R.W., M.W. Lechevallier, R.C. Svindland, and M.J. Friedman. 2004. Occurrence of transient low and negative pressures in distribution systems. *J. Am. Water Works Assoc.* 96:52–66.

Guo, M., H. Hu, J.R. Bolton, and M.G. El-Din. 2009. Comparison of low- and medium-pressure ultraviolet lamps: photoreactivation of *Escherichia coli* and total coliforms in secondary effluents of municipal wastewater treatment plants. *Water Res.* 43:815–821.

Guo, H., Y. Wyart, J. Perot, F. Nauleau, and P. Moulin. 2010. Low-pressure membrane integrity tests for drinking water treatment: a review. *Water Res.* 44:41–54.

Guo, M., J. Huang, H. Hu, W. Liu, and J. Yang. 2012. UV inactivation and characteristics after photoreactivation of *Escherichia coli* with plasmid: health safety concern about UV disinfection. *Water Res.* 46:4031.

Guttman, L. and J. van Rijn. 2011. Isolation of bacteria capable of growth with 2-methylisoborneol and geosmin as the sole carbon and energy sources. *Appl. Environ. Microbiol.* 78:363–370.

Haas, C.N. and R.S. Engelbrecht. 1980. Physiological alterations of vegetative microorganisms resulting from chlorination. *J. Water Pollut. Control Fed.* 52:1976–1989.

Haas, C.N. and J.N.S. Eisenberg. 2001. Risk assessment. In *World Health Organization (WHO). Water Quality: Guidelines, Standards and Health*, L. Fewtrell and J. Bartram (Eds). IW Publishing, London, UK. ISBN: 1 900222 28 0.

Haas, C.N., J.B. Rose, and C.P. Gerba. 1999. *Quantitative Microbial Risk Assessment*. John Wiley & Sons, New York, 449 pp.

Hageskal, G., A.K. Knutsen, P. Gaustad, G. Sybren de Hoog, and I. Skaar. 2006. Diversity and significance of mold species in Norwegian drinking water. *Appl. Environ. Microbiol.* 72:7586–7593.

Hageskal, G., P. Gaustad, B.T. Heier, and I. Skaar. 2007. Occurrence of moulds in drinking water. *J. Appl. Microbiol.* 102:774–780.

Hageskal, G., N. Lima, and I. Skaar. 2009. The study of fungi in drinking water. *Mycol. Res.* 113:165–172.

Hall, R.M. and M.D. Sobsey. 1993. Inactivation of hepatitis A virus (HAV) and MS-2 by ozone and ozone-hydrogen peroxide in buffered water. *Water Sci. Technol.* 27:371–378.

Hallam, N.B., J.R. West, C.F. Forster, and J. Simms. 2001. The potential for biofilm growth in water distribution systems. *Water Res.* 35:4063–4071.

Halle, C., P.M. Huck, S. Peldszus, J. Haberkamp, and M. Jekel. 2009. Assessing the performance of biological filtration as pretreatment to low pressure membranes for drinking water. *Environ. Sci. Technol.* 43:3878–3884.

Hallier-Soulier, S. and E. Guillot. 2003. An immunomagnetic separation-reverse transcription polymerase chain reaction (IMS-RT-PCR) test for sensitive and rapid detection of viable waterborne *Cryptosporidium parvum*. *Environ. Microbiol.* 5:592–598.

Hamelin, C., F. Sarhan, and Y.S. Chung. 1978. Induction of deoxyribonucleic acid degradation in *Escherichia coli* by ozone. *Experientia* 34:1578–1579.

Hamilton, W.A. 1985. Sulphate anaerobic bacteria and anaerobic corrosion. *Annu. Rev. Microbiol.* 39:195–217.

Hamilton, W.A. 1987. Biofilms: microbial interactions and metabolic activities. In *Ecology of Microbial Communities*, M. Fletcher, T.R.G. Gray, and J.G. Jones (Eds). Cambridge University Press, Cambridge, UK, pp. 361–385.

Hammer, M.J. and M.J. Hammer, Jr. 2008. *Water and Wastewater Technology*, 6th edition. Pearson/Prentice Hall, Columbus, OH, 553 pp.

Hammes, F.A. and T. Egli. 2005. New method for assimilable organic carbon determination using flow-cytometric enumeration and a natural microbial consortium as inoculum. *Environ. Sci. Technol.* 39:3289–3294.

Hammes, F., E. Salhi, O. Köster, H.-P. Kaiser, T. Egli, and U. von Gunten. 2006. Mechanistic and kinetic evaluation of organic disinfection by-products and assimilable organic carbon (AOC) formation during the ozonation of drinking water. *Water Res.* 40:2275–2286.

Hammes, F., S. Meylan, E. Salhi, O. Köster, T. Egli, and U. von Gunten. 2007. Formation of assimilable organic carbon (AOC) and specific natural organic matter (NOM) fractions during ozonation of phytoplankton. *Water Res.* 41(7):1447–1454.

Hammes, F., M. Vital, and T. Egli. 2010. Critical evaluation of the volumetric "bottle effect" on microbial batch growth. *Appl. Environ. Microbiol.* 76:1278–1281.

Handwerker, J., J.G. Fox, and D.B. Schauer. 1995. Detection of *Helicobacter pylori* in drinking water using polymerase chain reaction amplification (Abstr. O-203). In *Abstracts of the 95th General Meeting of the American Society of Microbiology*, Washington, DC, 435 pp.

Hang, J., A.K. Sundaram, P. Zhu, D.R. Shelto, J.S. Karns, P.A.W. Martin, S. Li, P. Amstutz, and C.-M. Tang. 2008. Development of a rapid and sensitive immunoassay for detection and subsequent recovery of *Bacillus anthracis* spores in environmental samples. *J. Microbiol. Methods* 73:242–246.

Hänninen, M.L., H. Haajanen, T. Pummi, K. Wermundsen, M.-L. Katila, H. Sarkkinen, I. Miettinen, and H. Rautelin. 2003. Detection and typing of *Campylobacter jejuni* and *Campylobacter coli* and analysis of indicator organisms in three waterborne outbreaks in Finland. *Appl. Environ. Microbiol.* 69:1391–1396.

Harb, O.S. and Y. Abu Kwaik. 2002. *Legionella* in the environment: persistence, evolution and pathogenicity. In *Encyclopedia of Environmental Microbiology*, G. Bitton (Editor-in-Chief). Wiley-Interscience, New York, pp. 1796–1806.

Harding, A.S. and K.J. Schwab. 2012. Using limes and synthetic psoralens to enhance solar disinfection of water (SODIS): a laboratory evaluation with norovirus, *Escherichia coli*, and MS2. *Am. J. Trop. Med. Health* 86:566–572.

Harrington, G.W., D.R. Noguera, A.I. Kandou, and D.J. Vanhoven. 2002. Pilot-scale evaluation of nitrification control strategies. *J. Am. Water Works Assoc.* 94(11):78–89.

Harris, G.D., V.D. Adams, D.L. Sorenson, and M.S. Curtis. 1987. Ultraviolet inactivation of selected bacteria and viruses with photoreactivation of the bacteria. *Water Res.* 21:687–692.

Hashimoto, A., S. Kunikane, and T. Hirata. 2002. Prevalence of *Cryptosporidium* oocysts and *Giardia* cysts in the drinking water supply in Japan. *Water Res.* 36:519–526.

Haufele, A. and H.V. Sprockhoff. 1973. Ozone for disinfection of water contaminated with vegetative and spore forms of bacteria, fungi and viruses. *Zentralbl. Bakteriol. Orig. B* 175:53–70.

Havelaar, A.H. and J.M. Melse. 2003. *Quantifying Public Health Risks in the WHO Guidelines for Drinking-Water Quality: A Burden of Disease Approach.* RIVM 734301022/2003. National Institute for Public Health and the Environment, Bilthoven.

Havelaar, A.H., M. van Olphen, and J.F. Schijven. 1995. Removal and inactivation of viruses by drinking water treatment processes under full-scale conditions. *Water Sci. Technol.* 31(5–6):55–68.

Hawley, R.L. and E.M. Eitzen, Jr. 2001. Biological weapons—a primer for microbiologists. *Annu. Rev. Microbiol.* 55:235–253.

Hayes, E.B., T.D. Matte, T.R. O'Brien, T.W. McKinley, G.S. Logsdon, J.B. Rose, B.L.P. Ungar, D.M. Word, P.F. Pinsky, M.L. Cummings, M.A. Wilson, E.G. Long, E.S. Hurwittz, and D.D. Juranek. 1989. Large community outbreak of cryptosporidiosis due to contamination of a filtered public water supply. *N. Engl. J. Med.* 320:1372–1376.

Haynes, S.L., K.M. White, and M.R. Rodgers. 2006. Assessment of the effectiveness of low-pressure UV light for inactivation of *Helicobacter pylori*. *Appl. Environ. Microbiol.* 72:3763–3765.

He, X., M. Pelaez, J.A. Westrick, K.E. O'Shea, A. Hiskia, T. Triantis, T. Kaloudis, M.I. Stefan, A.A. de la Cruz, and D.D. Dionysiou. 2012. Efficient removal of microcystin-LR by UV-C/H_2O_2 in synthetic and natural water samples. *Water Res.* 46:1501–1510.

Heaselgrave, W. and S. Kilvington. 2010. Antimicrobial activity of simulated solar disinfection against bacterial, fungal, and protozoan pathogens and its enhancement by riboflavin. *Appl. Environ. Microbiol.* 75:6010–6012.

van Heerden, J., M.M. Ehlers, A. Heim, and W.O.K. Grabow. 2005. Prevalence, quantification and typing of adenoviruses detected in river and treated drinking water in South Africa. *J. Appl. Microbiol.* 99:234–242.

Heffelfinger, J.D., J.L. Kool, S.K. Fridkin, V.J. Fraser, J.C. Carpenter, J. Hageman, J. Carpenter, and C.G. Whitney. 2003. Risk of hospital-acquired legionnaires disease in cities using monochloramine versus other water disinfectants. *Infect. Control Hosp. Epidemiol.* 24(8):569–574.

Heijnen, L. and G. Medena. 2009. Method for rapid detection of viable *Escherichia coli* in water using real-time NASBA. *Water Res.* 43:3124–3132.

Heim, T.H. and A.M. Dietrich. 2007. Sensory aspects and water quality impacts of chlorinated and chloraminated drinking water in contact with HDPE and cPVC pipe. *Water Res.* 41:757–764.

Helmi, K., S. Skraber, C. Gantzer, R. Willame, L. Hoffmann, and H.-M. Cauchie. 2008. Interactions of *Cryptosporidium parvum*, *Giardia lamblia*, vaccinal poliovirus Type 1, and bacteriophages ΦX174 and MS2 with a drinking water biofilm and a wastewater biofilm. *Appl. Environ. Microbiol.* 74:2079–2088.

Henderson, D.A., T.V. Inglesby, J.G. Bartlet, M.S. Ascher, and E. Eitzen. 1999. Smallpox as a biological weapon: medical and public health management. *J. Am. Med. Assoc.* 281:2127–2137.

Henne, K., L. Kahlisch, I. Brettar, and M.G. Höfle. 2012. Analysis of structure and composition of bacterial core communities in mature drinking water biofilms and bulk water of a citywide network in Germany. *Appl. Environ. Microbiol.* 78:3530–3538.

Hentzer, M., H. Wu, J.B. Andersen, K. Riedel, T.B. Rasmussen, N. Bagge, N. Kumar, M.A. Schembri, Z. Song, P. Kristoffersen, M. Manefield, J.W. Costerton, S. Molin, L. Eberl, P. Steinberg, S. Kjelleberg, N. Høiby, and M. Givskov. 2003. Attenuation of *Pseudomonas aeruginosa* virulence by quorum sensing inhibitors. *EMBO J.* 22(15):3803–3815.

Hentzer, M., M. Givskov, and L. Eberl. 2004. Quorum sensing in biofilms: gossip in slime city. In *Microbial Biofilms*, M. Ghannoum and G.A. O'Toole (Eds). ASM Press, Washington, DC, pp. 118–140.

Hernroth, B.E., A.-C. Conden-Hansson, A.-S. Rehnstam-Holm, R. Girones, and A.K. Allard. 2002. Environmental factors influencing human viral pathogens and their potential indicator organisms in the blue mussel, *Mytilus edulis*: the first Scandinavian report. *Appl. Environ. Microbiol.* 68:4523–4533.

Herrmann, J.E. and N.R. Blacklow. 1995. Enteric adenoviruses. In *Infections of the Gastrointestinal Tract*, M.J. Blaser, P.D. Smith, J.I. Ravdin, H.B. Greenberg, and R.L. Guerrant (Eds). Raven Press, New York, pp. 1047–1053.

Herson, D.S., B. McGonigle, M.A. Payer, and K.H. Baker. 1987. Attachment as a factor in the protection of *Enterobacter cloacae* from chlorination. *Appl. Environ. Microbiol.* 53:1178–1180.

Herwaldt, B.L. 2007. *Cyclospora cayetanensis*: a review, focusing on the outbreaks of cyclosporiasis in the 1990s. *Clin. Infect. Dis.* 31:1040–1057.

Herwaldt, B.L., G.F. Craun, S.L. Stokes, and D.D. Juranek. 1992. Outbreaks of waterborne disease in the United States: 1989–90. *J. Am. Water Works Assoc.* 84:129–135.

Herwaldt, B.L., M.-L. Ackers; the *Cyclospora* Working Group. 1997. An outbreak in 1996 of cyclosporiasis associated with imported raspberries. *N. Engl. J. Med.* 336:1548–1556.

Herzog, A.B., S.D. McLennan, A.K. Pandey, C.P. Gerba, C.N. Haas, J.B. Rose, and S.A. Hashsham. 2009. Implications of limits of detection of various methods for *Bacillus anthracis* in computing risks to human health. *Appl. Environ. Microbiol.* 75(19):6331–6339.

Hewitt, J., D. Bell, G.C. Simmons, M. Rivera-Aban, S. Wolf, and G.E. Greening. 2007. Gastroenteritis outbreak caused by waterborne norovirus at a New Zealand ski resort. *Appl. Environ. Microbiol.* 73:7853–7857.

Hibler, C.P. and C.M. Hancock. 1990. Waterborne giardiasis. In *Drinking Water Microbiology*, G.A. McFeters (Ed.). Springer Verlag, New York, pp. 271–293.

Hijnen, W.A.M. and D. van der Kooij. 1992. The effect of low concentrations of assimilable organic carbon (AOC) in water on biological clogging of sand beds. *Water Res.* 26:963–972.

Hijnen, W.A.M., E.F. Beerendonk, and G.J. Medema. 2006. Inactivation credit of UV radiation for viruses, bacteria and protozoan (oo)cysts in water. A review. *Water Res.* 40:3–22.

Hijnen, W.A.M., Y.J. Dullemont, J.F. Schijven, A.J. Hanzens-Brouwer, M. Rosielle, and G. Medema. 2007. Removal and fate of *Cryptosporidium parvum*, *Clostridium perfringens* and small-sized centric diatoms (*Stephanodiscus hantzschii*) in slow sand filters. *Water Res.* 41:2151–2162.

Hilborn, E.D., M.O. Royster, and D.J. Drabkowski. 2002. Survey of US public health laboratories: microbial pathogens on the CCL. *J. Am. Water Works Assoc.* 94:88–96.

Hilborn, E.D., T.C. Covert, M.A. Yakrus, S.I. Harris, S.F. Donnelly, E.W. Rice, S. Toney, S.A. Bailey, and G.N. Stelma, Jr. 2006. Persistence of nontuberculous mycobacteria in a drinking water system after addition of filtration treatment. *Appl. Environ. Microbiol.* 72:5864–5869.

Himberg, K., A.M. Keijola, L. Hiisvirta, H. Pyysalo, and K. Sivonen. 1989. The effect of water treatment processes on the removal of hepatoxins from *Microcystis* and *Oscillatoria* cyanobacteria: a laboratory study. *Water Res.* 23:979–984.

Hinzelin, F. and J.C. Block. 1985. Yeast and filamentous fungi in drinking water. *Environ. Lett.* 6:101–106.

Ho, L., T. Meyn, A. Keegan, D. Hoefel, J. Brookes, C.P. Saint, and G. Newcombe. 2006. Bacterial degradation of microcystin toxins within a biologically active sand filter. *Water Res.* 40:768–774.

Ho, L., D. Hoefel, C.P. Saint, and G. Newcombe. 2007. Isolation and identification of a novel microcystin-degrading bacterium from a biological sand filter. *Water Res.* 41:4685–4695.

Ho, L., P. Lambling, H. Bustamante, P. Duker, and G. Newcombe. 2011. Application of powdered activated carbon for the adsorption of cylindrospermopsin and microcystin toxins from drinking water supplies. *Water Res.* 45:2954–2964.

Ho, L., E. Sawade, and G. Newcombe. 2012a. Biological treatment options for cyanobacteria metabolite removal—a review. *Water Res.* 46:1536–1548.

Ho, L., K. Braun, R. Fabris, D. Hoefel, J. Morran, P. Monis, and M. Drikas. 2012b. Comparison of drinking water treatment process streams for optimal bacteriological water quality. *Water Res.* 46:3934–3942.

Hoefel, D., L. Ho, P.T. Monis, G. Newcombe, and C.P. Saint. 2009. Biodegradation of geosmin by a novel Gram-negative bacterium; isolation, phylogenetic characterization and degradation rate determination. *Water Res.* 43:2925–2935.

Hoff, J.C. 1978. The relationship of turbidity to disinfection of potable water. In *Evaluation of the Microbiology Standards for Drinking Water*, C.W. Hendricks (Ed.). EPA-570/9-78/00C, U.S. Environmental Protection Agency, Washington, DC, pp. 103–117.

Hoff, J.C. and E.W. Akin. 1986. Microbial resistance to disinfectants: mechanisms and significance. *Environ. Health Perspect.* 69:7–13.

Hong, P.-Y., C. Hwang, F. Ling, G.L. Andersen, M.W. LeChevallier, and W.-T. Liu. 2010. Pyrosequencing analysis of bacterial biofilm communities in meters of a drinking water distribution system. *Appl. Environ. Microbiol.* 76:5631–5635.

Hopkins, R.J., P.A. Vial, C. Ferreccio, J. Ovalle, P. Prado, V. Sotomayor, R.G. Russel, S.S. Wasserman, and J.G. Morris. 1993. Seroprevalence of *Helicobacter pylori* in Chile: vegetables may serve as one route of transmission. *J. Infect. Dis.* 168:222–226.

Hörman, A., R. Rimhanen-Finne, L. Maunula, C.-H. von Bonsdorff, J. Rapala, K. Lahti, and M.-L. Hänninen. 2004. Evaluation of the purification capacity of nine portable small-scale water purification devices. *Water Sci. Technol.* 50(1):179–183.

Horn, J.B., D.W. Hendricks, J.M. Scanlan, L.T. Rozelle, and W.C. Trnka. 1988. Removing *Giardia* cysts and other particles from low-turbidity waters using dual-stage filtration. *J. Am. Water Works Assoc.* 80:68–77.

Hozalski, R.M. and E.J. Bouwer. 1998. Deposition and retention of bacteria in backwashed filters. *J. Am. Water Works Assoc.* 90:71–85.

Hozalski, R.M., S. Goel, and E.J. Bouwer. 1992. Use of biofiltration for removal of natural organic matter to achieve biologically stable drinking water. *Water Sci. Technol.* 26:2011–2014.

Hrudey, S.E. 2009. Chlorination disinfection by-products, public health risk tradeoffs and me. *Water Res.* 43:2057–2092.

Hrudey, S.E., E.J. Hrudey, and S.J.T. Pollard. 2006. Risk management for assuring safe drinking water. *Environ. Int.* 32:948–957.

Hsu, S.C., R. Martin, and B.B. Wentworth. 1984. Isolation of *Legionella* species from drinking water. *Appl. Environ. Microbiol.* 48:830–832.

Huamanchay, O., L. Genzlinger, M. Iglesias, and Y.R. Ortega. 2004. Ingestion of *Cryptosporidium* oocysts by *Caenorhabditis elegans*. *J. Parasitol.* 90:1176–1178.

Huang, P., J.T. Weber, D.M. Sosin, P.M. Griffin, E.G. Long, J.J. Murphy, F. Kocka, C. Peters, and C. Kallick. 1996. The first reported outbreak of diarreal illness associated with *Cyclospora* in the United States. *Ann. Intern. Med.* 123:409–414.

Huang, W.J., B.L. Cheng, and Y.L. Cheng. 2007. Adsorption of microcystin-LR by three types of activated carbon. *J. Hazard. Mater.* 141:115–122.

Huang, J., N. Graham, M.R. Templeton, Y. Zhang, C. Collins, and M. Nieuwenhuijsen. 2009. A comparison of the role of two blue–green algae in THM and HAA formation. *Water Res.* 43:3009–3018.

Huang, Y.-F., Y.-F. Fang, and X.-P. Yan. 2010. Amine-functionalized magnetic nanoparticles for rapid capture and removal of bacterial pathogens. *Environ. Sci. Technol.* 44:7908–7913.

Huck, P.M. 1990. Measurement of biodegradable organic matter and bacterial growth potential in drinking water. *J. Am. Water Works Assoc.* 82:78–86.

Huertas, A., B. Barbeau, C. Desjardins, A. Galarza, M.A. Figueroa, and G.A. Toranzos. 2003. Evaluation of *Bacillus subtilis* and coliphage MS2 as indicators of advanced water treatment efficiency. *Water Sci. Technol.* 37(3):255–259.

Huffman, D.E., T.R. Slifko, K. Salisbury, and J.B. Rose. 2000. Inactivation of bacteria, virus and *Cryptosporidium* by a point-of-use device using pulsed broad spectrum white Light. *Water Res.* 34:2491–2498.

Huffman, D.E., A. Gennaccaroa, J.B. Rose, and B.W. Dussert. 2002. Low- and medium-pressure UV inactivation of microsporidia *Encephalitozoon intestinalis*. *Water Res.* 36:3161–3164.

Huisman, L. and W.E. Wood. 1974. *Slow Sand Filtration*. World Health Organization, Geneva.

Huk, A., R.R. Colwell, R. Rahman, A. Ali, M.A.R. Chowdhury, S. Parveen, D.A. Sack, and E. Russek-Cohen. 1990. Detection of *Vibrio cholerae* 01 in the aquatic environment by fluorescent-monoclonal antibody and culture methods. *Appl. Environ. Microbiol.* 56:2370–2373.

Huk, A., E. Lipp, and R. Colwell. 2002. Cholera. In *Encyclopedia of Environmental Microbiology*, G. Bitton (Editor-in-Chief). Wiley-Interscience, New York, pp. 853–861.

Humrighouse, B.W., N.J. Adcock, and E.W. Rice. 2011. Use of acid treatment and a selective medium to enhance the recovery of *Francisella tularensis* from water. *Appl. Environ. Microbiol.* 77(18):6729–6732.

Hunter, P.R. 1993. The microbiology of bottled natural mineral waters. *J. Appl. Bacteriol.* 74:345–352.

Hunter, P.R. 2009. Household water treatment in developing countries: comparing different intervention types using meta-regression. *Environ. Sci. Technol.* 43:8991–8997.

Hunter, P.R., P. Payment, N. Ashbolt, and J. Bartram. 2003. Assessment of risk. In *Assessing Microbial Safety of Drinking Water: Improving Approaches and Methods*, A. Dufour, M. Snozzi, W. Koster, J. Bartram, E. Ronchi and L. Fewtrell (Eds). Published on behalf of the World Health Organization by IWA Publishing. ISBN 92 4 154630 1(WHO) ISBN 1 84339 036 1 (IWA Publishing), pp. 79–103.

van Ingen, J., H. Blaak, J. de Beer, A.M. de Roda Husman, and D. van Soolingen. 2010. Rapidly growing nontuberculous mycobacteria cultured from home tap and shower water. *Appl. Environ. Microbiol.* 76:6017–6019.

Iqbal, S.S., M.W. Mayo, J.G. Bruno, B.V. Bronk, C.A. Batt, and J.P. Chambers. 2000. A review of molecular recognition technologies for detection of biological threat agents. *Biosens. Bioelectron.* 15:549–578.

REFERENCES **245**

Isaac-Renton, J., W.R. Bowie, A. King, G.S. Irwin, C.S. Ong, C.P. Fung, M.O. Shokeir, and J.P. Dubey. 1998. *Toxoplasma gondii* oocysts in drinking water. *Appl. Environ. Microbiol.* 63:2278–2280.

Ishizaki, K., N. Shinriki, and T. Ueda. 1984. Degradation of nucleic acids with ozone. V. Mechanism of action of ozone on deoxyribonucleoside 5′-monophosphates. *Chem. Pharm. Bull.* 32:3601–3606.

Ishizaki, K., K. Sawadaishi, K. Miura, and N. Shinriki. 1987. Effect of ozone on plasmid DNA of *Escherichia coli in situ. Water Res.* 21:823–827.

Ivnitski, D., D.J. O'neil, A. Gattuso, R. Schlicht, M. Calidonna, and R. Fisher. 2003. Nucleic acid approaches for detection and identification of biological warfare and infectious disease agents. *Biotechniques* 35(4):862–869.

Izaguirre, G., C.J. Hwang, S.W. Krasner, and M.J. McGuire. 1982. Geosmin and 2-methylisoborneol from cyanobacteria in three water supply systems. *Appl. Environ. Microbiol.* 43:708–714.

Izquierdo, F., J.A.C. Hermida, S. Fenoy, M. Mezo, M. Gonzalez-Warleta, and C. del Aguila. 2011. Detection of microsporidia in drinking water, wastewater and recreational rivers. *Water Res.* 45:4837–4843.

Izydorczyk, K., C. Carpentier, J. Mrówczyński, A. Wagenvoort, T. Jurczak, and M. Tarczyńska. 2009. Establishment of an alert level framework for cyanobacteria in drinking water resources by using the Algae Online Analyser for monitoring cyanobacterial chlorophyll *a. Water Res.* 43:989–996.

Jacongelo, J.G., S.A. Adham, and J.M. Laine. 1995. Mechanism of *Cryptosporidium, Giardia* and MS2 virus removal by MF and UF. *J. Am. Water Works Assoc.* 87(9):107–114.

Jagger, J. 1958. Photoreactivation. *Bacteriol. Rev.* 22:99–114.

Jain, P. and T. Pradeep. 2005. Potential of silver nanoparticle-coated polyurethane foam as an antibacterial water filter. *Biotechnol. Bioeng.* 90:59–63.

Jakubowski, W. and T.H. Ericksen. 1979. Methods of detection of *Giardia* cysts in water supplies. In *Waterborne Transmission of Giardiasis.* EPA-600/9-79-001. U.S. Environmental Protection Agency, Cincinnati, OH.

Jayasekara, N.Y., G.M. Heard, J.M. Cox, and G.H. Fleet. 1999. Association of microorganisms with the inner surfaces of bottles of non-carbonated mineral waters. *Food Microbiol.* 16:115–128.

Jenkins, M.W., S.K. Tiwari, and J. Darby. 2011. Bacterial, viral and turbidity removal by intermittent slow sand filtration for household use in developing countries: experimental investigation and modeling. *Water Res.* 45:6227–6239.

Jensen, S.E., C.L. Anders, L.J. Goatcher, T. Perley, S. Kenefick, and S.E. Hrudey. 1994. Actinomycetes as a factor in odour problems affecting drinking water from the North Saskatchewan River. *Water Res.* 28:1393–1401.

Jensen, P.K., J.H.J. Ensink, G. Jayasinghe, W. Van Der Hoek, S. Cairncross, and A. Dalsgaard. 2002. Domestic transmission routes of pathogens: the problem of in-house contamination of drinking water during storage in developing countries. *Trop. Med. Int. Health* 7:604–609.

Jernigan, J.A., D.S. Stephens, D.A. Ashford, C. Omenaca, M.S. Topiel, M. Galbraith, M. Tapper, T.L. Fisk, S. Zaki, T. Popovic, R.F. Meyer, C.P. Quinn, S.A. Harper, S.K. Fridkin, J.J. Sejvar, C.W. Shepard, M. McConnell, J. Guarner, W.J. Shieh, J.M. Malecki, J.L. Gerberding, J.M. Hughes, B.A. Perkins, and members of the Anthrax Bioterrorism Investigation Team. 2001. Bioterrorism-related inhalational anthrax: the first 10 cases reported in the United States. *Emerg. Infect. Dis.* 7:933–944.

Jiang, S.C. 2006. Human adenoviruses in water: occurrence and health implications: a critical review. *Environ. Sci. Technol.* 40:7132–7140.

Jiang, V., B. Jiang, J. Tate, U.D. Parashar, and M.M. Patel. 2010. Performance of rotavirus vaccines in developed and developing countries. *Hum. Vaccin.* 6:532–542.

Jiang, W., S. Xia, J. Liang, Z. Zhang, and S.W. Hermanowic. 2013. Effect of quorum quenching on the reactor performance, biofouling and biomass characteristics in membrane bioreactors. *Water Res.* 47:187–196.

Jo, C.H., A.M. Dietrich, and J.M. Tanko. 2011. Simultaneous degradation of disinfection byproducts and earthy-musty odorants by the UV/H_2O_2 advanced oxidation process. *Water Res.* 45:2507–2516.

Jofre, J., E. Olle, F. Ribas, A. Vidal, and F. Lucena. 1995. Potential usefulness of bacteriophages that infect *Bacteroides fragilis* as model organisms for monitoring virus removal in drinking water treatment plants. *Appl. Environ. Microbiol.* 61:3227–3231.

Johnson, D.W., N.J. Pienazek, D.W. Griffin, L. Misener, and J.B. Rose. 1995. Development of a PCR protocol for sensitive detection of *Cryptosporidium* ocysts in water samples. *Appl. Environ. Microbiol.* 61:3849–3855.

Johnson, A.M., G.D. Di Giovanni, and P.A. Rochelle. 2012. Comparison of assays for sensitive and reproducible detection of cell culture-infectious *Cryptosporium parvum* and *Cryptosporidium hominis* in drinking water. *Appl. Environ. Microbiol.* 78:156–162.

Jolis, D., C. Lam, and P. Pitt. 2001. Particle effects on ultraviolet disinfection of coliform bacteria in recycled water. *Water Environ. Res.* 73:233–236.

Jonassen, T.O., E. Kjeldsberg, and B. Grinde. 1993. Detection of human astrovirus serotype 1 by polymerase chain reaction. *J. Virol. Methods* 44:83–88.

Jones, C.R., M.R. Adams, P.A. Zhdan, and A.H.L. Chamberlain. 1999. The role of surface physicochemical properties in determining the distribution of the autochthonous microflora in mineral water bottles. *J. Appl. Microbiol.* 86:917–927.

Joret, J.-C. and Y. Levi. 1986. Methode rapide d'evaluation du carbone eliminable des eaux par voie biologique. *Trib. Cebedeau* 510:3–9.

Joret, J.-C., A. Hassen, M.M. Bourbigot, F. Agbalika, P. Hartmann, and J.M. Foliguet. 1986. Inactivation des virus dans l'eau sur une filiere de production a ozonation etagee. *Water Res.* 20:871–876.

Joret, J.C., Y. Levi, T. Dupin, and M. Gibert. 1988. Rapid method for estimating bioeliminable organic carbon in water. Presented at the *American Water Works Association Conference*, Orlando, FL, June 19–23, 1988.

Joret, J.C., Y. Levi, and C. Volk. 1990. Biodegradable dissolved organic carbon (BDOC) content of drinking water and potential regrowth of bacteria. Presented at the *International Symposium Health-Related Water Microbiology*, Tubingen, Germany, April 1–6, 1990.

Joret, J.-C., D. Perrine, and B. Langlais. 1992. Effect of temperature on the inactivation of *Cryptosporidium* oocysts by ozone. *IWPRC International Symposium*, Washington, DC, May 26–29, 1992.

Juhna, T., D. Birzniece, and J. Rubulis. 2007. Effect of phosphorus on survival of *Escherichia coli* in drinking water biofilms. *Appl. Environ. Microbiol.* 73:3755–3758.

Jung, Y.J., B.S. Oh, and J.-W. Kang. 2008. Synergistic effect of sequential or combined use of ozone and UV radiation for the disinfection of *Bacillus subtilis* spores. *Water Res.* 42:1613–1621.

Juttner, F. 1981. Detection of lipid degradation products in the water of a reservoir during a bloom of *Synura uvella. Appl. Environ. Microbiol.* 41:100–106.

Kahlisch, L., K. Henne, L. Groebe, J. Draheim, M.G. Hofle, and I. Brettar. 2010. Molecular analysis of the bacterial drinking water community with respect to live/dead status. *Water Sci. Technol.* 61(1):9–14.

Kalmbach, S., W. Manz, B. Bendinger, and U. Szewzyk. 2000. *In situ* probing reveals *Aquabacterium* commune as a widespread and highly abundant bacterial species in drinking water biofilms. *Water Res.* 34:575–581.

Kalscheur, K.N., C.E. Gerwe, J. Kweon, G.E. Speitel, Jr., and D.F. Lawler. 2006. Enhanced softening: effects of source water quality on NOM removal and DBP formation. *J. Am. Water Works. Assoc.* 98(11):93–105.

Kaminski, J.C. 1994. *Cryptosporidium* and the public water supply. *N. Engl. J. Med.* 331:1529.

Kane, S.R., S.E. Létant, G.A. Murphy, T.M. Alfaro, P.W. Krauter, R. Mahnke, T.C. Legler, and E. Raber. 2009. Rapid, high-throughput, culture-based PCR methods to analyze samples for viable spores of *Bacillus anthracis* and its surrogates. *J. Microbiol. Methods* 76:278–284.

Kaneko, M. 1989. Effect of suspended solids on inactivation of poliovirus and T2-phage by ozone. *Water Sci. Technol.* 21:215–219.

Kang, S.-I., M. Her, J.W. Kim, J.Y. Kim, K.Y. Ko, Y.-M. Ha, and S.C. Jung. 2011. Advanced multiplex PCR assay for differentiation of *Brucella* species. *Appl. Environ. Microbiol.* 77:6726–6728.

Kaplan, L.A., T.L. Bott, and D.J. Reasoner. 1993. Evaluation and simplification of the assimilable organic carbon nutrient bioassay for bacterial growth in drinking water. *Appl. Environ. Microbiol.* 59:1532–1539.

Kasuga, I., D. Shimazaki, and S. Kunikane. 2007. Influence of backwashing on the microbial community in a biofilm developed on biological activated carbon used in a drinking water treatment plant. *Water Sci. Technol.* 55(8–9):173–180.

Katz, A. and N. Narkis. 2001. Removal of chlorine dioxide disinfection by-products by ferrous salts. *Water Res.* 34:101–108.

Keegan, A., D. Daminato, C.P. Saint, and P.T. Monis. 2008. Effect of water treatment processes on *Cryptosporidium* infectivity. *Water Res.* 42:1805–1811.

Keevil, C.W. 2002. Pathogens in environmental biofilms. In *Encyclopedia of Environmental Microbiology*, G. Bitton (Editor-in-Chief). Wiley-Interscience, New York, pp. 2339–2356.

Kehoe, S.C., M.R. Barer, L.O. Devlin, and K.G. McGuigan. 2004. Batch process solar disinfection is an efficient means of disinfecting drinking water contaminated with *Shigella dysenteriae* type I. *Lett. Appl. Microbiol.* 38:410–414.

Keinänen, M.M., L.K. Korhonen, M.J. Lehtola, I.T. Miettinen, P.J. Martikainen, T. Vartiainen, and M.H. Suutari. 2002. The microbial community structure of drinking water biofilms can be affected byphosphorus availability. *Appl. Environ. Microbiol.* 68:434–439.

Kemmy, F.A., J.C. Fry, and R.A. Breach. 1989. Development and operational implementation of a modified and simplified method for determination of assimilable organic carbon (AOC) in drinking water. *Water Sci. Technol.* 21:155–159.

Keserue, H.-A., H.P. Fuchslin, and T. Egli. 2011. Rapid detection of *Giardia lamblia* cysts in water samples by immunomagnetic separation and flow cytometric analysis. *Appl. Environ. Microbiol.* 77:5420–5427.

Khan, A.L., D.L. Swerdlow, and D.D. Juranek. 2001. Precautions against biological and chemical terrorism directed at food and water supplies. *Public Health Rep.* 116:3–14.

Khan, M.M.T., B.H. Pyle, and A.K. Camper. 2010. Specific and rapid enumeration of viable but nonculturable and viable-culturable Gram-negative bacteria by using flow cytometry. *Appl. Environ. Microbiol.* 76:5088–5096.

Khiari, D. and S. Watson. 2007. Taste and odours compounds in dinking water. Where are we today? *Water Sci. Technol.* 55(5):365–366.

Kilvington, S. and J. Beeching. 1995a. Identification and epidemiological typing of *Naegleria fowleri* with DNA probes. *Appl. Environ. Microbiol.* 61:2071–2078.

Kilvington, S. and J. Beeching. 1995b. Development of a PCR for identification of *Naegleria fowleri* from the environment. *Appl. Environ. Microbiol.* 61:3764–3767.

Kilvington, S. and D.G. White. 1985. Rapid identification of thermophilic *Naegleria* including *Naegleria fowleri* using API ZYM system. *J. Clin. Pathol.* 38:1289–1292.

Kim, J. and B. Kang. 2008. DBPs removal in GAC filter-adsorber. *Water Res.* 42:145–152.

Kim, S., K. Ghafoor, J. Lee, M. Feng, J. Hong, D.-U. Lee, and J. Park. 2013a. Bacterial inactivation in water, DNA strand breaking, and membrane damage induced by ultraviolet-assisted titanium dioxide photocatalysis. *Water Res.* 47:4403–4411.

Kim, S.-R., H.-S. Oh, S.-J. Jo, K.-M. Yeon, C.-H. Lee, D.-J. Lim, C.-H. Lee, and J.-K. Lee. 2013b. Biofouling control with bead-entrapped quorum quenching bacteria in membrane bioreactors: physical and biological effects. *Environ. Sci. Technol.* 47:836–842.

King, D., V. Luna, A. Cannons, J. Cattani, and P. Amuso. 2003. Performance assessment of three commercial assays for direct detection of *Bacillus anthracis* spores. *J. Clin. Microbiol.* 41:3454–3455.

Klaine, S.J., P.J.J. Alvarez, G.E. Batley, T.F. Fernandes, R.D. Handy, D.Y. Lyon, S. Mahendra, M.J. McLaughlin, and J.R. Lead. 2008. Nanomaterials in the environment: behavior, fate, bioavailability, and effects. *Environ. Toxicol. Chem.* 27:1825–1851.

Klein, P.D., D.Y. Graham, A. Gailor, A.R. Opekun, and E. O'Brian Smith. 1991. Water source as risk factor for *Helicobacter pylori* infection in Peruvian children. *Lancet* 337:1503–1506.

Knudson, G.B. 1985. Photoreactivation of UV-irradiated *Legionella pneumophila* and other *Legionella* species. *Appl. Environ. Microbiol.* 49:975–980.

Koch, A.L. 2002. Viable but not culturable (VBNC) microorganisms. In *Encyclopedia of Environmental Microbiology*, G. Bitton (Editor-in-Chief). Wiley-Interscience, New York, pp. 3246–3255.

Koide, M., A. Saito, and M. Okazaki. 1999. Isolation of *Legionella longbeachae* serogroup 1 from potting soils in Japan. *Clin. Infect. Dis.* 29:943–944.

van der Kooij, D. 1983. Biological processes in carbon filters. In *Activated Carbon in Drinking Water Technology*. Research Report. AWWA Research Foundation, Denver, CO, pp. 119–152.

van der Kooij, D. 1990. Assimilable organic carbon (AOC) in drinking water. In *Drinking Water Microbiology*, G.A. McFeters (Ed.). Springer-Verlag, New York, pp. 57–87.

van der Kooij, D. 1992. Assimilable organic carbon as an indicator of bacterial regrowth. *J. Am. Water Works Assoc.* 84:57–65.

van der Kooij, D. 1995. Significance and assessment of the biological stability of drinking water. In *Water Pollution: Quality and Treatment of Drinking Water*, Springer-Verlag, New York, pp. 89–102.

Van der Kooij, D. 2000. Biological stability: a multidimensional quality aspect of treated water. *Water Air Soil Pollut.* 123:25–34.

van der Kooij, D. 2002. Assimilable organic carbon (AOC) in treated water: determination and significance. In *Encyclopedia of Environmental Microbiology*, G. Bitton (Editor-in-Chief). Wiley-Interscience, New York, pp. 312–327.

van der Kooij, D. and W.A.M. Hijnen. 1981. Utilization of low concentrations of starch by a *Flavobacterium* species isolated from tap water. *Appl. Environ. Microbiol.* 41:216–221.

van der Kooij, D. and W.A.M. Hijnen. 1984. Substrate utilization by an oxalate-consuming *Spirillum* species in relation to its growth in ozonated water. *Appl. Environ. Microbiol.* 47:551–559.

van der Kooij, D. and W.A.M. Hijnen. 1985. Determination of the concentration of maltose and starch-like compounds in drinking water by growth measurements with a well defined strain of *Flavobacterium* species. *Appl. Environ. Microbiol.* 49:765–771.

van der Kooij, D. and W.A.M. Hijnen. 1988. Multiplication of a *Klebsiella pneumonae* strain in water at low concentration of substrate. *Proceedings International Conference on Water and Wastewater Microbiology*, Newport Beach, CA, Feb. 8–11, 1988.

van der Kooij, D. and H.R. Veenendaal. 1994. Assessment of the biofilm formation potential of synthetic materials in contact with drinking water during distribution. In *Proceeding American Water Works Association Water Quality Technology Conference*, Miami, FL, pp. 1395–1407.

van der Kooij, D., A. Visser, and W.A.M. Hijnen. 1982a. Determining the concentration of easily assimilable organic carbon in drinking water. *J. Am. Water Works Assoc.* 74:540–545.

van der Kooij, D., J.P. Oranje, and W.A.M. Hijnen. 1982b. Growth of *Pseudomonas aeruginosa* in tap water in relation to utilization of substrates at concentrations of a few micrograms per liter. *Appl. Environ. Microbiol.* 44:1086–1095.

Kool, J.L., J.C. Carpenter, and B.S. Fields. 1999. Effect of monochloramine disinfection of municipal drinking water on risk of nosocomial Legionnaires disease. *Lancet* 353(9149):272–277.

Kool, J.L., J.C. Carpenter, and B.S. Fields. 2000. Monochloramine and Legionnaires's disease. *J. Am. Water Works Assoc.* 92(9):88–96.

Korich, D.G., J.R. Mead, M.S. Madore, and N.A. Sinclair. 1989. Effets of chlorine and ozone on *Cryptosporidium* oocyst viability (Abstr.). In *89th Annual Meeting of the American Society of Microbiology*, New Orleans, LA.

Korich, D.G., J.R. Mead, M.S. Madore, N.A. Sinclair, and C.R. Sterling. 1990. Effect of ozone, chlorine dioxide, chlorine, and monochloramine on *Cryptosporidium parvum* oocyst viability. *Appl. Environ. Microbiol.* 56:1423–1428.

Korsholm, E. and H. Søgaard. 1988. An evaluation of direct microscopical counts and endotoxin measurements as alternatives for total plate counts. *Water Res.* 22:783–788.

Kosek, M., C. Bern, and R. Guerrant. 2003. The global burden of diarrhoeal disease, as estimated from studies published between 1992 and 2000. *Bull. W.H.O.* 81:197–204.

Koskinen, R., T. Ali-Vehmas, P. Kämpfer, M. Laurikkala, I. Tsitko, E. Kostyal, F. Atroshi, and M. Salkinoja-Salonen. 2000. Characterization of *Sphingomonas* isolates from Finnish and Swedish drinking water distribution systems. *J. Appl. Microbiol.* 89:687–696.

Kramer, M.H., B.L. Herwaldt, G.F. Craun, R.L. Calderon, and D.D. Juranek. 1996. Waterborne disease: 1993 and 1994. *J. Am. Water Works. Assoc.* 88:66–80.

Kreft, P., M. Umphres, J.-M. Hand, C. Tate, M.J. McGuire, and R.R. Trussel. 1985. Converting from chlorine to chloramines: a case study. *J. Am. Water Works Assoc.* 77:38–45.

Krewski, D., J. Balbus, D. Butler-Jones, C. Haas, J. Isaac-Renton, K.J. Roberts, and M. Sinclair. 2002. Managing health risks from drinking water—a report to the Walkerton inquiry. *J. Toxicol. Environ. Health A* 65:1635–1823.

Krometis, L.-A.H., G.W. Characklis, and M.D. Sobsey. 2009. Identification of particle size classes inhibiting protozoan recovery from surface water samples via U.S. Environmental Protection Agency method 1623. *Appl. Environ. Microbiol.* 75:6619–6621.

Kuehn, W. and U. Mueller. 2000. Riverbank filtration: an overview. *J. Am. Water Works Assoc.* 92(12):60–69.

Kuhn, R.C. and K.H. Oshima. 2001. Evaluation and optimization of a reusable hollow fiber ultrafilter as a first step in concentrating *Cryptosporidium parvum* oocysts from water. *Water Res.* 35:2779–2783.

Kuhn, I., G. Allestam, G. Huys, P. Janssen, K. Kersters, K. Krovacek, and T.-A. Stenstrom. 1997. Diversity, persistence, and virulence of *Aeromonas* strains isolated from drinking water distribution systems in Sweden. *Appl. Environ. Microbiol.* 63:2708–2715.

Kukkula, M., L. Maunula, E. Silvennoinen, and C.-H. von Bonsdorff. 1999. Outbreak of viral gastroenteritis due to drinking water contaminated by Norwalk-like viruses. *J. Infect. Dis.* 180:1771–1776.

Kvist, M., V. Hancock, and P. Klemm. 2008. Inactivation of efflux pumps abolishes bacterial biofilm formation. *Appl. Environ. Microbiol.* 74:7376–7382.

Labatiuk, C.W., F.W. Schaefer, III, G.R. Finch, and M. Belosevic. 1991. Comparison of animal infectivity, excystation, and fluorogenic dye as measures of *Giardia muris* cyst inactivation by ozone. *Appl. Environ. Microbiol.* 57:3187–3192.

Lafrance, D. 2011. Delivering safe water to the moon. *J. Am. Water Works Assoc.* 103(7):6.

Lalezary-Craig, S., M. Pirbazari, M.S. Dale, T.S. Tanaka, and M.J. McGuire. 1988. Optimizing the removal of geosmin and 2-methylisoborneol by powdered activated carbon. *J. Am. Water Works Assoc.* 80:73–80.

Lam, A.K.-Y., E.E. Prepas, D. Spink, and S.E. Hrudey. 1995. Chemical control of hepatoxic phytoplankton blooms: implications for human health. *Water Res.* 29:1845–1854.

Lambertini, E., M.A. Borchardt, B.A. Kieke, Jr., S.K. Spencer, and F.J. Loge. 2012. Risk of viral acute gastrointestinal illness from nondisinfected drinking water distribution systems. *Environ. Sci. Technol.* 46:9299–9307.

Lamont, Y., A. Rze-Zutka, J.G. Anderson, S.J. MacGregor, M.J. Given, C. Deppe, and N. Cook. 2007. Pulsed UV-light inactivation of poliovirus and adenovirus. *Lett. Appl. Microbiol.* 45:564–567.

Lange, K.P., W.D. Bellamy, D.W. Hendricks, and G.S. Logsdon. 1986. Diatomaceous earth filtration of *Giardia* cysts and other substances. *J. Am. Water Works Assoc.* 78(1): 76–84.

Lantagne, D.S. 2001. *Investigation of the Potters for Peace Colloidal Silver Impregnated Ceramic Filter*. USAID, Washington, DC, 79 pp.

Lantagne, D.S. 2008. Sodium hypochlorite dosage for household and emergency water treatment. *J. Am. Water Works Assoc.* 100(8):106–118.

Lau, H.Y. and N.J. Ashbolt. 2009. The role of biofilms and protozoa in *Legionella* pathogenesis: implications for drinking water. *J. Appl. Microbiol.* 107:368–378.

Lauderdale, C.V., H.C. Aldrich, and A. Lindner. 2004. Isolation and characterization of a bacterium capable of removing taste- and odor-causing 2-methylisoborneol from water. *Water Res.* 38:4135–4142.

Lautenschlager, K., N. Boon, Y. Wang, T. Egli, and F. Hammes. 2010. Overnight stagnation of drinking water in household taps induces microbial growth and changes in community composition. *Water Res.* 44:4868–4877.

Lautenschlager, K., C. Hwang, W.-T. Liu, N. Boon, O. Köster, H. Vrouwenvelder, T. Egli, and F. Hammes. 2013. A microbiology-based multi-parametric approach towards assessing biological stability in drinking water distribution networks. *Water Res.* 47:3015–3025.

Lawton, L.A., A. Welgamage, P.M. Manage, and C. Edwards. 2011. Novel bacterial strains for the removal of microcystins from drinking water. *Water Sci. Technol.* 63(6):1137–1142.

Lazarova, V. and J. Manem. 1995. Biofilm characterization and activity analysis in water and wastewater treatment. *Water. Res.* 29:2227–2245.

Lechelt, M., W. Blohm, B. Kirschneit, M. Pfeiffer, E. Gresens, J. Liley, R. Holz, C. Luring, and C. Moldaenke. 2000. Monitoring of surface water by ultrasensitive *Daphnia* taximeter. *Environ. Toxicol.* 15:390–400.

LeChevallier, M.W. 2002. Microbial removal by pretreatment, coagulation and ion exchange. In *Encyclopedia of Environmental Microbiology*, G. Bitton (Editor-in-Chief). Wiley-Interscience, New York, pp. 2012–2019.

LeChevallier, M.W. 2004. Control, treatment, and disinfection of *Mycobacterium avium* complex in drinking water. In *Pathogenic Mycobacteria in Water*, S. Pedley, J. Bartram, G. Rees, A. Dufour, and J. Cotruvo (Eds). World Health Organization, Geneva.

LeChevallier, M.W. and K.K. Au. 2004. *Water Treatment and Pathogen Control: Process Efficiency in Achieving Safe Drinking Water.* IWA Publishing, London, UK. ISBN: 1 84339 069 8.

LeChevallier, M.W. and G.A. McFeters. 1985a. Enumerating injured coliforms in drinking water. *J. Am. Water Works Assoc.* 77:81–87.

LeChevallier, M.W. and G.A. McFeters. 1985b. Interactions between heterotrophic plate count bacteria and coliform organisms. *Appl. Environ. Microbiol.* 49:1138–1141.

LeChevallier, M.W. and G.A. McFeters. 1990. Microbiology of activated carbon. In *Drinking Water Microbiology*, G.A. McFeters (Ed.). Springer Verlag, New York, pp. 104–119.

LeChevallier, M.W., R.J. Seidler, and T.M. Evans. 1980. Enumeration and characterization of standard plate count bacteria in chlorinated and raw water supplies. *Appl. Environ. Microbiol.* 40:922–930.

LeChevallier, M.W., T.S. Hassenauer, A.K. Camper, and G.A. McFeters. 1984. Disinfection of bacteria attached to granular activated carbon. *Appl. Environ. Microbiol.* 48:918–923.

LeChevallier, M.W., C.D. Cawthon, and R.G. Lee. 1988. Inactivation of biofilm bacteria. *Appl. Environ. Microbiol.* 54:2492–2499.

LeChevallier, M.W., C.H. Lowry, and R.G. Lee. 1990. Disinfecting biofilm in a model distribution system. *J. Am. Water Works Assoc.* 82:85–99.

LeChevallier, M.W., W. Schulz, and R.G. Lee. 1991a. Bacterial nutrients in drinking water. *Appl. Environ. Microbiol.* 57:857–862.

LeChevallier, M.W., W.D. Norton, and R.G. Lee. 1991b. *Giardia* and *Cryptosporidium* spp. in filtered drinking water supplies. *Appl. Environ. Microbiol.* 57:2617–2621.

LeChevallier, M.W., W.C. Becker, P. Schorr, and R.G. Lee. 1992. Evaluating the performance of biologically active rapid filters. *J. Am. Water Works Assoc.* 84:136–146.

LeChevallier, M.W., N.E. Shaw, L.A. Kaplan, and T.L. Bott. 1993. Development of a rapid assimilable organic carbon method in water. *Appl. Environ. Microbiol.* 59:1526–1531.

LeChevallier, M.W., M.-C. Besner, M. Friedman, and V.L. Speight. 2011. Microbiological quality control in distribution systems. In *Water Quality and Treatment: A Hanbook on Drinking Water*, 6th edition, J.K. Edzwald (Ed.). AWWA, Denver, CO, pp. 21.1–21.84.

Lecican, A., A. Alkeskas, C. Gunter, S.J. Forsythe, and M.J. Figeras. 2013. Adherence to and invasion of human intestinal cells by *Arcobacter* species and their virulence genotypes. *Appl. Environ. Microbiol.* 79:4951–4957.

Leclerc, H. and A. Moreau. 2002. Microbiological safety of natural mineral water. *FEMS Mirobiol. Rev.* 26:207–222.

Leclerc, H., L. Schwartzbrod, and E. Dei-Cas. 2002. Microbial agents associated with waterborne disease. *Crit. Rev. Microbiol.* 28:371–409.

Le Dantec, C., J.P. Duguet, A. Montiel, N. Dumoutier, S. Dubrou, and V. Vincent. 2002a. Chlorine disinfection of atypical mycobacteria isolated from a water distribution system. *Appl. Environ. Microbiol.* 68:1025–1032.

Le Dantec, C., J.-P. Duguet, A. Montiel, N. Dumoutier, S. Dubrou, and V. Vincent. 2002b. Occurrence of mycobacteria in water treatment lines and in water distribution systems. *Appl. Environ. Microbiol.* 68:5318–5325.

Leduc, A., S. Gravel, J. Abikhzer, S. Roy, and J. Barbeau. 2012. Polymerase chain reaction detection of potentially pathogenic free-living amoebae in dental units. *Can. J. Microbiol.* 58:884–886.

Lee, S.-H. and S.-J. Kim. 2002. Detection of infectious enteroviruses and adenoviruses in tap water in urban areas in Korea. *Water Res.* 36:248–256.

Lee, J., C.S. Lee, K.M. Hugunin, C.J. Maute, and R.C. Dysko. 2010. Bacteria from drinking water supply and their fate in gastrointestinal tracts of germ-free mice: a phylogenetic comparison study. *Water Res.* 44:5050–5058.

Le Guyader, F., F.H. Neill, M.K. Estes, S.S. Monroe, T. Ando, and R.L. Atmar. 1996. Detection and analysis of a small round-structured virus strain in oysters implicated in an outbreak of acute gastroenteritis. *Appl. Environ. Microbiol.* 62:4268–4272.

Le Guyader, F., L. Haugarreau, L. Miossec, E. Dubois, and M. Pommepuy. 2000. Three-year study to assess human enteric viruses in shellfish. *Appl. Environ. Microbiol.* 66:3241–3248.

Le Guyader, F., S. Parnaudeau, J. Schaeffer, A. Bosch, F. Loisy, M. Pommepuy, and R.L. Atmar. 2009. Detection and quantification of noroviruses in shellfish. *Appl. Environ. Microbiol.* 75:618–624.

Lehtola, M.J., I.T. Miettinen, T. Vartiainen, and P.J. Martikainen. 2002. Changes in content of microbially available phosphorus, assimilable organic carbon and microbial growth potential during drinking water treatment processes. *Water Res.* 36:3681–3690.

Lehtola, M.J., E. Torvinen, I.T. Miettinen, and C.W. Keevil. 2006. Fluorescence *In Situ* hybridization using peptide nucleic acid probes for rapid detection of *Mycobacterium avium* subsp. *avium* and *Mycobacterium avium* subsp. *paratuberculosis* in potable-water biofilms. *Appl. Environ. Microbiol.* 72:848–853.

Lehtola, M.J., E. Torvinen, J. Kusnetsov, T. Pitkänen, L. Maunula, C.-H. von Bonsdorff, P.J. Martikainen, S.A. Wilks, C.W. Keevil, and I.T. Miettinen. 2007. Survival of *Mycobacterium avium, Legionella pneumophila, Escherichia coli*, and Caliciviruses in drinking water-associated biofilms grown under high-shear turbulent flow. *Appl. Environ. Microbiol.* 73:2854–2859.

Lénès, D., N. Deboosere, F. Ménard-Szczebara, J. Jossent, V. Alexandre, C. Machinal, and M. Vialette. 2010. Assessment of the removal and inactivation of influenza viruses H5N1 and H1N1 by drinking water treatment. *Water Res.* 44:2473–2486.

Leoni, E., P.P. Legnani, M.A. Bucci Sabattini, and F. Righi. 2001. Prevalence of *Legionella* spp. in swimming pool environment. *Water Res.* 35:3749–3753.

Letterman, R.D., A. Amirtharajah, and C.R. O'Melia. 1999. Coagulation and flocculation. In *Water Quality and Treatment*, R.D. Letterman (Ed.). McGraw-Hill, New York, pp. 6.1–6.66.

Levin, T.R., J.A. Schmittdiel, J.M. Henning, K. Kunz, C.J. Henke, C.J. Colby, and J.V. Selby. 1998. A cost analysis of a *Helicobacter pylori* eradication strategy in a large health maintenance organization. *Am. J. Gastroenterol.* 93:743–747.

Levin, R.B., P.R. Epsstein, T.E. Ford, W. Harrington, E. Olson, and E.G. Reichard. 2002. U.S. drinking water challenges in the twenty-first century. *Environ. Health Perspect.* 110(suppl. 1):43–52.

Levy, R.V. 1990. Invertebrates and associated bacteria in drinking water distribution lines. In *Drinking Water Microbiology*, G.A. McFeters (Ed.). Springer Verlag, New York, pp. 224–248.

Levy, R.V., F.L. Hart, and R.D. Cheetham. 1986. Occurrence and public health significance of invertebrates in drinking water systems. *J. Am. Water Works Assoc.* 78:105–110.

Li, J. and E.R. Blatchley, III. 2007. Volatile disinfection byproduct formation resulting from chlorination of organic-nitrogen precursors in swimming pools. *Environ. Sci. Technol.* 41:6732–6739.

Li, Q., R. Xie, Y.W. Li, E.A. Mintz, and J.K. Shang. 2007. Enhanced visible-light-induced photocatalytic disinfection of *E. coli* by carbon-sensitized nitrogen-doped titanium oxide. *Environ. Sci. Technol.* 41:5050–5056.

Li, Q., S. Mahendra, D.Y. Lyon, L. Brunet, M.V. Liga, D. Li, and P.J.J. Alvarez. 2008a. Antimicrobial nanomaterials for water disinfection and microbial control: potential applications and implications. *Water Res.* 42:4591–4602.

Li, Q., M.A. Page, B.J. Marinas, and J.K. Shang. 2008b. Treatment of coliphage MS2 with palladium-modified nitrogen-doped titanium oxide photocatalyst illuminated by visible light. *Environ. Sci. Technol.* 42:6148–6153.

Li, D., S.A. Craik, D.W. Smith, and M. Belosevic. 2009. Infectivity of *Giardia lamblia* cysts obtained from wastewater treated with ultraviolet light. *Water Res.* 43:3037–3046.

Li, L., N. Gao, Y. Deng, J. Yao, and K. Zhang. 2012. Characterization of intracellular & extracellular algae organic matters (AOM) of *Microcystis aeruginosa* and formation of AOM-associated disinfection byproducts and odor & taste compounds. *Water Res.* 46:1233–1240.

van Lieverloo, J.H.M., D. van der Kooij, and W. Hoogenboezem. 2002. Invertebrates and protozoa (free living) in drinking water distribution systems. In *Encyclopedia of Environmental Microbiology*, G. Bitton (Editor-in-Chief). Wiley-Interscience, New York, pp. 1718–1733.

van Lieverloo, J.H.M., D.W. Bosboom, G.L. Bakker, A.J. Brouwer, R. Voogt, and J.E.M. De Roos. 2004. Sampling and quantifying invertebrates from drinking water distribution mains. *Water Res.* 38:1101–1112.

van Lieverloo, J.H.M., W. Hoogenboezem, G. Veenendaal, and D. van der Kooij. 2012. Variability of invertebrate abundance in drinking water distribution systems in the Netherlands in relation to biostability and sediment volumes. *Water Res.* 46:4918–4932.

Lim, D.V., J.M. Simpson, E.A. Kearns, and M.F. Kramer. 2005. Current and developing technologies for monitoring agents of bioterrorism and biowarfare. *Clin. Microbiol. Rev.* 18(4):583–607.

Lin, S., R. Huang, Y. Cheng, J. Liu, B.L.T. Lau, and M.R. Wiesner. 2013. Silver nanoparticle-alginate composite beads for point-of-use drinking water disinfection. *Water Res.* 47:3959–3965.

Linden, K.G., J. Thurston, R. Schaefer, and J.P. Malley, Jr. 2007. Enhanced UV inactivation of adenoviruses under polychromatic UV lamps. *Appl. Environ. Microbiol.* 73:7571–7574.

Linden, K.G., G.-A. Shin, J.-K. Lee, K. Scheible, C. Shen, and P. Posy. 2009. Demonstrating 4-log adenovirus inactivation in a medium-pressure UV disinfection reactor. *J. Am. Water Works Assoc.* 101(4):90–99.

Lindström, M., R. Keto, A. Markkula, M. Nevas, S. Hielm, and H. Korkeala. 2001. Multiplex PCR assay for detection and identification of *Clostridium botulinum* types A, B, E, and F in food and fecal material. *Appl. Environ. Microbiol.* 67:5694–5699.

Linke, S., J. Lenz, S. Gemein, M. Exner, and J. Gebel. 2010. Detection of *Helicobacter pylori* in biofilms by real-time PCR. *Int. J. Hyg. Environ. Health* 213:176–182.

Linville, T.J. and K.A. Thompson. 2006. Protecting the security of our nation's water systems: challenges and successes. *J. Am. Water Works Assoc.* 98(3):234–241.

Liu, X. and R.M. Slawson. 2001. Factors affecting drinking water biofiltration. *J. Am. Water Works Assoc.* 93(12):90–101.

Liu, W., H. Wu, Z. Wang, S.L. Ong, J.Y. Hu, and W.J. Ng. 2002. Investigation of assimilable organic carbon (AOC) and bacterial regrowth in drinking water distribution system. *Water Res.* 36:891–898.

Liu, R., Z. Yu, H. Zhang, M. Yang, B. Shi, and X. Liu. 2012a. Diversity of bacteria and mycobacteria in biofilms of two urban drinking water distribution systems. *Can. J. Microbiol.* 58:261–270.

Liu, P., Q. Huang, and W. Chen. 2012b. Heterologous expression of bacterial nitric oxide syn-thase gene: a potential biological method to control biofilm development in the environment. *Can. J. Microbiol.* 58:336–344.

Liu, G., F.Q. Ling, A. Magic-Knezev, W.T. Liu, J.Q.J.C. Verberk, and J.C. Van Dijk. 2013. Quantification and identification of particle-associated bacteria in unchlorinated drinking water from three treatment lants by cultivation-independent methods. *Water Res.* 47:3523–3533.

Locas, A., B. Barbeau, and V. Gauthier. 2007. Nematodes as a source of total coliforms in a distribution system. *Can. J. Microbiol.* 53(5):580–585.

Locas, A., J. Demers, and P. Payment. 2008. Evaluation of photoreactivation of *Escherichia coli* and enterococci after UV disinfection of municipal wastewater. *Can. J. Microbiol.* 54:971–975.

Lodder, W.J., H.H.J.L. van den Berg, S.A. Rutjes, and A.M. de Roda Husman. 2010. Presence of enteric viruses in source waters for drinking water production in the Netherlands. *Appl. Environ. Microbiol.* 76:5965–5971.

Logsdon, G.S. and J.C. Hoff. 1986. Barriers to the transmission of waterborne disease. In *Waterborne Diseases in the United States*, G.F. Craun (Ed.). CRC Press, Boca Raton, FL, pp. 255–274.

Logsdon, G.S. and E.C. Lippy. 1982. The role of filtration in preventing waterborne disease. *J. Am. Water Works Assoc.* 74:649–655.

Logsdon, G.S., J.M. Symons, R.L. Hoye, Jr., and M.M. Arozarena. 1981. Alternative filtration methods for removal of *Giardia* cysts and cysts models. *J. Am. Water Works Assoc.* 73:111–118.

Logsdon, G.S., M.B. Hosley, S.D.N. Freeman, J.J. Neemann, and G.C. Budd. 2006. Filtration processes—a distinguished history and a promising future. *J. Am. Water Works Assoc.* 98(3):150–162.

Longley, K.E., B.E. Moore, and C.A. Sorber. 1980. Comparison of chlorine and chlorine dioxide as disinfectants. *J. Water Pollut. Control Fed.* 52:2098–2105.

Lonigro, A., A. Pollice, R. Spinelli, F. Berrilli, D. Di Cave, C. D'Orazi, P. Cavallo, and O. Brandonisio. 2006. *Giardia* cysts and *Cryptosporidium* oocysts in membrane-filtered municipal wastewater used for irrigation. *Appl. Environ. Microbiol.* 72:7916–7918.

Loo, S.-L., A.G. Fane, T.-T. Lim, W.B. Krantz. Y.-N. Liang, X. Liu, and X. Hu. 2013. Super-absorbent cryogels decorated with silver nanoparticles as a novel water technology for point-of-use disinfection. *Environ. Sci. Technol.* 47:9363–9371.

Lopez, A.S., D.R. Dodson, M.J. Arrowood, P.A. Orlandi, Jr., A.J. da Silva, J.W. Bier, S.D. Hanauer, R.L. Kuster, S. Oltman, M.S. Baldwin, K.Y. Won, E.M. Nace, M.L. Eberhard, and B.L. Herwaldt. 2001. Outbreak of cyclosporiasis associated with basil in Missouri in 1999. *Clin. Infect. Dis.* 32:1010–1017.

Lopez, G.U., C.P. Gerba, A.H. Tamimi, M. Kitajima, S.L. Maxwell, and J.B. Rose. 2013. Transfer efficiency of bacteria and viruses from porous and nonporous fomites to fin-gers under different relative humidity conditions. *Appl. Environ. Microbiol.* 79:5728–5734.

Lopman, B.A., D.W. Brown, and M. Koopmans. 2003. Caliciviruses: occurrence in Europe. In *Encyclopedia of Environmental Microbiology*, G. Bitton (Editor-in-Chief), Internet Edition. Wiley-Interscience, New York.

Lorch, W. 1987. *Handbook of Water Purification*, 2nd edition. Ellis Horwood Ltd, Chichester, UK.

Loret, J.-F. and G. Greub. 2010. Free-living amoebae: biological by-passes in water treatment. *Int. J. Hyg. Environ. Health* 213:167–175.

Loy, A., W. Beisker, and H. Meier. 2005. Diversity of bacteria growing in natural mineral water after bottling. *Appl. Environ. Microbiol.* 71:3624–3832.

Lu, W. and X.-J. Zhang. 2005. Factors affecting bacterial growth in drinking water distribution system. *Biomed. Environ. Sci.* 18:137–140.

Lui, Y.S., J.W. Qiu, Y.L. Zhang, M.H. Wong, and Y. Liang. 2011. Algal-derived organic matter as precursors of disinfection by-products and mutagens upon chlorination. *Water Res.* 45:1454–1462.

Lv, Y., H. Liu, Z. Wang, S. Liu, L. Hao, Y. Sang, D. Liu, J. Wang, and R.I. Boughton. 2009. Silver nanoparticle-decorated porous ceramic composite for water treatment. *J. Memb. Sci.* 331:50–56.

MacKenzie, W.R., N.J. Hoxie, M.E. Proctor, M.S. Gradus, K.A. Blair, D.E. Peterson, J.J. Kazmierezak, D.G. Addiss, K.R. Fox, J.B. Rose, and J.P. Davis. 1994. A massive outbreak in Milwaukee of *Cryptosporidium* infection transmitted through the public water supply. *N. Engl. J. Med.* 331:161–167.

Madaeni, S.S., A.G. Fane, and G.S. Grohmann. 1995. Virus removal from water and wastewater using membranes. *J. Memb. Sci.* 102:65–75.

Madge, B.A. and J.N. Jensen. 2002. Disinfection of wastewater using a 20-kHz ultrasound unit. *Water Environ. Res.* 74:159–169.

Madge, B.A. and J.N. Jensen. 2006. Ultraviolet disinfection of fecal coliform in municipal wastewater: effects of particle size. *Water Environ. Res.* 78(3):294–304.

Magic-Knezev, A. and D. Van der Kooij. 2004. Optimisation and significance of ATP analysis for measuring active biomass in granular activated carbon filters used in water treatment. *Water Res.* 38:3971–3979.

Magic-Knezev, A., B. Wullings, and D. Van der Kooij. 2009. *Polaromonas* and *Hydrogenophaga* species are the predominant bacteria cultured from granular activated carbon filters in water treatment. *J. Appl. Microbiol.* 107:1457–1467.

Mahbubani, M.H., A.K. Bej, M. Perlin, F.W. Schaefer, III, W. Jakubowski, and R.M. Atlas. 1991. Detection of *Giardia* cysts by using the polymerase chain reaction and distinguishing live from dead cysts. *Appl. Environ. Microbiol.* 57:3456–3461.

Mahoney, F.J., T.A. Farley, K.Y. Kelso, S.A. Wilson, J.M. Horan, and L.M. McFarland. 1992. An outbreak of hepatitis A associated with swimming in a public pool. *J. Infect. Dis.* 165:613–618.

Maier, R.M., I.L. Pepper, and C.P. Gerba. 2009. *Environmental Microbiology*, 2nd edition. Academic Press, Burlington, MA, 598 pp.

Makino, S.-I. and H.-I. Cheun. 2003. Application of the real-time PCR for the detection of airborne microbial pathogens in reference to the anthrax spores. *J. Microbiol. Methods* 53:141–147.

Malato, S., P. Fernández-Ibáñez, M.I. Maldonado, J. Blanco, and W. Gernjal. 2009. Decontamination and disinfection of water by solar photocatalysis: recent overview and trends. *Catal. Today* 147:1–59.

Malone, A.S., Y-K. Chung, and A.E. Youssef. 2006. Genes of *Escherichia coli* O157:H7 that are Involved in high-pressure resistance. *Appl. Environ. Microbiol.* 72:2661–2671.

Manage, P.M., C. Edwards, B.K. Singh, and L.A. Lawton. 2009. Isolation and identification of novel microcystin-degrading bacteria. *Appl. Environ. Microbiol.* 75:6924–6928.

Manem, J.A. and B.E. Rittmann. 1992. Removing trace-level organic pollutants in a biological filter. *J. Am. Water Works Assoc.* 84:152–157.

Manjon, F., L. Villen, D. Garcia-Fresnadillo, and G. Orellana. 2008. On the factors influencing the performance of solar reactors for water disinfection with photosensitized singlet oxygen. *Environ. Sci. Technol.* 42:301–307.

Marciano-Cabral, F. 1988. Biology of *Naegleria* spp. *Microbiol. Rev.* 52:114–133.

Marois-Fiset, J.-T., A. Carabin, A. Lavoie, and C.C. Dorea. 2013. Effects of temperature and pH on reduction of bacteria in a point-of-use drinking water treatment product for emergency relief. *Appl. Environ. Microbiol.* 79:2107–2109.

Marquez, B. 2005. Bacterial efflux systems and efflux pump inhibitors. *Biochimie* 87:1137–1147.

Marshall, J.A., L.D. Bruggink, K. Sturge, N. Subasinghe, A. Tan, and G.G. Hogg. 2007. Molecular features of astrovirus associated with a gastroenteritis outbreak in an aged-care centre. *Eur. J. Clin. Microbiol. Infect. Dis.* 26:67–71.

Marston, B.J., J.F. Plouffe, and T.M. File, Jr. 1997. Incidence of community acquired pneumonia requiring hospitalization: results of a population based active surveillance study in Ohio. *Arch. Internal Med.* 157:1709–1718.

Martin, N. and R. Gehr. 2007. Reduction of photoreactivation with the combined UV/peracetic acid process or delayed exposure to visible light. *Water Environ. Res.* 79(9):991–999.

Martinez, J.L., M.B. Sanchez, L. Martınez-Solano, A. Hernandez, L. Garmendia, A. Fajardo, and C. Alvarez-Ortega. 2009. Functional role of bacterial multidrug efflux pumps in microbial natural ecosystems. *FEMS Microbiol. Rev.* 33:430–449.

Martiny, A.C., T.M. Jorgensen, H.-J. Albrechtsen, E. Arvin, and S. Molin. 2003. Long-term succession of structure and diversity of a biofilm formed in a model drinking water distribution system. *Appl. Environ. Microbiol.* 69:6899–6907.

Masago, Y., H. Katayama, T. Watanabe, E. Haramoto, A. Hashimoto, T. Omura, T. Hirata, and S. Ohgaki. 2006. Quantitative risk assessment of noroviruses in drinking water based on qualitative data in Japan. Environ. *Sci. Technol.* 40:7428–7433.

Mateju, V., S. Cizinska, J. Krejei, and T. Janoch. 1992. Biological water denitrification: a review. *Enzyme Microb. Technol.* 14:170–183.

Matsui, S.M. 1995. Astroviruses. In *Infections of the Gastrointestinal Tract*, M.J. Blaser, P.D. Smith, J.I. Ravdin, H.B. Greenberg, and R.L. Guerrant (Eds). Raven Press, New York, pp. 1035–1045.

Matsuki, T., K. Watanabe, J. Fujimoto, Y. Miyamoto, T. Takada, K. Matsumoto, H. Oyaizu, and R. Tanaka. 2002. Development of 16S rRNA-gene-targeted group-specific primers for the detection and identification of predominant bacteria in human feces. *Appl. Environ. Microbiol.* 68:5445–5451.

Mattle, M.J. and T. Kohn. 2012. Inactivation and tailing during UV254 disinfection of viruses: contributions of viral aggregation, light shielding within viral aggregates, and recombination. *Environ. Sci. Technol.* 46:10022–10030.

Mayer, C.L. and C.J. Palmer. 1996. Evaluation of PCR, nested PCR, and fluorescent antibodies for detection of *Giardia* and *Cryptosporidium* species in wastewater. *Appl. Environ. Microbiol.* 62:2081–2085.

Mbonimpa, E.G., B. Vadheim, and E.R. Blatchley, III. 2012. Continuous-flow solar UVB disinfection reactor for drinking water. *Water Res.* 46:2344–2354.

McBride, G.B., D. Till, T. Ryan, A. Ball, G. Lewis, S. Palmer, and P. Weinstein. 2002. *Freshwater Microbiology Research Programme. Pathogen Occurrence and Human Health Risk Assessment Analysis.* Technical Publication. Ministry for the Environment, Wellington, New Zealand, 93 pp. http://www.mfe.govt.nz/publications/water/freshwater-microbiology-nov02/

McCuin, R.M. and J.L. Clancy. 2003. Modifications to United States Environmental Protection Agency methods 1622 and 1623 for detection of *Cryptosporidium* oocysts and *Giardia* cysts in water. *Appl. Environ. Microbiol.* 69:267–274.

McDowall, B., D. Hoefel, G. Newcombe, C.P. Saint, and L. Ho. 2009. Enhancing the biofiltration of geosmin by seeding sand filter columns with a consortium of geosmin-degrading bacteria. *Water Res.* 43:433–440.

McFeters, G.A. (Ed.). 1990. *Drinking Water Microbiology.* Springer-Verlag, New York, 502 pp.

McFeters, G.A. and A.K. Camper. 1988. Microbiological analysis and testing. In *Bacterial Regrowth in Distribution Systems*, W.G. Characklis, (Ed.). Research Report. AWWA Research Foundation, Denver, CO, pp. 73–95.

McGuire, M.J., X. Wu, N.K. Blute, D. Askenaizer, and G. Qin. 2009. Prevention of nitrification using chlorite ion: results of a demonstration project in Glendale, Calif. *J. Am. Water Works Assoc.* 101(10):41–59.

McLean, R.J., M. Whiteley, D.J. Stickler, and W.C. Fuqua. 1997. Evidence of autoinducer activity in naturally occurring biofilms. *FEMS Microbiol. Lett.* 154:259–263.

McLean, R.J.C., C.L. Bates, M.B. Barnes, C.L. McGowin, and G.M. Aron. 2004. Methods of studying biofilms. In *Microbial Biofilms*, M. Ghannoum and G.A. O'Toole (Eds). ASM Press, Washington, DC, pp. 379–413.

McLennan, S.D., L.A. Peterson, J.B. Rose. 2009. Comparison of point-of-use technologies for emergency disinfection of sewage-contaminated drinking water. *Appl. Environ. Microbiol.* 75:7283–7286.

McRae, B.M., T.M. LaPara, and R.M. Hozalski. 2004. Biodegradation of haloacetic acids by bacterial enrichment cultures. *Chemosphere* 55:915–922.

Meadows, C.A. and B.H. Snudden. 1982. Prevalence of *Yersinia enterocolitica* in waters at the lower Chippewa River basin, Wisconsin. *Appl. Environ. Microbiol.* 43:953–954.

Medema, G.J., P. Payment, A. Dufour, W. Robertson, M. Waite, P. Hunter, R. Kirby, and Y. Andersson. 2003. Safe drinking water: an ongoing challenge. In *Assessment of Risk Assessing Microbial Safety of Drinking Water: Improving Approaches and Methods*, A. Dufour, M. Snozzi, W. Koster, J. Bartram, E. Ronchi, and L. Fewtrell (Eds). Published on behalf of the World Health Organization by IWA Publishing. ISBN 92 4 154630 1(WHO) ISBN 1 84339 036 1(IWA Publishing), pp. 11–45.

Meier, J.R., R.B. Knohl, W.E. Coleman, H.R. Ringhand, J.W. Munch, W.H. Kaylor, R.P. Streicher, and F.C. Kopfler. 1987. Studies on the potent bacterial mutagen, 3-chloro-4-(dichloromethyl)-5-hydroxy-2(5H)-furanone: aqueous stability, XAD-recovery and analytical determination in drinking water and in chlorinated humic acid solution. *Mutat. Res.* 189:363–373.

Melamed, S., T. Elad, and S. Belkin. 2012. Microbial sensor cell arrays. *Curr. Opin. Biotechnol.* 23:2–8.

Mena, K.D., C.P. Gerba, C.N. Haas, and J.B. Rose. 2003. Risk assessment of waterborne coxsackievirus. *J. Am. Water. Works Assoc.* 95(7):122–129.

Meng, Q.S. and C.P. Gerba. 1996. Comparative inactivation of enteric adenoviruses, polioviruses and coliphages by ultraviolet irradiation. *Water Res.* 30:2665–2668.

Meschke, J.S. and M.D. Sobsey. 2002. Norwalk-like viruses: detection methodologies and environmental fate. In *Encyclopedia of Environmental Microbiology*, G. Bitton (Editor-in-Chief). Wiley-Interscience, New York, pp. 2221–2235.

Messner, M.J., C.L. Chappell, and P.C. Okhuysen. 2001. Risk assessment for *Cryptosporidium*: a hierarchical Bayesian analysis of human dose response data. *Watar Res.* 35:3934–3940.

Mezule, L., S. Larsson, and T. Juhna. 2013. Application of DVC-FISH method in tracking *Escherichia coli* in drinking water distribution networks. *Drink. Water. Eng. Sci. Discuss.* 6:25–31.

Michen, B., J. Fritsch, C. Aneziris, and T. Graule. 2013. Improved virus removal in ceramic depth filters modified with MgO. *Environ. Sci. Technol.* 47:1526–1533.

Mielke, R.E., J.H. Priester, R.A. Werlin, J. Gelb, A.M. Horst, E. Orias, and P.A. Holden. 2013. Differential growth of and nanoscale TiO$_2$ accumulation in *Tetrahymena thermophila* by direct feeding versus trophic transfer from *Pseudomonas aeruginosa*. *Appl. Environ. Microbiol.* 79:5616–5624.

Miettinen, I.T., T. Vartiainen, and P.J. Martikainen. 1997. Phophorus and bacterial growth in drinking water. *Appl. Environ. Microbiol.* 63:3242–3245.

Miettinen, I.T., K. Malaska, L. Korhonen, U. Lignell, P. Karkkainen, H. Rintala, E. Kauhanen, and A. Nevalainen. 2007. Occurrence of fungi and *Actinomyces* in Finnish drinking waters; In *Proceedings of the World Environmental and Water Resources Congress*, Tampa, FL. http://dx.doi.org/10.1061/40927(243)483.

Mikol, Y.B., W.R. Richardson, W.H. van der Schalie, T.R. Shedd, and M.W. Widder. 2007. An online real-time biomonitor for contaminant surveillance in water supplies. *J. Am. Water Works Assoc.* 99(2):107–115.

Milferstedt, K., M.-N. Pons, and E. Morgenroth. 2006. Optical method for long-term and large-scale monitoring of spatial biofilm development. *Biotechnol. Bioeng.* 94:773–782.

Miller, M.B. and B.L. Bassler. 2001. Quorum sensing in bacteria. *Annu. Rev. Microbiol.* 55:165–199.

Miltner, E.C. and L.E. Bermudez. 2000. *Mycobacterium avium* grown in *Acanthamoeba castellanii* is protected from the effects of antimicrobials. *Antimicrob. Agents Chemother.* 44:1990–1994.

Mintz, E., J. Bartram, P. Lochery, and M. Vegelin. 2001. Not Just a drop in the bucket: expanding access to point-of-use water treatment systems. *Am. J. Public. Health* 91:1565–1570.

Misund, A., B. Frengstad, U. Siewers, and C. Reimann. 1999. Variation of 66 elements in European bottled mineral waters. *Sci. Total Environ.* 244:21–41.

Mitchell, D.K., D.O. Matson, X. Jiang, T. Berke, S.S. Monroe, M.J. Carter, M.M. Willcocks, and L.K. Pickering. 1999. Molecular epidemiology of childhood astrovirus infection in child care centers. *J. Infect. Dis.* 180:514–517.

Moe, C.L. 1997. Waterborne transmission of infectious agents. In *Manual of Environmental Microbiology*, C.J. Hurst, G.R. Knudsen, M.J. McInerney, L.D. Stetzenbach, and M.V. Walter (Eds). ASM Press, Washington, DC, pp. 136–152.

Mofidi, A.A., H. Baribeau, P.A. Rochelle, R. De Leon, B.M. Coffey, and J.F. Green. 2001. Disinfection of *Cryptosporidium parvum* with polychromatic UV light. *J. Am. Water Works Assoc.* 93(6):95–109.

Mofidi, A.A., E.A. Meyer, P.M. Wallis, C.I. Chou, B.P. Meyer, S. Ramalingam, and B.M. Coffey. 2002. The effect of UV light on the inactivation of *Giardia lamblia* and *Giardia muris* cysts as determined by animal infectivity assay (P-2951-01). *Water Res.* 36:2098–2108.

Montgomery, M.A. and M. Elimelech. 2008. Water and sanitation in developing countries: including health in the equation. *Environ. Sci. Technol.* 41:17–24.

Moore, M.R., M. Pryor, B. Fields, C. Lucas, M. Phelan, and R.E. Besser. 2006. Introduction of monochloramine into a municipal water system: impact on colonization of buildings by *Legionella* spp. *Appl. Environ. Microbiol.* 72:378–383.

Morel, N., H. Volland, J. Dano, P. Lamourette, P. Sylvestre, M. Mock, and C. Creminion. 2012. Fast and sensitive detection of *Bacillus anthracis* spores by immunoassay. *Appl. Environ. Microbiol.* 78:6491–6498.

Moreno, Y., P. Piqueres, J.L. Alonso, A. Jiménez, A. González, and M.A. Ferrús. 2007. Survival and viability of *Helicobacter pylori* after inoculation into chlorinated drinking water. *Water Res.* 41:3490–3496.

Morin, P., A. Camper, W. Jones, D. Gatel, and J.C. Goldman. 1996. Colonization and disinfection of biofilms hosting coliform-colonized carbon fines. *Appl. Environ. Microbiol.* 62:4428–4432.

Morita, S., A. Namikoshi, T. Hirata, K. Oguma, H. Katayama, S. Ohgaki, N. Motoyama, and M. Fujiwara. 2002. Efficacy of UV irradiation in inactivating *Cryptosporidium parvum* Oocysts. *Appl. Environ. Microbiol.* 68:5387–5393.

Mota, P.Q.F. and S.C. Edberg. 2002. Nosocomial infections. In *Encyclopedia of Environmental Microbiology*, G. Bitton (Editor-in-Chief). Wiley-Interscience, New York, pp. 2235–2250.

Moyad, M.A. 2003. What do I tell my patients about drinking water and the risk of bladder cancer? *Urol. Nurs.* 23(5):371–377.

Muhammad, N., R. Sinha, E.R. Krishnan, H. Piao, and C.L. Patterson. 2008. Evaluating surrogates for *Cryptosporidium* removal in point-of-use systems. *J. Am. Water Works Assoc.* 100(12):98–106.

Muhsen, K. and D. Cohen. 2013. Letter to the editor: *Helicobacter pylori* infection and anemia. *Am. J. Trop. Med. Hyg.* 89:398.

Muniesa, M., J. Jofre, C. Garcia-Aljaro, and A.R. Blanch. 2006. Occurrence of *Escherichia coli* O157:H7 and other enterohemorrhagic *Escherichia coli* in the environment. *Environ. Sci. Technol.* 40:7141–7149.

Muniesa, M., J.A. Hammerl, S. Hertwig, B.A. Appel, and H. Brussow. 2012. Shiga toxin-producing *Escherichia coli* O104:H4: a new challenge for microbiology. *Appl. Environ. Microbiol.* 78:4065–4073.

Muraca, P., J.E. Stout, and V.L. Yu. 1987. Comparative assessment of chlorine, heat, ozone, and UV light for killing *Legionella pneumophila* within a model plumbing system. *Appl. Environ. Microbiol.* 53:447–453.

Muraca, P., V.L. Yu, and J.E. Stout. 1988. Environmental aspects of Legionnaires' Disease. *J. Am. Water Works Assoc.* 80:78–86.

Murray, G.E., R.S. Tobin, B. Junkins, and D.J. Kushner. 1984. Effect of chlorination on antibiotic resistance profiles of sewage-related bacteria. *Appl. Environ. Microbiol.* 48:73–77.

Myint, K.S.A., J.R. Campbell, and A.L. Corwin. 2002. Hepatitis viruses (HAV-HEV). In *Encyclopedia of Environmental Microbiology*, G. Bitton (Editor-in-Chief). Wiley-Interscience, New York, pp. 1530–1540.

Nadan, S., J.E. Walter, W.O.K. Grabow, D.K. Mitchell, and M.B. Taylor. 2003. Molecular characterization of astroviruses by reverse transcriptase PCR and sequence analysis: comparison of clinical and environmental isolates from South Africa. *Appl. Environ. Microbiol.* 69:747–753.

Nagy, L.A. and B.H. Olson. 1982. The occurrence of filamentous fungi in drinking water distribution systems. *Can. J. Microbiol.* 28:667–671.

Najm, I.N. and S.W. Krasner. 1995. Effects of bromide and natural organic matter on the formation of ozonation by-products. *J. Am. Water Works Assoc.* 87:1–15.

Najm, I. and R.R. Trussell. 2001. NDMA formation in water and wastewater. *J. Am. Water Works Assoc.* 93(2):92–99.

Najm, I.M., V.M. Snoeyink, B.W. Lykins, Jr., and J.Q. Adams. 1991. Using powdered activated carbon: a critical review. *J. Am. Water Works Assoc.* 83:65–76.

Namkung, E. and B.E. Rittmann. 1987. Removal of taste- and odor-causing compounds by biofims grown on humic substances. *J. Am. Water Works Assoc.* 79:107–112.

Naranjo, J.E., Cl. Chaidez, M. Quinõnez, C.P. Gerba, J. Olson, and J. Dekko. 1997. Evaluation of a portable water purification system for the removal of enteric pathogens. *Water Sci. Technol.* 35(11–12):55–59.

Narkis, N. and Y. Kott. 1992. Comparison between chlorine dioxide and chlorine for use as a disinfectant of wastewater effluents. *Water Sci. Technol.* 26:1483–1492.

Nataro, J.P. and J.B. Kaper. 1998. Diarrheagenic *Escherichia coli. Clin. Microbiol. Rev.* 11:142–201.

National Academy of Sciences (NAS). 1983. *Risk Assessment in the Federal Government: Managing the Process.* National Academy Press, Washington, DC.

National Research Council (NRC). 2006. *Drinking Water Distribution Systems: Assessing and Reducing Risks.* NRC, Washington, DC, 400 pp. http://www.nap.edu/catalog/11728.html (accessed May 5, 2014).

Neilson, A.H., A.S. Allard, and M. Remberger. 1985. Biodegradation and transformation of recalcitrant compounds. In *Handbook of Environmental Chemistry*, O. Hutzinger (Ed.). Springer, New York.

Nenonen, N.P., C. Hannoun, P. Horal, B. Hernroth, and T. Bergstrom. 2008. Tracing of norovirus outbreak strains in mussels collected near sewage effluents. *Appl. Environ. Microbiol.* 74:2244–2249.

Nerenberg, R., B.E. Rittmann, and W.J. Soucie. 2000. Ozone/biofiltration for removing MIB and geosmin. *J. Am. Water Works Assoc.* 92(12):85–95.

Neu, T.R. and J.R. Lawrence. 2002. Laser scanning microscopy in combination with fluorescence techniques for biofilm study. In *Encyclopedia of Environmental Microbiology*, G. Bitton (Editor-in-Chief). Wiley-Interscience, New York, pp. 1772–1788.

Neumann, U. and Weckesser, J. 1998. Elimination of microcystin peptide toxins from water by reverse osmosis. *Environ. Toxicol. Water Qual.* 13:143–148.

Neumeister, B., S. Schoniger, M. Faigle, M. Eichner, and K. Dietz. 1997. Multiplication of different *Legionella* species in mono Mac 6 cells and in *Acanthamoeba castellani. Appl. Environ. Microbiol.* 63:1219–1224.

Nichols, R.A.B., L. Connelly, C.B. Sullivan, and H.V. Smith. 2010. Identification of *Cryptosporidium* species and genotypes in Scottish raw and drinking waters during a one-year monitoring period. *Appl. Environ. Microbiol.* 76:5977–5986.

Nicholson, W.L. and B. Galeano. 2003. UV resistance of *Bacillus anthracis* spores revisited: validation of *Bacillus subtilis* spores as UV surrogates for spores of *B. anthracis* Sterne. *Appl. Environ. Microbiol.* 69:1327–1330.

Ndiongue, S., W.B. Anderson, A. Tadwalkar, J. Rudnickas, M. Lin, and P.M. Huck. 2006. Using pilot-scale investigations to estimate the remaining geosmin and MIB removal capacity of full-scale GAC-capped drinking water filter. *Water Qual. Res. J. Canada* 41(3):296–306.

Niemi, R.M., S. Knuth, and K. Lundstrom. 1982. Actinomycetes and fungi in surface waters and in potable water. *Appl. Environ. Microbiol.* 43:378–388.

Niquette, P., P. Servais, and R. Savoir. 2000. Impacts of pipe materials on densities of fixed bacterial biomass in a drinking water distribution system. *Water Res.* 34:1952–1956.

Nivens, D.E., J.Q. Chambers, T.R. Anderson, and D.C. White. 1993. Long-term, on-line monitoring of microbial biof ilms using a quartz crystal microbalance. *Anal. Chem.* 65:65–69.

Nivens, D.E., R.J. Palmer, and D.C. White. 1995. Continuous nondestructive monitoring of microbial biofilms: a review of analytical techniques. *J. Ind. Microbiol.* 15:263–276.

Njoroge, J. and V. Sperandio. 2009. Jamming bacterial communication: new approaches for the treatment of infectious diseases. *EMBO Mol. Med.* 1:201–210.

Noble, P.A., H.F. Ridgway, and B.H. Olson. 1994. Incorporation of the luciferase genes into *Pseudomonas fluorescens* strain P17: development of a bioluminescent sensor for assimilable organic carbon (Abstr. Q-351). In: *94th Meeting American Society Microbiology*, Las vegas, Nevada, May 23–27, 1994.

Norton, C.D. and M.W. LeChevallier. 1997. Chloramination: its effect on distribution system water quality. *J. Am. Water Works Assoc.* 89:66–77.

Norton, C.D. and M.W. LeChevallier. 2000. A pilot study of bacteriological population changes through potable treatment and distribution. *Appl. Environ. Microbiol.* 66:268–276.

Noss, C.I., F.S. Hauchman, and V.P. Olivieri. 1986. Chlorine dioxide reactivity with proteins. *Water Res.* 20:351–356.

Nuanualsuwan, S. and D.O. Cliver. 2003. Infectivity of RNA from inactivated poliovirus. *Appl. Environ. Microbiol.* 69:1629–1632.

Nuhoglu, A., T. Pekdemir, E. Yildiz, B. Keskinler, and G. Akay. 2002. Drinking water denitrification by a membrane bio-reactor. *Water Res.* 36:1155–1166.

Nwachcuku, N. and C.P Gerba. 2004. Emerging waterborne pathogens: can we kill them all? *Curr. Opin. Biotechnol.* 15:175–180.

Oates, P.M., P. Shanahan, and M.F. Polz. 2003. Solar disinfection (SODIS): simulation of solar radiation for global assessment and application for point-of-use water treatment in Haiti. *Water Res.* 37:47–54.

Odeymi, O. 1990. Use of solar radiation for drinking water disinfection in West Africa. (Abstr.). In *International Symposium Health-Related Microbiology*, Tubingen, Germany, April 1–6, 1990.

Oh, H.-S., K.-M. Yeon, C.-S. Yang, S.-R. Kim, C.-H. Lee, S.-Y. Park, J.Y. Han, and J.-K. Lee. 2012. Control of membrane biofouling in MBR for wastewater treatment by quorum quenching bacteria encapsulated in microporous membrane. *Environ. Sci. Technol.* 46:4877–4884.

Okinaka, R.T., K. Cloud, O. Hampton, A.R. Hoffmaster, K.K. Hill, P. Keim, T.M. Koehler, G. Lamke, S. Kumano, J. Mahillon, D. Manter, Y. Martinez, D. Ricke, R. Svensson, and P.J. Jackson. 1999. Sequence and organization of pXO1, the large *Bacillus anthracis* plasmid harboring the anthrax toxin genes. *J. Bacteriol.* 181:6509–6515.

Okoli, C., M. Boutonnet, S. Järås, and G. Rajarao-Kuttuva. 2012. Protein-functionalized magnetic iron oxide nanoparticles: time efficient potential-water treatment. *J. Nanopart. Res.* 14:1194–1203.

Oliveira, D.P., P.A. Carneiro, C.M. Rech, M.V.B. Zanoni, L.D. Claxton, and G.A. Umbuzeiro. 2006. Mutagenic compounds generated from the chlorination of disperse azo-dyes and their presence in drinking water. *Environ. Sci. Technol.* 40:6682–6689.

Oliver, J.D. 2005. The viable but nonculturable state in bacteria. *J. Microbiol.* 43:93–100.

Ollos, P.J., P.M. Huck, and R.M. Slawson. 2003. Factors affecting biofilm accumulation in model distribution systems. *J. Am. Water Works Assoc.* 95:87–97.

Olson, B.H. and L.A. Nagy. 1984. Microbiology of potable water. *Adv. Appl. Microbiol.* 30:73–132.

Olson, B.H., R. McCleary, and J. Meeker. 1991. Background and models for bacterial biofilm formation and function in water distribution systems. In *Modeling the Environmental Fate of Microorganisms*, C.J. Hurst (Ed.). American Society for Microbiology, Washington, DC, pp. 255–285.

Ongerth, J.E. and P.E. Hutton. 1997. DE filtration to remove *Cryptosporidium*. *J. Am. Water Works Assoc.* 89(12):39–46.

Ono, K., H. Tsuji, S.K. Rai, A. Yamamoto, K. Masuda, T. Endo, H. Hotta, T. Kawamura, and S.i. Uga. 2001. Contamination of river water by *Cryptosporidium parvum* oocysts in western Japan. *Appl. Environ. Microbiol.* 67:3832–3836.

Onstad, G.D., S. Strauch, J. Meriluoto, G.A. Codd, and U. von Gunten. 2007. Selective oxidation of key functional groups in cyanotoxins during drinking water ozonation. *Environ. Sci. Technol.* 41:4397–4404.

Onstad, G., H. Weinberg, and S.W. Krasner. 2008. Occurrence of halogenated furanones in U.S. drinking waters. *Environ. Sci. Technol.* 42:3341–3348.

Ortega, Y.R., V.A. Cama, and A.B. Prisma. 2002. Cyclospora: basic biology, occurrence, fate and methodologies. In *Encyclopedia of Environmental Microbiology*, G. Bitton (Editor-in-Chief). Wiley-Interscience, Hoboken, NJ, pp. 995–1001.

Oskam, G. 1995. Main principles of water quality improvement in reservoirs. *Aqua* 44:23–29.

Osswald, J., S. Rellan, A. Gago, and V. Vasconcelos. 2007. Toxicology and detection methods of the alkaloid neurotoxin produced by cyanobacteria, anatoxin-a. *Environ. Int.* 33:1070–1089.

Otterholt, E. and C. Charnock. 2011. Microbial quality and nutritional aspects of Norwegian brand waters. Int. *J. Food Microbiol.* 144:455–463.

Ovez, B., J. Mergaert, and M. Saglam. 2006. Biological denitrification in drinking water treatment using the seaweed *Gracilaria Verrucosa* as carbon source and biofilm carrier. *Water Environ. Res.* 78(4):430–434.

Ovrutsky, A.R., E.D. Chan, M. Kartalija, X. Bai, M. Jackson, S. Gibbs, J.O. Falkinham, III, M.D. Iseman, P.R. Reynolds, G. McDonnell, and V. Thomas. 2013. Cooccurrence of free-living amoebae and nontuberculous *Mycobacteria* in hospital water networks, and preferential growth of *Mycobacterium avium* in *Acanthamoeba lenticulata*. *Appl. Environ. Microbiol.* 79:3185–3192.

Oyanedel-Craver, V. and J.A. Smith. 2008. Sustainable colloidal-silver-impregnated ceramic filter for point-of-use water treatment. *Environ. Sci. Technol.* 42:927–933.

Oziol, L. and N. Bouaïcha. 2010. First evidence of estrogenic potential of the cyanobacterial heptotoxins the nodularin-R and the microcystin-LR in cultured mammalian cells. *J. Hazard. Mater.* 174(1–3):610–615.

Page, M.A., J.L. Shisler, and B.J. Marinas. 2009. Kinetics iof adenovirus type 2 with free chlorine. *Water Res.* 43:2916–2926.

Pal, S., J. Joardar, and J.M. Song. 2006. Removal of *E. coli* from water using surface-modified activated carbon filter media and its performance over an extended use. *Environ. Sci. Technol.* 40:6091–6097.

Palma, M., D. DeLuca, S. Worgall, and L.E. Quadri. 2004. Transcriptome analysis of the response of *Pseudomonas aeruginosa* to hydrogen peroxide. *J. Bacteriol.* 186:248–252.

Palmateer, G., D. Manz, A. Jurkovic, R. McInnis, S. Unger, K.K. Kwan, and B.J. Dutka. 1999. Toxicant and parasite challenge of Manz intermittent slow sand filter. *Environ. Toxicol.* 14:217–225.

Palmer, S.R., P.R. Gully, J.M. White, A.D. Pearson, W.G. Suckling, D.M. Jones, J.C.L. Rawes, and J.L. Penner. 1983. Waterborne outbreak of *Campylobacter* gastroenteritis. *Lancet* 1:287–290.

Papapetropoulou, M., A. Tsintzou, and A. Vantarakis. 1997. Environmental mycobacteria in bottled table waters in Greece. *Can. J. Microbiol.* 43:499–502.

Parag, Y. and J.T. Roberts. 2009. A battle against the bottles: building, claiming, and regaining tap-water trustworthiness. *Soc. Nat. Resour.* 22:625–636.

Paris, T., S. Skali-Lami, and J.C. Block. 2007. Effect of wall shear rate on biofilm deposition and grazing in drinking water flow chambers. *Biotechnol. Bioeng.* 97(6):1550–1560.

Paris, T., S. Skali-Lamia, and J.-C. Block. 2009. Probing young drinking water biofilms with hard and soft particles. *Water Res.* 43:117–126.

Park, S. 2002. *Campyloacter jejuni* and other enteric campyloacters. In *Encyclopedia of Environmental Microbiology*, G. Bitton (Editor-in-Chief). Wiley-Interscience, New York, pp. 803–810.

Park, S.R., W.G. Mackay, and D.C. Reid. 2001. *Helicobacter* sp. recovered from drinking water biofilm sampled from a water distribution system. *Water Res.* 35:1624–1626.

Parrish, S.C., J. Myers, and A. Lazarus. 2008. Nontuberculous mycobacterial pulmonary infections in non-HIV patients. *Postgrad. Med.* 120:78–86.

Parrotta, M.J. and F. Bekdash. 1998. UV disinfection of small groundwater supplies. *J. Am. Water Works Assoc.* 90:71–81.

Parshionikar, P.U., S. Willian-True, G.S. Fout, D.E. Robbins, S.A. Seys, J.D. Cassady, and R. Harris. 2003. Waterborne outbreak of gastroenteritis associated with a norovirus. *Appl. Environ. Microbiol.* 69:5263–5268.

Paterson, R.R.M. 2006. Fungi and fungal toxins as weapons. *Mycol. Res.* 110:1003–1010.

Pavagadhi, S., A.A.L. Tang, M. Sathishkumar, K.P. Loh, and R. Balasubramanian. 2013. Removal of microcystin-LR and microcystin-RR by graphene oxide: adsorption and kinetic experiments. *Water Res.* 47:4621–4629.

Payment, P. 1989. Elimination of viruses and bacteria during drinking water treatment: review of 10 years of data from the Montreal metropolitan area. In *Biohazards of Drinking Water Treatment*, R.A. Larson (Ed.). Lewis, Chelsea, MI, pp. 59–65.

Payment, P. 1991. Fate of human enteric viruses, coliphages, and *Clostridium perfringens* during drinking-water treatment. *Can. J. Microbiol.* 37:154–157.

Payment, P., F. Gamache, and G. Paquette. 1989. Comparison of microbiological data from two water filtration plants and their distribution system. *Water Sci. Technol.* 21:287–289.

Payment, P., L. Richardson, J. Siemiatycki, R. Dewar, M. Edwardes, and E. Franco. 1991a. A randomized trial to evaluate the risk of gastrointestinal disease due to consumption of drinking water meeting current microbiological standards. *Am. J. Public Health* 81:703–708.

Payment, P., E. Franco, L. Richardson, and J. Siemiatycki. 1991b. Gastrointestinal health effects associated with the consumption of drinking water produced by point-of-use domestic reverse-osmosis filtration units. *Appl. Environ. Microbiol.* 57:945–948.

Payment, P., E. Franco, and J. Siemiatycki. 1993. Absence of relationship between health effects due to tapwater consumption and drinking water quality parameters. *Water Sci. Technol.* 27:137–143.

Payment, P., J. Siemiatycki, L. Richardson, G. Renaud, E. Franco, and M. Prevost. 1997. A prospective epidemiological study of gastrointestinal health effects due to the consumption of drinking water. *Int. J. Environ. Health Res.* 7:5–31.

Peck, M.W. and S.C. Stringer. 2005. The safety of pasteurized in-pack chilled meat products with respect to the foodborne botulism hazard. *Meat Sci.* 70:461–475.

Pedersen, K. 1990. Biofilm development on stainless steel and PVC surfaces in drinking water. *Water Res.* 24:239–243.

Peeters, J.E., E.A. Mazas, W.J. Masschelein, I.V. Martinez de Maturana, and E. Debacker. 1989. Effect of disinfection of drinking water with ozone or chlorine dioxide on survival of *Cryptosporidium parvum* oocysts. *Appl. Environ. Microbiol.* 55:1519–1522.

Pejack, E. 2011. Solar pasteurization. In *Drinking Water Treatment, Strategies for Sustainability*, C. Ray and R. Jain (Eds). Springer, New York.

Percival, S.L. and J.G. Thomas. 2009. Transmission of *Helicobacter pylori* and the role of water and biofilms. *J. Water Health* 7(3):469–477.

Percival, S.L., J.T. Walker, and P.R. Hunter. 2000. *Microbiological Aspects of Biofilms and Drinking Water*. CRC Press, Boca Raton, FL, 229 pp.

Pereira, V.J., R. Marques, M. Marques, M.J. Benoliel, and M.T. Barreto Crespo. 2013. Free chlorine inactivation of fungi in drinking water sources. *Water Res.* 47:517–523.

Perelman, L. and A. Ostfeld. 2013. Operation of remote mobile sensors for security of drinking water distribution systems. *Water Res.* 47:4217–4226.

Perkins, S.D., J. Mayfield, V. Fraser, and L.T. Angenent. 2009. Potentially pathogenic bacteria in shower water and air of a stem cell transplant unit. *Appl. Environ. Microbiol.* 75:5363–5372.

Persson, P.E. 1979. Notes on muddy odour. III. Variability of sensory response to 2-methylisoborneol. *Aqua Fenn.* 9:48–52.

Peruski, A.H. and L.F. Peruski, Jr. 2003. Immunological methods for detection and identification of infectious disease and biological warfare agents. *Clin. Diagn. Lab. Immunol.* 10:506–513.

Pickup, R.W. 1991. Development of molecular methods for the detection of specific bacteria in the environment. *J. Gen. Microbiol.* 137:1009–1019.

Pigeot-Remy, S., F. Simonet, D. Atlan, J.C. Lazzaroni, and C. Guillard. 2012. Bactericidal efficiency and mode of action: a comparative study of photochemistry and photocatalysis. *Water Res.* 46:3208–3218.

Ping, T.S.T. 2010. Terrorism—a new perspective in the water management landscape. *Int. J. Water Resour. Dev.* 26:51–63.

Pintar, K.D.M. and R.M. Slawson. 2003. Effect of temperature and disinfection strategies on ammonia-oxidizing bacteria in a bench-scale drinking water distribution system. *Water Res.* 37:1805–1817.

Pintó, R.M., F.X. Abad, R. Gajardo, and A. Bosch. 1996. Detection of infectious astroviruses in water. *Appl. Environ. Microbiol.* 62:1811–1813.

Pintó, R.M., M.I. Costafreda, and A. Bosch. 2009. Risk assessment in shellfish-borne outbreaks of Hepatitis A. *Appl. Environ. Microbiol.* 75:7350–7355.

Pip, E. 2000. Survey of bottled drinking water available in Manitoba, Canada. *Environ. Health Perspect.* 108:863–866.

Pizzi, N.G. 2002. *Water Treatment Operator Handbook*. American Water Works Association, Denver, CO, 241 pp.

Plummer, J.D. and J.K. Edzwald. 2001. Effect of ozone on algae as precursors for trihalomethane and haloacetic acid production. *Environ. Sci. Technol.* 35:3661–3668.

Poitelon, J.-B., M. Joyeux, B. Welte, J.-P. Duguet, E. Prestel, and M.S. Dubow. 2010. Variations of bacterial 16S rDNA phylotypes prior to and after chlorination for drinking water production from two surface water treatment plants. *J. Ind. Microbiol. Biotechnol.* 37:117–128.

Pond, K., J. Rueedi, and S. Pedley. 2004. *Microrisk: Pathogens in Drinking Water Sources*. Robens Centre for Public and Environmental Health, University of Surrey, UK.

Pontius, F.W. 2002. Regulatory compliance planning to ensure water supply safety. *J. Water Works. Assoc.* 94:52–64.

Pontius, F.W. 2003. Update on USEPA's drinking water regulations. *J. Water Works. Assoc.* 95:57–68.

Popovtzer, R., T. Neufeld, D. Biran, E.Z. Ron, J. Rishpon, and Y. Shacham-Diamand. 2005. Novel integrated electrochemical nano-biochip for toxicity detection in water. *Nano Lett.* 5:1023–1027.

Porco, J. 2010. Municipal water distribution system security study: recommendations for science and technology investments. *J. Am. Water Works Assoc.* 102(4):30–32.

Post, G.B., T.B. Atherholt, and P.D. Cohn. 2011. Health and aesthetic aspects of drinking water. In *Water Quality and Treatment_A Handbook on Drinking Water*, 6th edition, J.K. Edzwald (Ed.). AWWA, Denver, CO.

Pougnard, C., P. Catala, J.-L. Drocourt, S. Legastelois, P. Pernin, E. Pringuez, and P. Lebaron. 2002. Rapid detection and enumeration of *Naegleria fowleri* in surface waters by solid-phase cytometry. *Appl. Environ. Microbiol.* 68:3102–3107.

Powers, E.M., C. Hernandez, S.N. Boutros, and B.G. Harper. 1994. Biocidal efficacy of a flocculating emergency water purification tablet. *Appl. Environ. Microbiol.* 60:2316–2323.

Poynter, S.F.B. and J.S. Slade. 1977. The removal of viruses by slow sand filtration. *Prog. Water Technol.* 9:75–78.

du Preez, M., R.M. Conroy, J.A. Wright, S. Moyo, N. Potgieter, and S.W. Gundry. 2008. Short report: use of ceramic water filtration in the prevention of diarrheal disease: a randomized controlled trial in rural South Africa and Zimbabwe. *Am. J. Trop. Med. Hyg.* 79:696–701.

du Preez, M., R.M. Conroy, S. Ligondo, J. Hennessy, M. Elmore-Meegan, A. Soita, and K.G. McGuigan. 2011. Randomized intervention study of solar disinfection of drinking water in the prevention of dysentery in Kenyan children aged under 5 years. *Environ. Sci. Technol.* 45:9315–9323.

Prévost, M., A. Rompré, H. Baribeau, J. Coallier, and P. Lafrance. 1997. Service lines: their effect on microbiological quality. *J. Am. Water Works Assoc.* 89:78–91.

Prévost, M.P., P. Laurent, P. Servais, and J.C. Joret (Eds). 2005. *Biodegradable Organic Matter in Drinking Water Treatment and Distribution*. American Water Works Association, Denver, CO.

Prevots, D.R., P.A. Shaw, D. Strickland, L.A. Jackson, M.A. Raebel, M.A. Blosky, R. Montes de Oca, Y.R. Shea, A.E. Seitz, S.M. Holland, and K.N. Olivier. 2010. Nontuberculous mycobacterial lung disease prevalence at four integrated health care delivery systems. *Am. J. Respir. Crit. Care Med.* 182:970–976.

Psutka, R., R. Peletz, S. Michelo, P. Kelly, and T. Clasen. 2011. Assessing the microbiological performance and potential cost of boiling drinking water in urban Zambia. *Environ. Sci. Technol.* 45:6095–6101.

Puzon, G.J., J.A. Lancaster, J.T. Wylie, and J.J. Plumb. 2009. Rapid detection of *Naegleria fowleri* in water distribution pipeline biofilms and drinking water samples. *Environ. Sci. Technol.* 43:6691–6696.

Qi, F., B. Xu, Z. Chen, J. Ma, D. Sun, and L. Zhang. 2009. Efficiency and products investigations on the ozonatilon of 2-Methylisoborneol in drinking water. *Water Environ. Res.* 81:2411–2419.

Qu, X., P.J.J. Alvarez, and Q. Li. 2013. Applications of nanotechnology in water and wastewater treatment. *Water Res.* 47:3931–3946.

Qualls, R.G., M.P. Flynn, and J.D. Johnson. 1983. The role of suspended particles in ultraviolet irradiation. *J. Water Pollut. Control Fed.* 55:1280–1285.

Quek, P.H., J.Y. Hu, X.N. Chu, Y.Y. Feng, and X.L. Tan. 2006. Photoreactivation of *Escherichia coli* following medium-pressure ultraviolet disinfection and its control using chloramination. *Water Sci. Technol.* 53(6):123–129.

Quick, R.E., A. Kimura, A. Thevos, M. Tembo, I. Shamputa, L. Hutwagner, and E. Mintz. 2002. Diarrhea prevention through household-level water disinfection and safe storage in Zambia. *Am. J. Trop. Med. Hyg.* 66:584–589.

Quinlivan, P.A., L. Li, and R.U.D. Knappe. 2005. Effects of activated carbon characteristics on the simultaneous adsorption of aqueous organic micropollutants and natural organic matter. *Water Res.* 39:1663–1673.

Raber, E. and A. Burklund. 2010. Decontamination options for *Bacillus anthracis*-contaminated drinking water determined from spore surrogate studies. *Appl. Environ. Microbiol.* 76:6631–6638.

Raber, E. and R. McGuire. 2002. Oxidative decontamination of chemical and biological warfare agents using L-Gel. *J. Hazard. Mater.* 93:339–352.

Rabold, J.G., C.W. Hoge, D.R. Shlim, C. Kefford, R. Rajah, and P. Echeverria. 1994. *Cyclospora* outbreak associated with chlorinated drinking water (letter). *Lancet* 344:1360–1361.

Radziminski, C., L. Ballantyne, J. Hodson, R. Creason, R.C. Andrews, and C. Chauret. 2002. Disinfection of *Bacillus subtilis* spores with chlorine dioxide: a bench-scale and pilot-scale study. *Water Res.* 36:1629–1639.

Ragull, S., M. Garcia-Nunez, M.L. Pedro-Botet, N. Sopena, M. Esteve, R. Montenegro, and M. Sabria. 2007. *Legionella pneumophila* in cooling towers: fluctuations in counts, determination of genetic variability by pulsed-field gel electrophoresis (PFGE), and persistence of PFGE patterns. *Appl. Environ. Microbiol.* 73:5382–5384.

Rahaman, M.S., C.D. Vecitis, and M. Elimelech. 2012. Electrochemical carbon-nanotube filter performance toward virus removal and inactivation in the presence of natural organic matter. *Environ. Sci. Technol.* 46:1556–1564.

Rahman, M.S., G. Encarnacion, and A.K. Camper. 2011. Nitrification and potential control mechanisms in simulated premises plumbing. *Water Res.* 45:5511–5523.

Rai, M., A. Yadav, and A. Gade. 2009. Silver nanoparticles as a new generation of antimicrobials. *Biotechnol. Adv.* 27(1):76–83.

Rainey, R.C. and A.K. Harding. 2005. Acceptability of solar disinfection of drinking water treatment in Kathmandu Valley, Nepal. *Int. J. Environ. Health Res.* 15(5):361–372.

Ram, P.K., E. Kelsey, Rasoatiana, R.R. Miarintsoa, O. Rakotomalala, C. Dunston, and R.E. Quick. 2007. Bringing safe water to remote populations: an evaluation of a portable point-of-use intervention in rural Madagascar. *Am. J. Public Health* 97:398–400.

Rand, J.L., R. Hofmann, M.Z.B. Alam, C. Chauret, R. Cantwell, R.C. Andrews, and G.A. Gagnon. 2007. A field study evaluation for mitigating biofouling with chlorine dioxide or chlorine integrated with UV disinfection. *Water Res.* 41:1939–1948.

Rao, V.C., J.M. Symons, A. Ling, P. Wang, T.G. Metcalf, J.C. Hoff, and J.L. Melnick. 1988. Removal of hepatitis A virus and rotavirus by drinking water treatment. *J. Am. Water Works Assoc.* 80:59–67.

Rapala, J., K. Lahtia, L.A. Räsänen, A.-L. Esala, S.I. Niemelä, and K. Sivonen. 2002. Endotoxins associated with cyanobacteria and their removal during drinking water treatment. *Water Res.* 36:2625–2637.

Rapala, J., M. Niemelä, K.A. Berg, L. Lepistö, and K. Lahti. 2006. Removal of cyanobacteria, cyanotoxins, heterotrophic bacteria and endotoxins at an operating surface water treatment plant. *Water Sci. Technol.* 54(3):23–28.

Ray, C., T. Grischek, J. Schubert, J.Z. Wang, and T.F. Speth. 2002. A perspective of riverbank filtration. *J. Am. Water Works Assoc.* 94(4):49–159.

Ray, C., J. Jasperse, and T. Grischek. 2011. Bank filtration as natural filtration. In *Drinking Water Treatment, Strategies for Sustainability*, C. Ray and R. Jain (Eds). Springer Science+Business Media, pp. 93–158.

Raynaud, M. 2004. A view of the European plastic pipes market in a global scenario. In *Proceedings of Plastics Pipes XII, April 19–22, 2004*, Milan, Italy.

Reasoner, D.J. 2002. Home treatment devices—microbiology of point of use and point of entries devices. In *Encyclopedia of Environmental Microbiology*, G. Bitton (Editor-in-Chief). Wiley-Interscience, New York, pp. 1563–1575.

Reasoner, D.J., J.C. Blannon, and E.E. Geldreich. 1987. Microbiological characteristics of third-faucet point-of-use devices. *J. Am. Water Works Assoc.* 79:60–66.

Regan, J.M., G.W. Harrington, H. Baribeau, R.D. Leon, and D.R. Noguera. 2003. Diversity of nitrifying bacteria in full-scale chloraminated distribution systems. *Water Res.* 37:197–205.

Regli, S., J.B. Rose, C.N. Haas, and C.P. Gerba. 1991. Modeling the risk from Giardia and viruses in drinking water. *J. Am. Water Works Assoc.* 83(11):76–84.

Reipa, V., J. Almeida, and K.D. Cole. 2006. Long-term monitoring of biofilm growth and disinfection using a quartz crystal microbalance and reflectance measurements. *J. Microbiol. Methods* 66:449–459.

Reller, M., C.E. Mendoza, M.B. Lopez, M. Alvarez, R.M. Hoekstra, C.A. Olson, K.G. Baier, B. Keswick, and S.P. Luby. 2003. A randomized controlled trial of household-based flocculant-disinfectant drinking water treatment for diarrhea prevention in rural Guatemala. *Am. J. Trop. Med. Hyg.* 69:411–419.

Ren, D. and J.A. Smith. 2013. Retention and transport of silver nanoparticles in a ceramic porous medium used for point-of-use water treatment. *Environ. Sci. Technol.* 47:3825–3832.

Rendtorff, R.C. 1979. The experimental transmission of *Giardia lamblia* among volunteer subjects. In *Waterborne Transmission of Giardiasis*, W.W. Jakubowski, and J.C. Hoff (Eds). EPA-600/9-79-001. Environmnetal Protection Agency, Office of Research and Development, Environmental Research Center, Cincinnati, OH, pp. 64–81.

Rendueles, O., J.B. Kaplan, and J.-M. Ghigo. 2013. Antibiofilm polysaccharides. *Environ. Microbiol.* 15:334–346.

Rennecker, J.L., A.M. Driedger, S.A. Rubin, and B.J. Mariñas. 2000. Synergy in sequential inactivation of *Cryptosporidium parvum* with ozone/free chlorine and ozone/monochloramine. *Water Res.* 34:4121–4130.

Revetta, R.P., A. Pemberton, R. Lamendella, B. Iker, and J.W. Santo Domingo. 2010. Identification of bacterial populations in drinking water using 16S rRNA-based sequence analyses. *Water Res.* 44:1353–1360.

Ribas, F., J. Frias, and F. Lucena. 1991. A new dynamic method for the rapid determination of the biodegradable dissolved organic carbon in drinking water. *J. Appl. Bacteriol.* 71:371–378.

Ribau-Teixeira, M. and M.J. Rosa. 2005. Microcystins removal by nanofiltration membranes. *Sep. Purif. Technol.* 46:192–201.

Ribau-Teixeira, M. and M.J. Rosa. 2006. Neurotoxic and hepatotoxic cyanotoxins removal by nanofiltration. *Water Res.* 40:2837–2848.

Rice, E.W., D.J. Reasoner, and P.V. Scarpino. 1988. Determining biodegradable organic matter in drinking water: a progress report. In *Water Quality Technology Conference of the American Water Works Association*, St. Louis, MO, Nov. 13–17.

Ridgway, H.F. and B.H. Olson. 1981. Scanning electron microscope evidence for bacterial colonization of a drinking-water distribution system. *Appl. Environ. Microbiol.* 41:274–287.

Ridgway, H.F., E.G. Means, and B.H. Olson. 1981. Iron bacteria in drinking-water distribution systems: elemental analysis of *Gallionella* stalks, using x-ray energy-dispersive microanalysis. *Appl. Environ. Microbiol.* 41:288–297.

Ried, A., J. Mielcke, A. Wieland, S. Schaefer, and M. Sievers. 2007. An overview of the integration of ozone systems in biological treatment steps. *Water Sci. Technol.* 55(12):253–258.

Rigal, S. and J. Danjou. 1999. Tastes and odors in drinking water distribution systems related to the use of synthetic materials. *Water Sci. Technol.* 40:203–208.

Rittmann, B.E. 1987. Aerobic biological treatment. *Environ. Sci. Technol.* 21:128–136.

Rittmann, B.E. 1989. Biodegradation processes to make drinking water biologically stable. In *Biohazards of Drinking Water Treatment*, R.A. Larson (Ed.). Lewis, Chelsea, MI, pp. 257–263.

Rittmann, B.E. 1995a. Transformations of organic micropollutants by biological processes. In *Water Pollution: Quality and Treatment of Drinking Water.* Springer, New York, pp. 31–42.

Rittmann, B.E. 1995b. Fundamentals and application of biofilm processes in drinking-water treatment. In *Water Pollution: Quality and Treatment of Drinking Water.* Springer, New York, pp. 61–87.

Rittmann, B.E. 2004. Biofilms in the water industry. In *Microbial Biofilms*, M. Ghannoum and G.A. O'Toole (Eds). ASM Press, Washington, DC, pp. 359–378.

Rittmann, B.E. and C.S. Laspidou. 2002. Biofilm detachment. In *Encyclopedia of Environmental Microbiology*, G. Bitton (Editor-in-Chief). Wiley-Interscience, New York, pp. 544–550.

Rittmann, B.E. and V.L. Snoeyink. 1984. Achieving biologically stable drinking water. *J. Am. Water Works Assoc.* 76:106–114.

Roberson, J.A. 2006. From common cup to *Cryptosporidium*: a regulatory evolution. *J. Am. Water Works Assoc.* 98(3):198–207.

Robertson, J.B. and S.C. Edberg. 1997. Natural protection of spring and well drinking water against surface microbial contamination. I. Hydrogeological parameters. *Crit. Rev. Microbiol.* 23:143–178.

Robinson, B.S., P.T. Monis, and P.J. Dobson. 2006. Rapid, sensitive, and discriminating identification of *Naegleria* spp. by real-time PCR and melting-curve analysis. *Appl. Environ. Microbiol.* 72:5857–5863.

Rochelle, P.A. 2002. *Giardia*: detection and occurrence in the environment. In *Encyclopedia of Environmental Microbiology*, G. Bitton (Editor-in-Chief). Wiley-Interscience, New York, pp. 1477–1489.

Rochelle, P.A. and J. Clancey. 2006. The evolution of microbiology in the drinking water treatment industry. *J. Am. Water Works Assoc.* 98(3):163–191.

Rochelle, P.A., A.M. Johnson, R. De Leon, and G.D. Di Giovanni. 2012. Assessing the risk of infectious *Cryptosporidium* in drinking water. *J. Am. Water Works Assoc.* 104(5):79–80.

Rodgers, J. and C.W. Keevil. 1995. Survival of *Cryptosporidium parvum* oocysts in biofilms and planktonic samples in a model system. In *Protozoan Parasites and Water*, W.B. Betts, D. Casemore, C. Fricker, H. Smith and J. Watkins (Eds). The Royal Society of Chemistry, Cambridge, UK, pp. 209–213.

Rodgers, F.G., P. Hufton, E. Kurzawska, C. Molloy, and S. Morgan. 1985. Morphological response of human rotavirus to ultraviolet radiation, heat and disinfectants. *J. Med. Microbiol.* 20:123–130.

Rodgers, M.R., B.J. Blackstone, A.L. Reyes, and T.C. Covert. 1999. Colonisation of point of use water filters by silver resistant non-tuberculous mycobacteria. *J. Clin. Pathol.* 52:629–630.

Rodrigo, S., M. Sinclair, A. Forbes, D. Cunliffe, and K. Leder. 2011. Drinking rainwater: a double-blinded, randomized controlled study of water treatment filters and gastroenteritis incidence. *Am. J. Public Health* 101:842–847.

Rodriguez, E., G.D. Onstad, T.P.J. Kull, J.S. Metcalf, J.L. Acero, and U. von Gunten. 2007a. Oxidative elimination of cyanotoxins: comparison of ozone, chlorine, chlorine dioxide and permanganate. *Water Res.* 41:3381–3393.

Rodriguez, M.J., J. Serodes, and D. Roy. 2007b. Formation and fate of haloacetic acids (HAAs) within the water treatment plant. *Water Res.* 41:4222–4232.

Rodriguez, E.M., J.L. Acero, L. Spoof, and J. Meriluoto. 2008. Oxidation of MC-LR and -RR with chlorine and potassium permanganate: Toxicity of the reaction products. *Water Res.* 42:1744–1752.

Roessler, P.F., and B.F. Severin. 1996. Ultraviolet light disinfection of water and wastewater. In *Modeling Disease Transmission and Its Prevention by Disinfection*, C. Hurst (Ed.). Cambridge University Press, Cambridge, UK, pp. 313–368.

Rogers, E.D., T.B. Henry, M.J. Twiner, J.S. Gouffon, J.T. McPherson, G.L. Boyer, G.S. Sayler, and S.W. Wilhelm. 2011. Global gene expression profiling in larval zebrafish exposed to microcystin-LR and *Microcystis* reveals endocrine disrupting effects of cyanobacteria. *Environ. Sci. Technol.* 45:1962–1969.

Rollinger, Y. and W. Dott. 1987. Survival of selected bacterial species in sterilized activated carbon filters and biological activated carbon filters. *Appl. Environ. Microbiol.* 53:777–781.

Rollins, D.M. and R.R. Colwell. 1986. Viable but non culturable stage of *Campylobacter jejuni* and its role in survival in the aquatic environment. *Appl. Environ. Microbiol.* 52:531–538.

Rooy, T.B.V. 2003. Bottling up our natural resources: the fight over bottled water extraction in the United States. *J. Land Use Environ. Law* 18:267–298.

Rosa, G. and T. Clasen. 2010. Estimating the scope of household water treatment in low- and middle-income countries. *Am. J. Trop. Med. Hyg.* 82:289–300.

Rosa, G., L. Miller, and T. Clasen. 2010. Microbiological effectiveness of disinfecting water by boiling in rural Guatemala. *Am. J. Trop. Med. Hyg.* 82(3):473–477.

Rose, J.B. 1990. Occurrence and control of *Cryptosporidium* in drinking water. In *Drinking Water Microbiology*, G.A. McFeters (Ed.). Springer Verlag, New York, pp. 294–321.

Rose, J.B. and C.P. Gerba. 1991. Use of risk assessment for development of microbial standards. *Water Sci. Technol.* 24(2):29–34.

Rose, L.J. and H. O'Connell. 2009. UV light inactivation of bacterial biothreat agents. *Appl. Environ. Microbiol.* 75:2987–2990.

Rose, J.B., L.K. Landeen, K.R. Riley, and C.P. Gerba. 1989. Evaluation of immunofluoresecence techniques for detection of *Cryptosporidium* oocysts and *Giardia* cysts from environmental samples. *Appl. Environ. Microbiol.* 55:3189–3196.

Rose, J.B., C.N. Haas, and S. Regli. 1991a. Risk assessment and control of waterborne giardiasis. *Am. J. Public Health* 81:709–713.

Rose, J.B., C.P. Gerba, and W. Jakubowski. 1991b. Survey of potable water supplies for *Cryptosporidium* and *Giardia*. *Environ. Sci. Technol.* 25:1393–1399.

Rose, L.J., E.W. Rice, B. Jensen, R. Murga, A. Peterson, R.M. Donlan, and M.J. Arduino. 2005. Chlorine inactivation of bacterial bioterrorism agents. *Appl. Environ. Microbiol.* 71:566–568.

Rose, L.J., E.W. Rice, L. Hodges, A. Peterson, and M.J. Arduino. 2007. Monochloramine inactivation of bacterial select agents. *Appl. Environ. Microbiol.* 73:3437–3439.

Rosegrant, M.W., X. Cal, and S.A. Cline. 2002. *Global Water Outlook to 2025*: *Averting an Impending Crisis*. Food Policy Report. Inter. Food Policy Res. Inst. (IFPRI) and Intern. Water Management Inst. (IWMI), 28 pp.

Rosenberg, F.A. 2002. Bottled water, microbiology. In *Encyclopedia of Environmental Microbiology*, G. Bitton (Editor-in-Chief). Wiley-Interscience, New York, pp. 795–802.

Rosenfeldt, E.J., B. Melcher, and K.G. Linden. 2005. Treatment of taste and odor causing compounds in water by UV and UV/H$_2$O$_2$ processes. *J. Water Supply Res. Technol.* 54(7):423–434.

Rosenszweig, W.D. and W.O. Pipes. 1989. Presence of fungi in drinking water. In *Biohazards of Drinking Water Treatment*, R.A. Larson (Ed.). Lewis, Chelsea, MI, pp. 85–93.

Rosenszweig, W.D., H.A. Minnigh, and W.O. Pipes. 1983. Chlorine demand and inactivation of fungal propagules. *Appl. Environ. Microbiol.* 45:182–186.

Rosenszweig, W.D., H.A. Minnigh, and W.O. Pipes. 1986. Fungi in potable distribution systems. *J. Am. Water Works Assoc.* 78:53–55.

Roszak, D.B. and R.R. Colwell. 1987. Survival strategies of bacteria in the natural environment. *Microbiol. Rev.* 51:365–379.

Rotz, L.D., A.S. Khan, S.R. Lillibridge, S.M. Ostroff, and J.M. Hughes. 2002. Public health assessment of potential biological terrorism agents. *Emerg. Infect. Dis.* 8(2):225–229.

Rowe-Taitt, C.A., J.P. Golden, M.J. Feldstein, J.J. Cras, K.E. Hoffman, and F.S. Ligler. 2000. Array biosensor for detection of biohazards. *Biosens. Bioelectron.* 14:785–794.

Rowland, B. and N.Y Voorheesville. 2003. Bacterial contamination of dental unit waterlines: what is your dentist spraying into your mouth. *Clin. Microbiol. Newslett.* 25:73–77.

Roy, D., P.K.Y. Wong, R.S. Engelbrecht, and E.S.K. Chian. 1981. Mechanism of enteroviral inactivation by ozone. *Appl. Environ. Microbiol.* 41:718–723.

de la Rubia, A., M. Rodriguez, V.M. Leon, and D. Prats. 2008. Removal of natural organic matter and THM formation potential by ultra- and nanofiltration of surface water. *Water Res.* 42:714–722.

Rubulis, J. and R. Juhna. 2007. Evaluating the potential of biofilm control in water supply systems by removal of phosphorus from drinking water. *Water Sci. Technol.* 55(8–9):211–217.

Rufener, S., D. Mäusezahl, H.-J. Mosler, and R. Weingartner. 2010. Quality of drinking water at source and point-of-consumption—drinking cup as a high potential recontamination risk: a field study in Bolivia. *J. Health Popul. Nutr.* 28(1):34–41.

Russel, A.D., J.R. Furr, and J.-Y. Maillard. 1997. Microbial susceptibility and resistance to biocides. *ASM News* 63:481–487.

Ryan, M.O., P.L. Gurian, C.N. Haas, J.B. Rose, and P.J. Duzinski. 2013. Acceptable microbial risk: cost-benefit analysis of a boil water order for *Cryptosporidium*. *J. Am. Water Works Assoc.* 105(4):189–192.

Ryu, H., D. Gerrity, J.C. Crittenden, and M. Abbaszadegan. 2008. Photocatalytic inactivation of *Cryptosporidium parvum* with TiO$_2$ and low-pressure ultraviolet irradiation. *Water Res.* 42:1523–1530.

Saikaly, P.E., M.A. Barlaz, and F.L. de los Reyes, III. 2007. Development of quantitative real-time PCR assays for detection and quantification of surrogate biological warfare agents in building debris and leachate. *Appl. Environ. Microbiol.* 73:6557–6565.

Saleh, M.A., F.H. Abdel-Rahman, B.B. Woodard, S. Clark, C. Wallace, A. Aboaba, W. Zhang, and J. Nance. 2008. Chemical, microbial and physical evaluation of commercial bottled waters in greater Houston area of Texas. *J. Environ. Sci. Health A Tox. Hazard. Subst. Environ. Eng.* 43:335–347.

Sakai, H., H. Katayama, K. Oguma, and S. Ohgaki. 2011. Effect of photoreactivation on ultraviolet inactivation of *Microcystis aeruginosa*. *Water Sci. Technol.* 63(6):1224–1229.

San Martin, M.F., G.V. Barbosa-Cánovas, and B.G. Swanson. 2002. Food processing by high hydrostatic pressure. *Crit. Rev. Food Sci. Nutr.* 42:627–645.

Sano, D., Y. Ueki, T. Watanabe, and T. Omura. 2006. Membrane separation of indigenous noroviruses from sewage sludge and treated wastewater. *Water Sci. Technol.* 54(3):77–82.

Sanz, E.N., I. Salcedo Dávila, J.A. Andrade Balao, and J.M. Quiroga Alonso. 2007. Modelling of reactivation after UV disinfection: effect of UV-C dose on subsequent photoreactivation and dark repair. *Water Res.* 41:3141–3151.

Sarkar, P. and C.P. Gerba. 2012. Inactivation of *Naegleria fowleri* by chlorine and ultraviolet light. *J. Am. Water Works Assoc.* 104(3):51–52.

Sathasivan, A. and S. Ohgaki. 1999. Application of new bacterial regrowth potential method for water distribution systems: a clear evidence of phosphorus limitation. *Water Res.* 33:137–144.

Sauch, J.F., D. Flanigan, M.L. Galvin, D. Berman, and W. Jakubowski. 1991. Propidium iodide as an indicator of *Giardia* cyst viability. *Appl. Environ. Microbiol.* 57:3243–3247.

Sauer, K., A.H. Rickard, and D.G. Davies. 2007. Biofilms and diversity. *Microbe* 2(7):347–353.

Sautour, M., V. Edel-Hermann, C. Steinberg, N. Sixt, J. Laurent, F. Dalle, S. Aho, P. Hartemann, C. L'Ollivier, M. Goyer, and A. Bonnin. 2012. *Fusarium* species recovered from the water distribution system of a French university hospital. *Int. J. Hyg. Environ. Health* 215:286–292.

Savage, N. and M.S. Diallo. 2005. Nanomaterials and water purification: opportunities and challenges. *J. Nanopart. Res.* 7:331–342.

Schachter, B. 2003. Slimy business: the biotechnology of biofilms. *Nat. Biotechnol.* 21:361–365.

Schaefer, F.W. 1997. Detection of protozoan parasites in source and finished drinking waters. In *Manual of Environmental Microbiology*, C.J. Hurst, G.R. Knudsen, M.J. McInerney, L.D. Stetzenbach, and M.V. Walter, (Eds). ASM Press, Washington, DC, pp. 153–175.

Schaudinn, C., P. Stoodley, A. Kainovic, T. O'keefe, B. Costerton, D. Robinson, M. Baum, G. Ehrlich, and P. Webster. 2007. Bacterial biofilms, other structures seen as mainstream concepts. *Microbe* 2(5):231–237.

Schaule, G., T. Griebe, and H.-C. Flemming. 2000. Steps in biofilm sampling and characterization in biofouling cases. In *Biofilms-Investigative Methods & Applications*, H.-C. Flemming, U. Szewzyk and T. Griebe (Eds). Technomic Publishing Co., Lancaster-Basel, pp. 1–21.

Schaule, G., D. Moschnitschka, S. Schulte, A. Tamachkiarow, and H.-C. Flemming. 2007. Biofilm growth in response to various concentrations of biodegradable material in drinking water. *Water Sci. Technol.* 55(8–9):191–195.

Schiemann, D.A. 1990. *Yersinia enterocolitica* in drinking water. In *Drinking Water Microbiology*, G.A. McFeters (Ed.). Springer Verlag, New York, pp. 322–339.

Schijven, J.F., P.F.M. Teunis, S.A. Rutjes, M. Bouwknegt, A.M. de Roda Husman. 2011. QMRAspot: a tool for Quantitative Microbial Risk Assessment from surface water to potable water. *Water Res.* 45:5564–5576.

Schijven, J.F., H.H.J.L. van den Berg, M. Colin, Y. Dullemont, W.A.M. Hijnen, A. Magic-Knezev, W.A. Oorthuizen, and G. Wubbels. 2013. A mathematical model for removal of human pathogenic viruses and bacteria by slow sand filtration under variable operational conditions. *Water Res.* 47:2592–2602.

Schlosser, O., C. Robert, C. Bourderioux, M. Rey, and M.R. de Roubin. 2001. Bacterial removal from inexpensive portable water treatment systems for travelers. *J. Travel Med.* 8(1):12–18.

Schlumpf, M., B. Cotton, M. Conscience, V. Haller, B. Steinmann, and W. Lichtensteiner. 2001. *In vitro* and *in vivo* estrogenicity of UV screens. *Environ. Health Perspect.* 109:239–244.

Schmeisser, C., C. Stockigt, C. Raasch, J. Wingender, K.N. Timmis, D.F. Wenderoth, H.-C. Flemming, H. Liesegang, R.A. Schmitz, K.-E. Jaeger, and W.R. Streit. 2003. Metagenome survey of biofilms in drinking-water networks. *Appl. Environ. Microbiol.* 69:7298–7309.

Schmidt, C. and H.-J. Brauch. 2008. N,N-Dimethylsulfamide as precursor for N-Nitrosodimethylamine (NDMA) formation upon ozonation and its fate during drinking water treatment. *Environ. Sci. Technol.* 42:6340–6346.

Schmidt, W.-P. and S. Cairncross. 2008. Household water treatment in poor populations: is there enough evidence for scaling up now? *Environ. Sci. Technol.* 43:986–992.

Schmidt, M., U. Panne, J. Adams, and R. Niessner. 2004. Investigation of biocide efficacy by photoacoustic biofilm monitoring. *Water Res.* 38:1189–1196.

Schmidt, W., H. Petzoldt, K. Bornmann, L. Imhof, and C. Moldaenke. 2009. Use of cyanopigment determination as an indicator of cyanotoxins in drinking water. *Water Sci. Technol.* 59(8):1531–1540.

Schneider, R.P. and A. Leis. 2002. Conditioning films in aquatic environments. In *Encyclopedia of Environmental Microbiology*, G. Bitton (Ed.). Wiley-Interscience, New York, pp. 928–941.

Schoen, M.E. and N.J. Ashbolt. 2011. An in-premise model for *Legionella* exposure during showering events. *Water Res.* 45:5826–5836.

Schofield, D.A., I.J. Molineux, and C. Westwater. 2012. Rapid identification and antibiotic susceptibility testing of *Yersinia pestis* using bioluminescent reporter phage. *J. Microbiol. Methods* 90:80–82.

Schooling, S.R., U.K. Charaf, D.G. Allison, and P. Gilbert. 2004. A role for rhamnolipid in biofilm dispersion. *Biofilms* 1:91–99.

Schramm, A., L.H. Larsen, N.P. Revsbech, N.B. Ramsing, R. Amann, and K.-H. Schleifer. 1996. Structure and function of a nitrifying biofilm determined by *in situ* hybridization and use of microelectrodes. *Appl. Environ. Microbiol.* 62:4641–4647.

Schuler, P.F. and M.M. Ghosh. 1990. Diatomaceous earth filtration of cysts and other particulates using chemical additives. *J. Am. Water Works Assoc.* 82(12):67–75.

Schulze-Robbecke, R., B. Janning, and R. Fischeder. 1992. Occurrence of mycobacteria in biofilm samples. *Tuber. Lung Dis.* 73:141–144.

Schupp, D.G. and S.L. Erlandsen. 1987. A new method to determine *Giardia* cyst viability: correlation of fluorescein diacetate and propidium iodide staining with animal infectivity. *Appl. Environ. Microbiol.* 53:704–707.

Schuster, C.J., A.G. Ellis, W.J. Robertson, D.F. Charron, J.J. Aramini, B.J. Marshall, and D.T. Medeiros. 2005. Infectious disease outbreaks related to drinking water in Canada, 1974–2001. *Can. J. Public Health* 96:254–258.

Schwab, K.J., F.H. Neill, F. Le Guyader, M.K. Estes, and R.L. Atmar. 2001. Development of a reverse transcription-PCR-DNA enzyme immunoassay for detection of "Norwalk-Like" viruses and hepatitis A virus in stool and shellfish. *Appl. Environ. Microbiol.* 67:742–749.

Schwartz, T., W. Kohnen, B. Jansen, and U. Obst. 2003. Detection of antibiotic-resistant bacteria and their resistance genes in wastewater, surface water, and drinking water biofilms. *FEMS Microbiol. Ecol.* 43:325–335.

Schwartzbrod, L. (Ed.). 1991. *Virologie des Milieux Hydriques*. TEC & DOC Lavoisier, Paris, France, 304 pp.

Searcy, K.E., A.I. Packman, E.R. Atwill, and T. Harter. 2006. Capture and retention of *Cryptosporidium parvum* oocysts by *Pseudomonas aeruginosa* biofilms. *Appl. Environ. Microbiol.* 72:6242–6247.

Servais, P. 1996. Rôle du chlore et de la matière organique biodégradable dans le contrôle de la croissance bactérienne en réseaux de distribution de l'eau potable. *Tribune Eau* 96:3–12.

Servais, P., G. Billen, and M.-C. Hascoet. 1987. Determination of the biodegradable fraction of dissolved organic matter in waters. *Water Res.* 21:445–452.

Servais, P., G. Billen, C. Ventresque, and G.P. Bablon. 1991. Microbial activity in GAC filters at the Choisy-le-Roi treatment plant. *J. Am. Water Works Assoc. J.* 83:62–68.

Servais, P., P. Laurent, and G. Randon. 1995. Comparison of the bacterial dynamics in various French distribution systems. *J. Water Supply Res. Technol.* 44(1):10–17.

Setlow, P. 2000. Resistance of bacterial spores. In *Bacterial Stress Responses*, G. Storz and R. Hengge-Aronis (Eds). American Society of Microbiology, Washington, DC, pp. 217–230.

Seymour, I.J. and H. Appleton. 2001. Foodborne viruses and fresh produce. *J. Appl. Microbiol.* 91:759–773.

Shands, K.N., J.L. Ho, R.D. Meyer, G.W. Gorman, P.H. Edelstein, G.F. Mallison, S.M. Finegold, and D.W. Fraser. 1985. Potable water as a source of Legionnaire's disease. *J. Am. Med. Assoc.* 253:1412–1416.

Shannon, M.A., P.W. Bohn, M. Elimelech, J.G. Georgiadis, B.J. Mariñas, and A.M. Mayes. 2007. Science and technology for water purification in the coming decades. *Nature* 452:301–310.

Sharp, R.R., A.K. Camper, J.J. Crippen, O.D. Schneider, and S. Leggiero. 2001. Evaluation of drinking water biostability using biofilm methods. *J. Environ. Eng.* 127:403–410.

Shaw, K., S. Walker, and B. Koopman. 2000. Improving filtration of *Cryptosporium*. *J. Am. Water Works Assoc.* 92(11):103–111.

Sheehan, T. 2002. Bioterrorism. In *Encyclopedia of Environmental Microbiology*, G. Bitton (Editor-in-Chief). Wiley-Interscience, New York, pp. 771–782.

Shields, J.M. and B.H. Olson. 2003. PCR-restriction fragment length polymorphism method for detection of *Cyclospora cayetanensis* in environmental waters without microscopic confirmation. *Appl. Environ. Microbiol.* 69:4662–4669.

Shin, G.-A., K.G. Linden, M.J. Arrowood, and M.D. Sobsey. 2001. Low-pressure UV inactivation and DNA repair potential of *Cryptosporidium parvum* oocysts. *Appl. Environ. Microbiol.* 67:3029–3032.

Shin, G.-A., K.G. Linden, and G. Faubert. 2009. Inactivation of *Giardia lamblia* cysts by polychromatic UV. *Lett. Appl. Microbiol.* 48:790–792.

Shirasaki, N., T. Matsushita, Y. Matsui, A. Oshiba, and K. Ohno. 2010. Estimation of norovirus removal performance in a coagulation-rapid sand filtration process by using recombinant norovirus VLPs. *Water Res.* 44:1307–1316.

Shon, H.K., S. Phuntsho, D.S. Chaudhary, S. Vigneswaran, and J. Cho. 2013. Nanofiltration for water and wastewater treatment—a mini review. *Drink. Water Eng. Sci.* 6:47–53.

Shotyk, W. and M. Krachler. 2007. Lead in bottled waters: contamination from gass and comparison with pristine groundwater. *Environ. Sci. Technol.* 41(10):3508–3513.

Shotyk, W., M. Krachler, and B. Chen. 2006. Contamination of Canadian and European bottled waters with antimony leaching from PET containers. *J. Environ. Monit.* 8:288–292.

Shrivastava, R., R.K. Upreti, S.R. Jain, K.N. Prasad, P.K. Seth, and U.C. Chaturvedi. 2004. Suboptimal chlorine treatment of drinking water leads to selection of multidrug resistant *Pseudomonas aeruginosa*. *Ecotoxicol. Environ. Saf.* 58:277–283.

Shrout, J.D. and R. Nerenberg. 2012. Monitoring bacterial twitter: does quorum sensing determine the behavior of water and wastewater treatment biofilms? *Environ. Sci. Technol.* 46:1995–2005.

Sibille, I., T. Sime-Ngando, L. Mathieu, and J.C. Block. 1998. Protozoan bacterivory and *Escherichia coli* survival in drinking water distribution systems. *Appl. Environ. Microbiol.* 64:197–202.

Siebel, E., Y. Wang, T. Egli, and F. Hammes. 2008. Correlations between total cell concentration, total adenosine tri-phosphate concentration and heterotrophic plate counts during microbial monitoring of drinking water. *Drink. Water Eng. Sci.* 1:71–86.

Silbaq, F.S. 2009. Viable ultramicrocells in drinking water. *J. Appl. Microbiol.* 106:106–117.

Silhan, J., C.B. Corfitzen, and H.J. Albrechtsen. 2006. Effect of temperature and pipe material on biofilm formation and survival of *Escherichia coli* in used drinking water pipes: a laboratory-based study. *Water Sci. Technol.* 54(3):49–56.

Simoes, L.C., M. Simoes, and M.J. Vieira. 2010. Influence of the diversity of bacterial isolates from drinking water on resistance of biofilms to disinfection. *Appl. Environ. Microbiol.* 76:6673–6679.

Simonet, J. and C. Gantzer. 2006. Inactivation of poliovirus 1 and F-specific RNA phages and degradation of their genomes by UV irradiation at 254 nanometers. *Appl. Environ. Microbiol.* 72:7671–7677.

Simpson, D.R. 2008. Biofilm processes in biologically active carbon water purification. *Water Res.* 42:2839–2848.

Singer, P.C. 2006. DBPs in drinking water: additional scientific and policy considerations for public health protection. *J. Am. Water Works Assoc.* 98(10):73–79.

Singh, A. and G.A. McFeters. 1986. Recovery, growth, and production of heat-stable entero-toxin by *Escherichia coli* after copper-induced injury. *Appl. Environ. Microbiol.* 51:738–742.

Singh, A. and G.A. McFeters. 1990. Injury of enteropathogenic bacteria in drinking water. In *Drinking Water Microbiology*, G.A. McFeters (Ed.). Springer Verlag, New York, pp. 368–379.

Siqueira, V.M., H.M.B. Oliveira, C. Santos, R.R.M. Paterson, N.B. Gusmão, and N. Lima. 2013. Biofilms from a Brazilian water distribution system include filamentous fungi. *Can. J. Microbiol.* 59:183–188.

Sisson, A.J., P.J. Wampler, R.R. Rediske, J.N. McNair, and D.J. Frobish. 2013. Long-term field performance of Biosand filters in the Artibonite Valley, Haiti. *Am. J. Trop. Med. Hyg.* 88:862–867.

Sivonen, K. and T. Borner. 2008. Biactive compounds produced by cyanobacteria. In *The Cyanobacteria: Molecular Biology, Genetics and Evolution*, A. Herrero and E. Flores (Eds). Caister Academic Press, Great Britain, pp. 159–198.

Skjevrak, I., A. Due, K.O. Gjerstad, and H. Herikstad. 2003. Volatile organic components migrating from plastic pipes (HDPE, PEX and PVC) into drinking water. *Water Res.* 37:1912–1920.

Skraber, S., L. Ogorzaly, K. Helmi, A. Maul, L. Hoffmann, H.-M. Cauchie, and C. Gantzer. 2009. Occurrence and persistence of enteroviruses, noroviruses and F-specific RNA phages in natural wastewater biofilms. *Water Res.* 43:4780–4789.

Slezak, L.A. and R.C. Sims. 1984. The application and effectiveness of slow sand filtration in the United States. *J. Am. Water Works Assoc.* 76:38–43.

Smith, H.V. 1996. Detection of *Giardia* and *Cryptosporidium* in water: current status and future prospects. In *Molecular Approaches to Environmental Microbiology*, R.W. Pickup and J.R. Saunders (Eds). Ellis Horwood, Chichester, UK, pp. 195–225.

Smith, D.B. 2002. Coliform bacteria: control in drinking water distribution systems. In *Encyclopedia of Environmental Microbiology*, G. Bitton (Editor-in-Chief). Wiley-Interscience, New York, pp. 905–914.

Smith, A.M., W. Reacher, W. Smerdon, G.K. Adak, G. Nichols, and R.M. Chalmers. 2006. Outbreaks of waterborne infectious intestinal disease in England and Wales, 1992–2003. *Epidemiol. Infect.* 134:1141–1149.

Smith-Somerville, H.E., V.B. Huryn, C. Walker, and A.L. Winters. 1991. Survival of *Legionella pneumophila* in the cold-water ciliate *Tetrahymena vorax*. *Appl. Environ. Microbiol.* 57:2742–2749.

Snoeyink, V.L. and D. Jenkins. 1980. *Water Chemistry*. John Wiley & Sons, New York.

Snyder, J.W., Jr., C.N. Mains, R.E. Anderson, and G.K. Bissonnette. 1995. Effect of point-of-use, activated carbon filters on the bacteriological quality of rural groundwater supplies. *Appl. Environ. Microbiol.* 61:4291–4295.

Snyder, S.A., B.J. Vanderford, and D.J. Rexing. 2005. Trace analysis of bromate, chlorate, iodate, and perchlorate in natural and bottled Waters. *Environ. Sci. Technol.* 39:4586–4593.

Soave, R. 1996. *Cyclospora*: an overview. *Clin. Infect. Dis.* 23:429–437.

Sobsey, M.D. 1989. Inactivation of health-related microorganisms in water by disinfection processes. *Water Sci. Technol.* 21:179–195.

Sobsey, M.D. 2002. *Managing Water in the Home: Accelerated Health Gains from Improved Water Supply.* WHO WHO/SDE/WSH/02.07. Geneva, Switzerland.

Sobsey, M.D. and B. Olson. 1983. Microbial agents of waterborne disease. In *Assessment of Microbiology and Turbidity Standards For Drinking Water*, P.S. Berger and Y. Argaman (Eds). EPA-570-9-83-001. U.S. Environmental Protection Agency, Office of Drinking Water, Washington, DC.

Sobsey, M.D., C. Stauber, L. Casanova, J. Brown, and M.A. Elliott. 2008. Point of use household drinking water filtration: a practical, effective solution for providing sustained access to safe drinking water in the developing world. *Environ. Sci. Technol.* 42:4261–4267.

Solheim, H.T., C. Sekse, A.M. Urdahl, Y. Wasteson, and L.L. Nesse. 2013. Biofilm as an environment for dissemination of *stx* genes by transduction. *Appl. Environ. Microbiol.* 79:896–900.

Sommer, R., G. Weber, A. Cabaj, J. Wekerle, and G. Schauberger. 1989. UV-inactivation of microorganisms in water. *Zentralbl. Hyg. Umweltmed* 189:214–224.

Sommer, R., A. Cabaj, G. Hirschmann, and T. Haider. 2008. Disinfection of drinking water by UV irradiation: basic principles – specific requirements—international implementations. *Ozone Sci. Eng.* 30:43–48.

Son, W.K., J.H. Youk, T.S. Lee & W.H. Park. 2004. Preparation of antimicrobial ultrafine cellulose acetate fibers with silver nanoparticles. *Macromol. Rapid Commun.* 25:1632–1637.

Sonnenberg, A. and J.E. Everhart. 1997. Health impact of peptic ulcer in the United States. *Am. J. Gastroenterol.* 92:614–620.

Spotts, W.E.A., M.E. Beatty, T.H. Taylor, Jr., R. Veyant, J. Sobel, M.J. Arduino, and D.A. Ashford. 2003. Inactivation of *Bacillus anthracis* spores. *Emerg. Infect. Dis.* 9:623–627.

Sproul, O.J., R.M. Pfister, and C.K. Kim. 1982. The mechanism of ozone inactivation of waterborne viruses. *Water Sci. Technol.* 14:303–314.

Srinivasan, S. and G.W. Harrington. 2007. Biostability analysis of drinking water distribution systems. *Water Res.* 41:2127–2138.

Srinivasan, S., G.W. Harrington, I. Xagoraraki, and R. Goel. 2008. Factors affecting bulk to total bacteria ratio in drinking water distribution systems. *Water Res.* 42:3393–3404.

Soracco, R.J. and D.H. Pope. 1983. Bacteriostatic and bactericidal modes of action of bis(tributyltin)oxide on *Legionella pneumophila*. *Appl. Environ. Microbiol.* 45:48–57.

Stamper, D.M., E.R. Holm, and R.A. Brizzolara. 2008. Exposure times and energy densities for ultrasonic disinfection of *Escherichia coli*, *Pseudomonas aeruginosa*, *Enterococcus avium*, and sewage. *J. Environ. Eng. Sci.* 7:139–146.

Stanfield, G. and P.H. Jago. 1987. *The Development and Use of a Method for Measuring the Concentration of Assimilable Organic Carbon in Water*. Report PRU 1628-M. Water Research Centre, Medmenham, UK.

Starkey, M., K.A. Gray, S.I. Chang, and M.R. Parsek. 2004. A sticky business: the extracellular polymeric substance matrix of bacterial biofilms. In *Microbial Biofilms*, M. Ghannoum and G.A. O'toole (Eds). ASM Press, Washington, DC, pp. 174–191.

States, S.J., L.F. Conley, M. Ceraso, T.E. Stephenson, R.S. Wolford, R.M. Wadosky, A.M. McNamara, and R.B. Yee. 1985. Effect of metals on *Legionella pneumophila* growth in drinking water plumbing systems. *Appl. Environ. Microbiol.* 50:1149–1154.

States, S.J., L.F. Conley, S.G. Towner, R.S. Wolford, T.E. Stephenson, A.M. McNamara, R.M. Wadowsky, and R.B. Yee. 1987. An alkaline approach to treating cooling waters for control of *Legionella pneumophila*. *Appl. Environ. Microbiol.* 53:1775–1779.

States, S.J., J.M. Kuchta, L.F. Conley, R.S. Wolford, R.M. Wadowsky, and R.B. Yee. 1989. Factors affecting the occurrence of the legionnaires' disease bacterium in public water supplies. In *Biohazards of Drinking Water Treatment*, R.A. Larsen (Ed.). Lewis, Chelsea, MI, pp. 67–83.

States, S.J., R.M. Wadowsky, J.M. Kuchta, R.S. Wolford, L.F. Conley, and R.B. Yee. 1990. *Legionella* in drinking water. In *Drinking Water Microbiology*, G.A. McFeters (Ed.). Springer Verlag, New York, pp. 340–367.

States, S., M. Scheuring, R. Evans, E. Buzza, B. Movahed, T. Gigliotti, and L. Casson. 2000. Membrane filtration as posttreatment. *J. Am. Water Works Assoc.* 92(8):59–68.

States, S., J. Newberry, J. Wichterman, J. Kuchta, M. Scheuring, and L. Casson. 2004. Rapid analytical techniques for drinking water security investigations. *J. Am. Water Works Assoc.* 96(1):52–64.

Stathopoulos, G.A. and T. Vayonas-Arvanitidou. 1990. Detection of *Campylobacter* and *Yersinia* species in waters and their relatioship to indicator microorganisms. In *International Symposium on Health-Related Water Microbiology*, Tubingen, Germany, April 1–6, 1990.

Stauber, C.E., M.A. Elliott, F. Koksal, G.M. Ortiz, F.A. DiGiano, and M.D. Sobsey. 2006. Characterisation of the biosand filter for *E. coli* reductions from household drinking water under controlled laboratory and field use conditions. *Water Sci. Technol.* 54(3): 1–7.

Stauber, C.E., G.M. Ortiz, D.P. Loomis, and M.D. Sobsey. 2009. A randomized controlled trial of the concrete biosand filter and its impact on diarrheal disease in Bonao, Dominican Republic. *Am. J. Trop. Med. Hyg.* 80:286–293.

Stauber, C.E., E.R. Printy, F.A. McCarty, K.R. Liang, and M.D. Sobsey. 2012. Cluster randomized controlled trial of the plastic BioSand water filter in Cambodia. *Environ. Sci. Technol.* 46:722–728.

Steed, K.A. and J.O. Falkinham, III. 2006. Effect of growth in biofilms on chlorine susceptibility of *Mycobacterium avium* and *Mycobacterium intracellulare*. *Appl. Environ. Microbiol.* 72:4007–4011.

Steele, M., S. Unger, and J. Odumeru. 2003. Sensitivity of PCR detection of *Cyclospora cayetanensis* in raspberries, basil, and mesclun lettuce. *J. Microbiol. Methods* 54:277–280.

Stewart, M.H., R.L. Wolfe, and E.G. Means. 1990. Assessment of the bacteriological activity associated with granular activated carbon treatment of drinking water. *Appl. Environ. Microbiol.* 56:3822–3829.

Stoimenov, P.K., R.L. Klinger, G.L. Marchin, and K.J. Klabunde. 2002. Metal oxide nanoparticles as bactericidal agents. *Langmuir* 18:6679–6686.

Stoodley, P., K. Sauer, D.G. Davies, and J.W. Costerton. 2002. Biofilms as complex differentiated communities. *Annu. Rev. Microbiol.* 56:187–209.

Storey, M.V., B. van der Gaag, and B.P. Burns. 2011. Advances in on-line drinking water quality monitoring and early warning systems. *Water Res.* 45:741–747.

Stott, R., E. May, E. Ramirez, and A. Warren. 2003. Predation of *Cryptosporidium* oocysts by protozoa and rotifers: implications for water quality and public health. *Water Sci. Technol.* 47(3):77–83.

Stout, J.E., V.L. Yu, and M.G. Best. 1985. Ecology of *Legionella pneumophila* within water distribution systems. *Appl. Environ. Microbiol.* 49:221–228.

Stout, J.E., M.G. Best, and V.L. Yu. 1986. Susceptibility of members of the family *Legionellaceae* to thermal stress: implications for heat eradication methods in water distribution systems. *Appl. Environ. Microbiol.* 52:396–399.

St-Pierre, K., S. Levesque, E. Frost, N. Carrier, R.D. Arbeit, and S. Michaud. 2009. Thermotolerant coliforms are not a good surrogate for *Campylobacter* spp. in environmental waters. *Appl. Environ. Microbiol.* 75:6736–6744.

Straub, T.M., R.A. Bartholomew, C.O. Valdez, N.B. Valentine, A. Dohnalkova, R.M. Ozanich, C.J. Bruckner-Lea, and D.R. Call. 2011. Human norovirus infection of Caco-2 cells grown as a 3-dimensional tissue structure. *J. Water Health* 9(2):225–240.

Straub, T.M., J.R. Hutchison, R.A. Bartholomew, C.O. Valdez, N.B. Valentine, A. Dohnalkova, R.M. Ozanich, and C.J. Bruckner-Lea. 2013. Defining cell culture conditions to improve human norovirus infectivity assays. *Water Sci. Technol.* 67(4):863–868.

Suffet, I.H., A. Corado, D. Chou, M.J. McGuire, and S. Butterworth. 1996. Taste and odor survey. *J. Am. Water Works Assoc.* 88(4):168–180.

Sullivan, J.P., Jr. 2004. The WateISAC—one year later. *J. Am. Water Works Assoc.* 96(1):29–32.

Summers, R.S., S.M. Kim, K. Shimabuku, S.-H. Chae, and C.J. Corwin. 2013. Granular activated carbon adsorption of MIB in the presence of dissolved organic matter. *Water Res.* 47:3507–3513.

Sunada, K., Y. Kikuchi, K. Hashimoto, and A. Fujishima. 1998. Bactericidal and detoxification effects of TiO_2 thin film photocatalysis. *Environ. Sci. Technol.* 32:726–728.

Sutton, S.D. 2002. Quatification of microbial biomass. In *Encyclopedia of Environmental Microbiology*, G. Bitton (Ed.). Wiley-Interscience, New York, pp. 2652–2660.

Swerdlow, D.L., B.A. Woodruff, R.C. Brady, P.M. Griffin, S. Tippen, H.D. Donnell, Jr., E.E. Geldreich, B.J. Payne, A. Meyer, Jr., J.S. Wells, K.D. Greene, M. Bright, N.H. Bean, and P.A. Blake. 1992. A waterborne outbreak in Missouri of *Escherichia coli* 0157:H7 associated with bloody diarrhea and death. *Ann. Intern. Med.* 117:812–819.

Sykes, G. and F.A Skinner (Eds). 1973. *Actinomycetales: Characteristics and Practical Importance*. Academic Press, London.

Sykora, J.L., G. Keleti, R. Roche, D.R. Volk, G.P. Kay, R.A. Burgess, M.A. Shapiro, and E.C. Lippy. 1980. Endotoxins, algae and *Limulus* amoebocyte lysate test in drinking water. *Water Res.* 14:829–839.

Sylvestry-Rodriguez, N., K.R. Bright, D.C. Slack, D.R. Uhlmann, and C.P. Gerba. 2008. Silver as a residual disinfectant to prevent biofilm formation in water distribution systems. *Appl. Environ. Microbiol.* 74:1639–1641.

Symons, G.E. 2006. Water treatment through the ages. *J. Am. Water Works Assoc.* 98(3):87–98.

Takaara, T., D. Sano, Y. Masago, and T. Omura. 2010. Surface-retained organic matter of *Microcystis aeruginosa* inhibiting coagulation with polyaluminum chloride in drinking water treatment. *Water Res.* 44:3781–3786.

Tang, K.W., C. Dziallas, and H.-P. Grossart. 2011. Zooplankton and aggregates as refuge for aquatic protection from UV, heat and ozone stresses used for water treatment. *Environ. Microbiol.* 13:378–390.

Tatchou-Nyamsi-König, J.-A., E. Dague, M. Mullet, J.F.L. Duval, F. Gaboriaud, and J.-C. Block. 2008. Adhesion of *Campylobacter jejuni* and *Mycobacterium avium* onto polyethylene terephtalate (PET) used for bottled waters. *Water Res.* 42:4751–4760.

Taylor, G.R. and M. Butler. 1982. A comparison of the virucidal properties of chlorine, chlorine dioxide, bromine chloride and iodine. *J. Hyg.* 89:321–328.

Taylor, R.H., M.J. Allen, and E.E. Geldreich. 1979. Testing of home carbon filters. *J. Am. Water Works Assoc.* 71:577–581.

Taylor, D.N., K.T. McDermott, J.R. Little, J.G. Wells, and M.J. Blaser. 1983. *Campylobacter* enteritis from untreated water in the Rocky Mountains. *Ann. Intern. Med.* 99:38–40.

Taylor, R.H., J.O. Falkinham, III, C.D. Norton, and M.W. LeChevallier. 2000. Chlorine, chloramine, chlorine dioxide, and ozone susceptibility of *Mycobacterium avium*. *Appl. Environ. Microbiol.* 66:1702–1705.

Taylor, S.J., L.J. Ahonen, F.A.A.M. de Leij, and J.W. Dale. 2003. Infection of *Acanthamoeba castellanii* with *Mycobacterium bovis* and *M. bovis* BCG and survival of *M. bovis* within the amoebae. *Appl. Environ. Microbiol.* 69:4316–4319.

Templeton, M.R., R.C. Andrews, and R. Hoffmann. 2007. Removal of particle-associated bacteriophages by dual media filtration at different filter cycle stages and impacts on subsequent UV disinfection. *Water Res.* 41:2393–2406.

Templeton, M.R., R.C. Andrews, and R. Hofmann. 2008. Particle-associated viruses in water: impacts on disinfection processes. *Crit. Rev. Environ. Sci. Technol.* 38:137–164.

Teunis, P.F., C.L. Chappell, and P.C. Okhuysen. 2002. *Cryptosporidium* dose-response studies: variation between isolates. *Risk Anal.* 22(1):175–183.

Teunis, P.F.M., I.D. Ogden, N.J.C. Strachan. 2008. Hierarchical dose response of *E. coli* O157:H7 from human outbreaks incorporating heterogeneity in exposure. *Epidemiol. Infect.* 136(6):761–770.

Theron, J. and T.E. Cloete. 2002. Emerging waterborne infections: contributing factors, agents, and detection tools. *Crit. Rev. Microbiol.* 28:1–26.

Theron, J., J.A. Walker, and T.E. Cloete. 2008. Nanotechnology and water treatment: applications and emerging opportunities. *Crit. Rev. Microbiol.* 34:43–69.

Thomas, J.M. and N.J. Ashbolt. 2011. Do free-living amoebae in treated drinking water systems present an emerging health risk? *Environ. Sci. Technol.* 45:860–869.

Thomas, J.G., G. Ramage, and J.L. Lopez-Ribot. 2004. Biofilms and implants infections. In *Microbial Biofilms*, M. Ghannoum and G.A. O'Toole (Eds). ASM Press, Washington, DC, pp. 269–293.

Thomas, V., K. Herrera-Rimann, D.S. Blanc, and G. Greub. 2006. Biodiversity of amoebae and amoeba-resisting bacteria in a hospital water network. *Appl. Environ. Microbiol.* 72:2428–2438.

Thomas, V., J.-F. Loret, M. Jousset, and G. Greub. 2008. Biodiversity of amoebae and amoeba-resisting bacteria in a drinking water treatment plant. *Environ. Microbiol.* 10:2728–2745.

Thomas, V., G. McDonnell, S.P. Denyer, and J.-Y. Maillard. 2010. Free-living amoebae and their intracellular pathogenic microorganisms: risks for water quality. *FEMS Microbiol. Rev.* 34:231–259.

Thompson, S.S., J.L. Jackson, M. Suva-Castillo, W.A. Yanko, Z. El Jack, J. Kuo, C.-L. Chen, F.P. Williams, and D.P. Schnurr. 2003. Detection of infectious adenoviruses in tertiary-treated and untraviolet-disinfected wastewater. *Water Environ. Res.* 75:163–170.

Tian, J.-Y., H. Liang, X. Li, S.J. You, S. Tian, and G.-B. Li. 2008. Membrane coagulation bioreactor (MCBR) for drinking water treatment. *Water Res.* 42:3910–3920.

Tiwari, S.K., W.P. Schmidt, J. Darby, Z.G. Kariuki, and M.W. Jenkins. 2009. Intermittent slow sand filtration for preventing diarrhoea among children in Kenyan households using unimproved water sources: randomized controlled trial. *Trop. Med. Int. Health* 14(11):1374–1382.

Tobin, R.S., D.K. Smith, and J.A. Lindsay. 1981. Effects of activated carbon and bacteriostatic filters on microbiological quality of drinking water. *Appl. Environ. Microbiol.* 41:646–651.

Tobin, R.S., P. Ewan, K. Walsh, and B. Dutka. 1986. A survey of *Legionella pneumophila* in water in 12 Canadian cities. *Water Res.* 20:495–501.

Tomboulian, P., L. Schweitzer, K. Mullin, J. Wilson, and D. Khiari. 2004. Materials used in drinking water distribution systems: contribution to taste and odor. *Water Sci. Technol.* 49:219–226.

Torok, T.J., R.V. Tauxe, R.P. Wise, J.R. Livengood, R. Sokolow, and S. Mauvais. 1997. A large community outbreak of Salmonellosis caused by intentional contamination of restaurant salad bars. *J. Am. Med. Assoc.* 278:389–395.

Torres, A.G. 2002. *Shigella*. In *Encyclopedia of Environmental Microbiology*, G. Bitton (Editor-in-Chief). Wiley-Interscience, New York, pp. 2865–2871.

Torvinen, E., M.J. Lehtola, P.J. Martikainen, and I.T. Miettinen. 2007. Survival of *Mycobacterium avium* in drinking water biofilms as affected by water flow velocity, availability of phosphorus, and temperature. *Appl. Environ. Microbiol.* 73:6201–6207.

Touron-Bodilis, A., C. Pougnard, H. Frenkiel-Lebosse, and S. Hallier-Soulier. 2011. Usefulness of real-time PCR as a complementary tool to the monitoring of *Legionella* spp. and *Legionella pneumophila* by cultire in industrial cooling systems. *J. Appl. Microbiol.* 111:499–510.

Trulear, M.G. and W.G. Characklis. 1982. Dynamics of biofilm processes. *J. Water Pollut. Control Fed.* 54:1288–1301.

Trussell, R.R. 2006. Water treatment: the past 30 years. *J. Am. Water Works Assoc.* 98(3):100–108.

Tsai, W.-L., C.E. Miller, and E.R. Richter. 2000. Determination of the sensitivity of a rapid *Escherichia coli* O157:H7 assay for testing 375-gram composite samples. *Appl. Environ. Microbiol.* 66:4149–4151.

Tung, H.-H., R.F. Unz, and Y.F. Xie. 2006. HAA removal by GAC adsorption. *J. Am. Water Works Assoc.* 98(6):107–112.

Tuovinen, O.H. and J.C. Hsu. 1982. Aerobic and anaerobic microorganisms in tubercles of the Columbus, Ohio, water distribution system. *Appl. Environ. Microbiol.* 44:761–764.

Ueda, T. and N.J. Horan. 2000. Fate of indigenous bacteriophage in a membrane bioreactor. *Water Res.* 34:2151–2159.

Unger, M. and M.R. Collins. 2008. Assessing *Escherichia coli* removal in the schmutzdecke of slow-rate biofilters. *J. Am. Water Works Assoc.* 100(12):60–72.

United Nations Educational, Scientific and Cultural Organization (UNESCO). 2009. *World Water Report. 3. Water in a Changing World.* UNESCO, Paris, France.

Upadhyayula, V.K.K., S. Deng, M.C. Mitchell, and G.B. Smith. 2009. Application of carbon nanotube technology for removal of contaminants in drinking water: a review. *Sci. Total Environ.* 408(1):1–13.

Urfer, D., P.M. Huck, S.D. Booth, and B.M. Bradley. 1997. Biological filtration for BOM and particle removal: a critical review. *J. Am. Water Works Assoc.* 89(12):83–96.

U.S. EPA. 1989. Drinking water; national primary drinking water regulations; filtration, disinfection; turbidity, *Giardia lamblia*, viruses, *Legionella*, and heterotrophic bacteria, final rule. *Fed. Regist.* 54(124):27486–27541.

U.S. EPA. 1990. *Technologies for Upgrading Existing or Designing New Drinking Water Treatment Facilities.* EPA/625/4-89/023.

U.S. EPA. 1998. United Announcement of the drinking water contaminant candidate list: notice. *Fed. Regist.* 63:10274–10287.

U.S. EPA. 2001. *Method 1623: Cryptosporidium and Giardia in water by filtration/IMS/FA.* EPA-821-R-99-006.

U.S. EPA. 2003. *Small Drinking Water Systems Handbook: A Guide to Packaged Filtration and Disinfection Technologies with Remote Monitoring and Control Tools.* EPA-600-R-03-041.

U.S. EPA. 2005. *Drinking Water and Ground Water Statistics for 2004.* EPA 816-K-05-001.

U.S. EPA. 2006. *Ultraviolet Disinfection Guidance Manual for the Final Long Term 2 Enhanced Surface Water Treatment Rule*. EPA 815-R-06-007.

U.S. EPA. 2007a. Anthrax spore decontamination using chlorine dioxide. http://www.epa. gov/pesticides/factsheets/chemicals/chlorinedioxidefactsheet.htm (accessed May 5, 2014).

U.S. EPA. 2007b. *Small Drinking Water Systems: State of the Industry and Treatment Technologies to Meet the Safe Drinking Water Act Requirements*. EPA/600/R-07/110.

U.S. EPA. 2012. *Microbial Risk Assessment Guideline: Pathogenic Microorganisms with Focus on Food and Water*. EPA/100/J-12/001 and USDA/FSIS/2012-001.

Vaerewijck, M.J.M., G. Huys, J.C. Palomino, J. Swings, and F. Portaels. 2005. Mycobacteria in drinking water distribution systems: ecology and significance for human health. *FEMS Microbiol. Rev.* 29:911–934.

Valster, R.M., B.A. Wullings, G.B. Bakker, H. Smidt, and D. van der Kooij. 2009. Free living protozoa in two unchlorinated drinking water supplies, identified by phylogenic analysis of 18S rRNA gene sequences. *Appl. Environ. Microbiol.* 75:4736–4746.

Vandenberg, O., A. Dediste, K. Houf, S. Ibekwem, H. Souayah, S. Cadranel, N. Douat, G. Zissis, J.-P. Butzler, and P. Vandamme. 2004. *Arcobacter* species in humans. *Emerg. Infect. Dis.* 10:1863–1867.

Vanden Bossche, G. 1994. Detergent conditioning of environmental samples: the most sensitive method for the detection of viral activity. Paper presented at the *IAWQ 17th Biennial International Conference Health-Related Water Microbiology Symposium*, Budapest, Hungary.

Van Hallem, D., H. van der Laan, S.G.J. Heijman, J.C. van Dijk, and G.L. Amy. 2009. Assessing the sustainability of the silver-impregnated ceramic pot filter for low-cost household drinking water treatment. *Phys. Chem. Earth* 34(1–2):36–42.

Van Reis, R. and A. Zydney. 2007. Bioprocess membrane technology. *J. Memb. Sci.* 297:16–50.

Varma, M., J.D. Hestera, F.W. Schaefer, III, M.W. Warea, and H.D.A. Lindquist. 2003. Detection of *Cyclospora cayetanensis* using a quantitative real-time PCR assay. *J. Microbiol. Methods* 53:27–36.

Vaughn, J.M. and J.F. Novotny. 1991. Virus inactivation by disinfectants. In *Modeling the Environmental Fate of Microorganisms*, C.J. Hurst (Ed.). American Society of Microbiology, Washington, DC, pp. 217-241.

Vaz-Moreira, I., O.C. Nunes, and C.M. Manaia. 2011. Diversity and antibiotic resistance patterns of Sphingomonadaceae isolates from drinking water. *Appl. Environ. Microbiol.* 77:5697–5706.

Velimirov, B. 2001. Nanobacteria, ultramicrobacteria and starvation forms: a search for the smallest metabolizing bacterium. *Microb. Environ.* 16:67–77.

Venkobachar, C., Z. Invegar, and A.V.S. Prabhakara Raj. 1977. Mechanism of disinfection: effect of chlorine on cell membrane functions. *Water. Res.* 11:727–729.

de Vet, W.W.J.M., I.J.T. Dinkla, L.C. Rietveld, and M.C.M. van Loosdrech. 2011. Biological iron oxidation by *Gallionella* spp. in drinking water production under fully aerated conditions. *Water Res.* 45:5389–5398.

Villanueva, C.M., K.P. Cantor, J.O. Grimalt, N. Malats, D. Silverman, A. Tardon, R. Garcia-Closas, C. Serra, A. Carrato, G. Castano-Vinyals, R. Marcos, N. Rothman, F.X. Real, M. Dosemeci, and M. Kogevinas. 2007. Bladder cancer and exposure to water disinfection byproducts through ingestion, bathing, showering, and swimming in pools. *Am. J. Epidemiol.* 165:148–156.

Visvesvara, G.S., M.J. Peralta, F.H. Brandt, M. Wilson, C. Aloisio, and E. Franko. 1987. Production of monoclonal antibodies to *Naegleria fowleri*, agent of primary amoebic meningoencephalitis. *J. Clin. Microbiol.* 25:1629–1634.

Vital, M., F. Hammes, and T. Egli. 2008. *Escherichia coli* O157 can grow in natural freshwater at low carbon concentrations. *Environ. Microbiol.* 10:2387–2396.

Vital, M., M. Dignum, A. Magic-Knezev, P. Ross, L. Rietveld, and F. Hammes. 2012. Flow cytometry and adenosine tri-phosphate analysis: alternative possibilities to evaluate major bacteriological changes in drinking water treatment and distribution systems. *Water Res.* 46:4665–4676.

Volk, C.J. and M.W. LeChevallier. 2000. Assessing biodegradable organic matter. *J. Am. Water Works Assoc.* 92(5):64–76.

Wadowsky, R.M. and R.B. Yee. 1985. Effect on non-legionellaceae bacteria on the multiplication of *Legionella pneumophila* in potable water. *Appl. Environ. Microbiol.* 49:1206–1210.

Wagner, M. and J. Oehlmann. 2009. Endocrine disruptors in bottled mineral water: total estrogenic burden and migration from plastic bottles. *Environ. Sci. Pollut. Res.* 16:278–286.

Walker, J.T. and P.D. Marsh. 2007. Microbial biofilm formation in DUWS and their control using disinfectants. *J. Dent.* 35:721–730.

Walker, G.S., F.P. Lee, and E.M. Aieta. 1986. Chlorine dioxide for taste and odor control. *J. Am. Water Works Assoc.* 78:84–93.

Walker, R.E., J.M. Petersen, K.W. Stephens, and L.A. Dauphin. 2010. Optimal swab processing recovery method for detection of bioterrorism-related *Francisella tularensis* by real-time PCR. *J. Microbiol. Methods* 83:42–47.

Wallis, C., C.H. Stagg, and J.L. Melnick. 1974. The hazards of incorporating charcoal filters into domestic water systems. *Water Res.* 8:111–113.

Wallis, P.M., S.L. Erlandsen, J.L. Isaac-Renton, M.E. Olson, W.J. Robertson, and H. van Keulen. 1996. Prevalence of *Giardia* cysts and *Cryptosporidium* oocysts and characterization of *Giardia* spp. Isolated from drinking water in Canada. *Appl. Environ. Microbiol.* 62:2789–2797.

Walse, S.S. and W.A. Mitch. 2008. Nitrosamine carcinogens also swim in chlorinated pools. *Environ. Sci. Technol.* 42:1032–1037.

Walski, T.M. 2006. A history of water distribution. *J. Am. Water Works Assoc.* 98(3):110–121.

Walter, R.K., P.-O. Lin, M. Edwards, and R.E. Richardson. 2011. Investigation of factors affecting the accumulation of vinyl chloride in polyvinyl chloride piping used in drinking water distribution systems. *Water Res.* 45:2607–2615.

Wang, X. and T.-T. Lim. 2013. Highly efficient and stable Ag-AgBr/TiO2 composites for destruction of *Escherichia coli* under visible light irradiation. *Water Res.* 47:4148–4158.

Wang, D., Q. Wu, L. Yao, M. Wei, X. Kou, and J. Zhang. 2008. New target tissue for food-borne virus detection in oysters. *Lett. Appl. Microbiol.* 47:405–409.

Wang, Y., F. Hammes, K. De Roy, W. Verstraete, and N. Boon. 2010. Past, present and future applications of flow cytometry in aquatic microbiology. *Trends Biotechnol.* 28:416–424.

Wang, H., S. Masters, Y. Hong, J. Stallings, J.O. Falkinham, III, M.A. Edwards, and A. Priden. 2012a. Effect of disinfectant, water age, and pipe material on occurrence and persistence of *Legionella*, mycobacteria, *Pseudomonas aeruginosa*, and two amoebas. *Environ. Sci. Technol.* 46:11566–11574.

Wang, H., M. Edwards, J.O. Falkinham, III, and A. Pruden. 2012b. Molecular survey of the occurrence of *Legionella* spp., *Mycobacterium* spp., *Pseudomonas aeruginosa*, and amoeba hosts in two chloraminated drinking water distribution systems. *Appl. Environ. Microbiol.* 78:6285–6294.

Warburton, D.W. 1993. A review of the microbiological quality of bottled water sold in Canada. Part 2. The need for more stringent standards and regulations. *Can. J. Microbiol.* 39:158–168.

Ware, M.W., L. Wymer, H.D. Lindquist, and F.W. Schaefer, III, 2003. Evaluation of an alternative IMS dissociation procedure for use with Method 1622: detection of *Cryptosporidium* in water. *J. Microbiol. Methods* 55:575–583.

Watson, S.B., F. Jüttner, and O. Köster. 2007. Daphnia behavioural responses to taste and odour compounds: ecological significance and application as an inline treatment plant monitoring tool. *Water Sci. Technol.* 55(5):23–31.

Webb, S. 2012. Harvesting photons to kill microbes: ES&T's top environmental technology article 2011. *Environ. Sci. Technol.* 46:3609–3610.

Weber, W.J., M. Pirbazari, and G.L. Melson. 1978. Biological growth on activated carbon: an investigation by scanning electron microscopy. *Environ. Sci. Technol.* 12:817–819.

Wegelin, M. and B. Sommer. 1998. Solar water disinfection (SODIS)—destined for worldwide use? *Waterlines* 16:30–32.

Wegelin, M., S. Canonica, K. Mechsner, F. Pesaro, and A. Metzler. 1994. Solar water disinfection: scope of the process and analysis of radiation experiments. *J. Water Supply Res. Technol.* 43:154–169.

Weinrich, L.A., E. Giraldo, and M.W. LeChevallier. 2009. Development and application of a bioluminescence-based test for assimilable organic carbon in reclaimed waters. *Appl. Environ. Microbiol.* 75:7385–7390.

Weinrich, L.A., O.D. Schneider, and M.W. LeChevallier. 2011. Bioluminescence-based method for measuring assimilable organic carbon in pretreatment water for reverse osmosis membrane desalination. *Appl. Environ. Microbiol.* 77:1148–1150.

Wellman, N., S.M. Fortun, and B.R. McLeod. 1996. Bacterial biofilms and the bioelectric effect. *Antimicrob. Agents Chemother.* 40:2012–2014.

van der Wende, E. and W.G. Characklis. 1990. Biofilms in potable water distribution systems. In *Drinking Water Microbiology*, G.A. McFeters (Ed.). Springer Verlag, New York, pp. 249–268.

Werner, P. 1985. Eine methode zur bestimmung der veirkemungsneigung von trinkwasser. *Von Wasser* 65:257–262.

Westerhoff, P., P. Prapaipong, E.T. Shock, and A. Hillaireau. 2008. Antimony leaching from polyethylene terephthalate (PET) plastic used for bottled drinking water. *Water Res.* 42:551–556.

Westrell, I., O. Bergstedt, T.A. Stenstrom, and N.J. Ashbolt. 2003. A theoretical approach to assess microbial risks due to failures in drinking water systems. Intern. *J. Environ. Health Res.* 13:181–197.

Westrick, J.A., D.C. Szlag, B.J. Southwell, and J. Sinclair. 2010. A review of cyanobacteria and cyanotoxins removal/inactivation in drinking water treatment. *Anal. Bioanal. Chem.* 397:1705–1714.

Wheeler, D., J. Bartram, and B.J. Lloyd. 1988. The removal of viruses by filtration through sand. In *Slow Sand Filtration: Recent Developments in Water Treatment Technology*, N.J.D. Graham (Ed.). Ellis Horwood, Chichester, UK, pp. 207–229.

Whitby, G.E., G. Palmateer, W.G. Cook, J. Maarschalkerweerd, D. Huber, and K. Flood. 1984. Ultraviolet disinfection of secondary effluent. *J. Water Pollut. Control Fed.* 56:844–850.

Whitler, J. 2007. Emergency preparedness for drinking water and wastewater systems. *J. Am. Water Works Assoc.* 99(3):36–38.

WHO/UNICEF. 2000. *Global Water Supply and Sanitation Assessment 2000*. http://www. who.int/water_sanitation_health (accessed May 5, 2014).

WHO/UNICEF. 2004. *Meeting the MDG Drinking Water and Sanitation Target: A Mid-Term Assessment of Progress*. WHO, Geneva.

WHO/UNICEF. 2009. *Diarrhoea: Why Children are Still Dying and What Can Be Done.* World Health Organization, Geneva, Switzerland.

WHO/UNICEF. 2010. *Progress on Sanitation and Drinking Water. 2010 Update.* World Health Organization, Geneva, Switzerland. http://whqlibdoc.who.int/publications/2010/9789241563956_eng_full_text.pdf

WHO/UNICEF. 2012. *Progess on Drinking Water and Sanitation. 2012 Update.* WHO/UNICEF Joint Monitoring Programme for Water Supply and Sanitation, Geneva, Switzerland. http://www.who.int/water_sanitation_health/publications/2012/jmp_report/en/index .html (accessed May 5, 2014).

Wickramanayake, G.B., A.J. Rubin, and O.J. Sproul. 1985. Effect of ozone and storage temperature on *Giardia* cysts. *J. Am. Water Works Assoc.* 77:74–77.

van der Wielen, P.W.J.J. and D. van der Kooij. 2010. Effect of water composition, distance and season on the adenosine triphosphate concentration in unchlorinated drinking water in the Netherlands. *Water Res.* 44:4860–4867.

van der Wielen, P.W.J.J. and D. van der Kooij. 2013. Nontuberculous mycobacteria, fungi, and opportunistic pathogens in unchlorinated drinking water in the Netherlands. *Appl. Environ. Microbiol.* 79:825–834.

Wigginton, K.R., B.M. Pecson, T. Sigstam, F. Bosshard, and T. Kohn. 2012. Virus inactivation mechanisms: impact of disinfectants on virus function and structural integrity. *Environ. Sci. Technol.* 46:12069–12078.

Wilcox, D.P., E. Chang, K.L. Dickson, and K.R. Johansson. 1983. Microbial growth associated with granular activated carbon in a pilot water treatment facility. *Appl. Environ. Microbiol.* 45:406–416.

Wilczak, A., J.G. Jacangelo, J.P. Marcinko, L.H. Odell, G.J. Kirmeyer, and R.L. Wolfe. 1996. Occurrence of nitrification in chloraminated distribution systems. *J. Am. Water Works Assoc.* 88:74–85.

Wilkinson, F.H. and K.G. Kerr. 1998. Bottled water as a source of multi-resistant *Stenotrophomonas* and *Pseudomonas* species for neutropenic patients. *Eur. J. Cancer Care* 7:12–14.

Willcomb, G.E. 1923. Twenty years of filtration practice at Albany. *J. Am. Water Works Assoc.* 10:97–103.

Williams, F.P. and E.W. Akin. 1986. Waterborne viral gastroenteritis. *J. Am. Water Works Assoc.* 78:34–39.

Williams, R.B. and G.L. Culp. 1986. *Handbook of Public Water Systems.* Van Nostrand Reinhold, New York.

Williams, M.M., J.W. Domingo, M.C. Meckes, C.A. Kelty, and H.S. Rochon. 2004. Phylogenetic diversity of drinking water bacteria in a distribution system simulator. *J. Appl. Microbiol.* 96:954–964.

Williams, P., K. Winzer, W.C. Chan, and M. Cámara. 2007. Look who's talking: communication and quorum sensing in the bacterial world. *Phil. Trans. R. Soc. B* 362:1119–1134.

Wingender, J. and H.C. Flemming. 2004. Contamination potential of drinking water distribution network biofilms. *Water Sci. Technol.* 49(11):277–285.

Wingender, J. and H.-C. Flemming. 2011. Biofilms in drinking water and their role as reservoir for pathogens. *Int. J. Hyg. Environ. Health* 214:417–423.

Winzer, K., K.R. Hardie, and P. Williams. 2002. Bacterial cell-to-cell communication: sorry, can't talk now—gone to lunch!. *Curr. Opin. Microbiol.* 5(2):216–222.

Witherell, L.E., R.W. Duncan, K.M. Stone, L.J. Stratton, L. Orciani, S. Kappel, and D.A. Jillson. 1988. Investigation of *Legionella pneumophila* in drinking water. *J. Am. Water Works Assoc.* 80:87–93.

Withers, H., S. Swift, and P. Williams. 2001. Quorum sensing as an integral component of gene regulatory networks in Gram-negative bacteria. *Curr. Opin. Microbiol.* 4:186–193.

Wolfe, R.L. 1990. Ultraviolet disinfection of potable water. *Env. Sci. Technol.* 24:768–773.

Wolfe, R.L. and N.I. Lieu. 2002. Nitrifying bacteria in drinking water. In *Encyclopedia of Environmental Microbiology*, G. Bitton (Editor-in-Chief). Wiley-Interscience, New York, pp. 2167–2176.

Wolfe, R.L., N.R. Ward, and B.H. Olson. 1984. Inorganic chloramines as drinking water disinfectants: a review. *J. Am. Water Works Assoc.* 76:74–88.

Wolfe, R.L., E.G. Means, III, M.K. Davis, and S.E. Barrett. 1988. Biological nitrification in covered reservoirs containing chloraminated water. *J. Am. Water Works Assoc.* 80(9):109–114.

Wolf, S., W.M. Williamson, J. Hewitt, M. Rivera-Aban, S. Lin, A. Ball, P. Scholes, and G.E. Greening. 2007. Sensitive multiplex real-time reverse transcription-PCR assay for the detection of human and animal noroviruses in clinical and environmental samples. *Appl. Environ. Microbiol.* 73:5464–5470.

Wong, M.-S., W.-C. Chu, D.-S. Sun, H.-S. Huang, J.-H. Chen, P.-J. Tsai, N.-T. Lin, M.-S. Yu, S.-F. Hsu, S.-L. Wang, and H.H. Chang. 2006. Visible-light-induced bactericidal activity of a nitrogen-doped titanium photocatalyst against human pathogens. *Appl. Environ. Microbiol.* 72:6111–6116.

Wood, D.J., K. Bijlsma, J.C. de Jong, and C. Tonkin. 1989. Evaluation of a commercial monoclonal antibody-based immunoassay for detection of adenovirus types 40 and 41 in stool specimens. *J. Clin. Microbiol.* 27:1155–1158.

Woolhouse, M.E.J. 2006. Where do emerging pathogens come from? *Microbe* 1(11):511–515.

World Health Organization. 1979. *WHO International Reference Center for Community Water Supply Annual Report*. Rijswijk, The Netherlands.

World Health Organization (WHO). 1996a. *Guidelines for Drinking Water Quality, Vol. 2: Recommendations*. WHO, Geneva, Switzerland.

World Health Organization (WHO). 1996b. *Water and Sanitation: WHO fact sheet no. 112*. WHO, Geneva.

World Health Organization (WHO). 1999. *Toxic Cyanobacteria in Water: A Guide to Their Public Health Consequences*. ISBN 0-419-23930-8. World Health Organization, Geneva, Switzerland.

World Health Organization (WHO). 2003. *The Right to Water*, WHO, Geneva, Switzerland.

World Health Organization (WHO). 2005. *Trihalomethanes in Drinking Water: Background Document for Development of WHO Guidelines for Drinking Water Quality*. WHO, Geneve, Switzerland.

World Health Organization (WHO). 2008. *Fuel for Life: Household Energy and Health (2006)*. http://www.who.int/indoorair/publications/fuelforlife.pdf (accessed May 20, 2008).

World Health Organization (WHO). 2011a. *Cholera*. Fact sheet no. 107. WHO. http://www.who.int/mediacentre/factsheets/fs107/en/index.html (accessed May 5, 2014). (accessed May 5, 2014).

World Health Organization (WHO). 2011b. *Evaluating Household Water Treatment Options: Health-Based Targets and Microbiological Performance Specifications*. WHO, Geneva, Switzerland.

World Health Organization (WHO). 2011c. *Guidelines for Drinking-Water Quality*, 4th edition. WHO, Geneva, Switzerland. http://www.who.int/water_sanitation_health/publications/2011/dwq_guidelines/en/. (accessed May 5, 2014).

Wright, H., D. Gaithuma, D. Greene, and M. Aieta. 2006. Integration of validation, design, and operation provides optimal implementation of UV disinfection. *J. Am. Water Works Assoc.* 98(10):81–92.

Wright, S.J., J.D. Semrau, and D.R. Keeney. 2008. Microbial fouling of a reverse osmosis municipal water treatment system. *Water Environ. Res.* 80:703–707.

Wu, X., E.M. Joyce, and T.J. Mason. 2012. Evaluation of the mechanisms of the effect of ultrasound on *Microcystis aeruginosa* at different ultrasonic frequencies. *Water Res.* 46:2851–2858.

Wuertz, S. 2002. Gene exchange in biofilms. In *Encyclopedia of Environmental Microbiology*, G. Bitton (Editor-in-Chief). Wiley-Interscience, New York, pp. 1408–1420.

Wyn-Jones, A.P. and J. Sellwood. 2001. Enteric viruses in the aquatic environment. *J. Appl. Microbiol.* 91:945–962.

Xagoraraki, I., G.W. Harrington, P. Assavasilavasukul, and J.H. Standridge. 2004. Removal of emerging waterborne pathogens and pathogen indicators. *J. Am. Water Works Assoc.* 96(5):102–113.

Xi, C., Y. Zhang, C.F. Marrs, W. Ye, C. Simon, B. Foxman, and J. Nriagu. 2009. Prevalence of antibiotic resistance in drinking water treatment and distribution systems. *Appl. Environ. Microbiol.* 75:5714–5718.

Xiao, S., W. An, Z. Chen, D. Zhang, J. Yu, and M. Yang. 2012. The burden of drinking water-associated cryptosporidiosis in China: the large contribution of the immunodeficient population identified by quantitative microbial risk assessment. *Water Res.* 46:4272–4280.

Xue, B., M. Jin, D. Yang, X. Guo, Z. Chen, Z. Shen, X. Wang, Z. Qiu, J. Wang, B. Zhang, and J. Li. 2013a. Effects of chlorine and chlorine dioxide on human rotavirus infectivity and genome stability. *Water Res.* 47:3329–3338.

Xue, Z., V.R. Sendamangalam, C.L. Gruden, and Y. Seo. 2013b. Multiple roles of extracellular polymeric substances on resistance of biofilm and detached clusters. *Environ. Sci. Technol.* 46:13212–13219.

Yamamura, H., K. Kimura, and Y. Watanabe. 2007. Mechanism involved in the evolution of physically irreversible fouling in microfiltration and ultrafiltration membranes used for drinking water treatment. *Environ. Sci. Technol.* 41:6789–6794.

Yang, H., J.A. Wright, and S.W. Gundry. 2012. Letter to the editor: household water treatment in China. *Am. J. Trop. Med. Hyg.* 86:554–555.

Yao, C., X. Li, K.G. Neoh, Z. Shi, and E.T. Kang. 2008. Surface modification and antibacterial activity of electrospun polyurethane fibrous membranes with quaternary ammonium moieties. *J. Memb. Sci.* 320:259–267.

Yates, M.V., J. Malley, P. Rochelle, and R. Hoffman. 2006. Effect of adenovirus resistance on UV disinfection requirements: a report on the state of adenovirus science. *J. Am. Water Works Assoc.* 98(6):93–106.

Yee, R.B. and R.M. Wadowsky. 1982. Multiplication of *Legionella pneumophila* in unsterilized tap water. *Appl. Environ. Microbiol.* 43:1130–1134.

Yeh, H.-Y., Y.-C. Hwang, M.V. Yates, A. Mulchandani, and W. Chen. 2008. Detection of hepatitis A virus by using a combined cell culture-molecular beacon assay. *Appl. Environ. Microbiol.* 74:2239–2243.

Yen, H.-K., T.-F. Lin, and P.-C. Liao. 2011. Simultaneous detection of nine cyanotoxins in drinking water using dual solid-phase extraction and liquid chromatography-mass spectrometry. *Toxicon* 58:209–218.

Yeon, K.M., W.S. Cheong, H.S. Oh, W.N. Lee, B.K. Hwang, C.H. Lee, H. Beyenal, and Z. Lewandowski. 2009a. Biofouling control paradigm in a membrane bioreactor for advanced wastewater treatment. *Environ. Sci. Technol.* 43:380–385.

Yeon, K.M., C.H. Lee, and J. Kim. 2009b. Magnetic enzyme carrier for effective biofouling control in the membrane bioreactor based on enzymatic quorum quenching. *Environ. Sci. Technol.* 43:7403–7409.

Yongsi, H.B.N. 2010. Suffering for water, suffering from water: access to drinking water and associated health risks in Cameroon. *J. Health Popul. Nutr.* 28(5):424–435.

Yoo, R.S., W.W. Carmichael, R.C. Hoehn, and S.E. Hrudey. 1995. *Cyanobacterial (Blue-Green Algal) Toxins: A Resource Guide*. American Water Works Association Research.

Young, B. and P. Setlow. 2004. Mechanisms of killing of *Bacillus subtilis* spores by Decon and Oxone, two general decontaminants for biological agents. *J. Appl. Microbiol.* 96:289–301.

Yu, J.C., W. Ho, J. Yu, H. Yip, P. Wong, and J. Zhao. 2005. Efficient visible-light-induced photocatalytic disinfection on sulfur-doped nanocrystalline titania. *Environ. Sci. Technol.* 39:1175–1179.

Yu, J., D. Kim, and T. Lee. 2010. Microbial diversity in biofilms on water distribution pipes of different materials. *Water Sci Technol.* 61(1):163–171.

Zaitlin, B. and S.B. Watson. 2006. Actinomycetes in relation to taste and odour in drinking water: myths, tenets and truths. *Water Res.* 40:1741–1753.

Zamyadi, A., L. Ho, G. Newcombe, H. Bustamante, and M. Prevost. 2012. Fate of toxic cyanobacterial cells and disinfection by-products formation after chlorination. *Water Res.* 46:1534–1535.

Zamyadi, A., S. Dorner, S. Sauvé, D. Ellis, A. Bolduc, C. Bastien, and M. Prévost. 2013. Species dependence of cyanobacteria removal efficiency by different drinking water treatment processes. *Water Res.* 47:2689–2700.

Zhang, Y. and M. Edwards. 2009. Accelerated chloramine decay and microbial growth by nitrification in premise plumbing. *J. Am. Water Works Assoc.* 101(11):51–62.

Zhang, P., T.M. LaPara, E.H. Goslan, Y. Xie, S.A. Parsons, and R.M. Hozalski. 2009a. Biodegradation of haloacetic acids by bacterial isolates and enrichment cultures from drinking water systems. *Environ. Sci. Technol.* 43:3169–3175.

Zhang, X., J.H. Pan, A.J. Du, W. Fu, D.D. Sun, and J.O. Leckie. 2009b. Combination of one-dimensional TiO_2 nanowire photocatalytic oxidation with microfiltration for water treatment. *Water Res.* 43:1179–1186.

Zhang, R., H.X. Li, X.F. Wu, F.C. Fan, B.Y. Sun, Z.S. Wang, Q. Zhang, and Y. Tao. 2009c. Current situation analysis on China rural drinking water quality [in Chinese]. *J. Environ. Health* 26:3–5.

Zhang, Z., C. McCann, J. Hanrahan, A. Jencson, D. Joyce, S. Fyffe, S. Piesczynski, R. Hawks, J.E. Stout, V.L. Yu, and R.D. Vidic. 2009d. *Legionella* control by chlorine dioxide in hospital water systems. *J. Am. Water Works Assoc.* 101(5):117–127.

Zhang, Y., N. Love, and M. Edwards. 2009e. Nitrification in drinking water systems. *Crit. Rev. Environ. Sci. Technol.* 39:153–208.

Zheng, M., X. Wang, L.J. Templeton, D.R. Smulski, R.A. LaRossa, and G. Storz. 2001. DNA microarray-mediated transcriptional profiling of the *Escherichia coli* response to hydrogen peroxide. *J. Bacteriol.* 183:4562–4570.

Zhu, I.Z., T. Getting, and D. Bruce. 2010. Review of biologically active filters in drinking water applications. *J. Am. Water Works Assoc.* 102(12):67–77.

Zierler, S., R.A. Danley, and L. Feingold. 1987. Type of disinfectant in drinking water and patterns of mortality in Massachussetts. *Environ. Health Perspect.* 68:275–287.

Zilinskas, R.A. 1997. Iraq's biological weapons. The past as future? *J. Am. Med. Assoc.* 278:418–424.

Zimmer, J.L. and R.M. Slawson. 2002. Potential repair of *Escherichia coli* DNA following exposure to UV radiation from both medium- and low-pressure UV sources used in drinking water treatment. *Appl. Environ. Microbiol.* 68:3293–3299.

Zimmer, J.L., R.M. Slawson, and P.M. Huck. 2003. Inactivation and potential repair of *Cryptosporidium parvum* following low- and medium-pressure ultraviolet irradiation. *Water Res.* 37:3517–3523.

Zodrow, K., L. Brunet, S. Mahendra, D. Li, A. Zhang, Q. Li, and P.J.J. Alvarez. 2009. Polysulfone ultrafiltration membranes impregnated with silver nanoparticles show improved biofouling resistance and virus removal. *Water Res.* 43:715–723.

Zwiener, C., S.D. Richardson, D.M. De Marini, T. Grummt, T. Glauner, and F.H. Frimmel. 2006. Drowning in disinfection byproducts? Assessing swimming pool water. *Environ. Sci. Technol.* 41(2):363–372.

van Zyl, W.B., N.A. Page, W.O.K. Grabow, A.D. Steele, and M.B. Taylor. 2006. Molecular epidemiology of group A rotaviruses in water sources and selected raw vegetables in Southern Africa. *Appl. Environ. Microbiol.* 72:4554–4560.

INDEX

Note: Page numbers followed by *f* and *t* indicate figures and tables, respectively.

Acanthamoeba, 11, 84, 130, 131
Acanthamoeba spp., 34
Acidithiobacillus ferrooxidans, 134
Acidithiobacillus thiooxidans, 134
Acidovorax, 98, 99
Acidovorax facilis, 199
Acinetobacter, 13, 37, 49, 99
Acinetobacter baumannii, 13
Actinomycetes, 129–130
Activated carbon, 48–51
Acute gastrointestinal illness (AGI), 92
Acyl homoserine lactones (AHL), 101
Adenoviruses, 4, 37
Aeromonas, 13, 34, 49, 99
Aeromonas hydrophila, 35, 37, 199
Afipia, 49, 200
AGI, *see* Acute gastrointestinal illness
Ag-impregnated blotting paper, 60–61
Ag-impregnated carbon filters, 59–60
AHL, *see* Acyl homoserine lactones
Airborne transmission (infectious agents), 4
Alcaligenes, 37, 49, 99
Algae and cyanobacteria
    cyanotoxins, 121–122
    impact on water treatment plants, 121
    water treatment options for, 122–127
Alternaria, 127
Amoxilline, 108
Anabaena, 121, 122

Anabaena flos-aquae, 121
Anacystis, 121
Anatoxin, 122
Antibiotic resistance genes (ARG), 108
Antibiotic-resistant bacteria (ARB), 108, 131
AOC, *see* Assimilable organic carbon
Aphanizomenon, 122
AQUAPOT Project, 188
ARB, *see* Antibiotic-resistant bacteria
Arcobacter, 10–11
Arcobacter butzleri, 10
Arcobacter cryaerophilus, 10
Arcobacter skirrowii, 10
ARG, *see* Antibiotic resistance genes
Arthrobacter, 49
Arthropods, 5
Ascaris suum, 84
Aspergillus, 127, 160
Aspergillus flavus, 161
Aspergillus terreus, 68
Assimilable organic carbon (AOC), 38, 39*f*, 144, 147–148
Astroviruses, 4, 20
Aureobasidium, 127
Autoinducers, 101

BAC, *see* Biological activated carbon
Bacillus, 13, 49, 99

Bacillus anthracis, 157–158
Bacillus cereus, 101
Bacillus fusiformis, 120
Bacillus sphaericus, 120
Bacillus subtilis, 75, 84, 182
Bacillus thuringiensis, 6
Bacterial pathogens, 7
　Campylobacter, 10–11
　Escherichia coli, 9
　Helicobacter pylori, 14
　Legionella pneumophila, 11–12
　Leptospira, 14–15
　opportunistic bacterial pathogens, 13–14
　Salmonella, 8
　Shigella, 8–9
　Vibrio cholerae, 8
　web resources, 27
　Yersinia, 9
Bacterial twitter, 101
Bacteroides, 7
Balantidium coli, 130
BAR, *see* Biofilm annular reactor
BDOC, *see* Biodegradable dissolved
　　organic carbon
Beta-Poisson model, 211–212
BFR, *see* Biofilm formation rate
Bifidobacterium, 7
Bioassays, 145–151
Biodegradable dissolved organic carbon
　　(BDOC), 52, 142, 148–150, 149*f*,
　　150*f*
Biodegradation, 120
Biofilm annular reactor (BAR), 150
Biofilm formation rate (BFR), 142
Biofilms, 93
　accumulation factors, 97, 99
　　concentration of assimilable organic
　　　carbon, 97
　　flow velocity and regimen, 98
　　limiting nutrients, 97
　　quorum sensing, 99
　　trace elements, 97
　　type of pipe material, 97–98
　advantages, 109–110
　control and prevention, 112–114
　detachment from surfaces, 101–102
　development processes, 95*f*
　　adhesion of microorganisms to
　　　surfaces, 96

cell anchoring to surfaces, 96
　cell growth and biofilm
　　accumulation, 97
　surface conditioning, 94–95
　transport of microorganisms to
　　conditioned surfaces, 95
　disadvantages, 110–112
　ecology, 99–100
　gene exchange and quorum sensing,
　　100–101
　study, 102–105, 103*t*
Biological activated carbon (BAC), 50
Biological and ToxinWeapons Convention
　　(BTWC) of 1972, 168
Biological treatment (drinking water),
　　141–143
　advantages, 142–143
Biological warfare (BW), 153–154
Biological weapons, 153
Biosand filter, 177–179, 178*f*
Biostability assessment (drinking water)
　bioassays
　　biomass-based methods, 145–148
　　determination of biofilm growth,
　　　149–150
　　DOC-based methods, 148–149
　　impact of leachates, 150–151
　　multiparametric approach to
　　　determine biostability, 150
　introduction, 143–145
Bioterrorism
　biotoxins, 159–161
　BW microbial agents, 154–159
　contamination of water supplies with
　　BW agents/biotoxins, 161–165
　disinfection of BW-contaminated
　　drinking water, 169–170
　history of biological warfare, 153–154
　introduction, 153
　protection of drinking water supplies,
　　168–169
　warning systems for assessing
　　contamination, 165–168
Bioterrorism Act, 169
Biotoxins, 159–161, 160*t*
Bisphenol A (BPA), 197
Bodo, 130
Boiling water, 176–177
Bosea, 200

Bottled water microbiology, 196*f*
  introduction, 195–197
  microorganisms, 199–203
  regulations, 203–204
  sources and categories
    distilled or purified water, 198
    mineral water, 198
    sparkling water, 198
    spring water, 198
    tap water, 198
Bottled water microorganisms
  microorganisms found in bottled water,
    199–203
  source water, 199
  water treatment before bottling, 199
Bromate, 75–76, 197
Brucella, 158
Burkholderia, 99
Burkholderia mallei, 159
BW, *see* Biological warfare
BW microbial agents, 115*t*
  categories of, 156–157, 156*t*
  features, 154–155, 155*t*
  major agents, 157–159

Caenorhabditis elegans, 133
Caliciviruses, 37
Campylobacter, 10–11
Campylobacter fetus, 10
Campylobacter jejuni, 4, 10, 35, 208
Canadian Food and Drug Act, 203
Candida, 128
Candida albicans, 111
Carboxyfluorescein diacetate (CFDA), 104
Cephalosporium, 127
Ceramic water filters, 179–180, 179*f*, 180*t*
Cercariae, 3
CFDA, *see* Carboxyfluorescein diacetate
Chlamydomonas, 121
Chloramination (drinking water), 71–72
Chloramphenicol, 108
Chlor-Floc tablets, 183–184
Chlorine
  chemistry of, 66–67
  chloramination of drinking water, 71–72
  disinfection by-products, 69–71
  inactivation of microorganisms, 67–68,
    68*t*
  on pathogens, adverse effects, 69

Chlorine dioxide, 72–74, 73*f*
Chromobacterium, 49
Ciprofloxacin, 108
Citrobacter, 13, 99
Cladosporium, 127, 128
Cladosporium tenuissimum, 68
Clostridium, 7
Clostridium baratii, 159–160
Clostridium botulinum, 159–160
Clostridium butyricum, 159
Clostridium perfringens, 35
CLSM, *see* Confocal laser scanning
  microscopy
Coagulation, 40, 40*f*, 183–184
Confocal laser scanning microscopy
  (CLSM), 49
Contaminant candidate list 3 (CCL3), 7*t*
Corynebacterium, 49
Coxsackieviruses, 37
Crenothrix, 134
Cryptosporidium, 3, 4, 37, 84, 130, 208
Cryptosporidium andersoni, 24
Cryptosporidium hominis, 23, 78
Cryptosporidium oocysts, 23–24, 40, 46, 68
Cryptosporidium parvum, 22–24, 23*f*, 75
Cryptosporidium spp., 20
Cryptosporidium ubiquitum, 24
Crystal violet assay, 105
CTC, *see* Cyanoditolyl tetrazolium chloride
Curvibacter, 200
Cyanoditolyl tetrazolium chloride (CTC),
  104
Cyanotoxins, 121–122
  water treatment options for, 122–127
Cyclospora, 24
Cyclospora cayetanensis, 24
Cylindrospermopsin, 122
Cylindrospermopsis, 122
Cytotoxins, 122

DAF, *see* Dissolved air flotation
DALY, *see* Disability-adjusted life year
Dark repair, 78
DBP, *see* Disinfection by-products
Delftia sp., 199
Denaturing gradient gel electrophoresis
  (DGGE), 202
Dermatotoxins, 122
Desulfovibrio desulfuricans, 134

DGGE, *see* Denaturing gradient gel
    electrophoresis
6-Diamidino-2-phenylindol (DAPI), 104
Diatomaceous earth filtration, 47–48
Disability-adjusted life year (DALY), 207
Disinfection by-products (DBP), 69–70
Dissolved air flotation (DAF), 124
Dissolved organic carbon (DOC), 143,
    148–149
Distilled/purified water, 198
DLVO theory (Derjaguin, Landau, Verwey,
    and Overbeek theory), 96
DOC, *see* Dissolved organic carbon
Drinking water
    biological treatment and biostability of,
        141–151
    and bioterrorism, 153–171
    bottled water microbiology, 195–204
    disinfection, 65–89
    distribution systems, 91–114
    esthetic and other concerns, 117–137
    microbial contaminants in, 1–27
    microbial risk assessment for, 207–216
    risk assessment for, 207–216
    treatment (microbiological aspects),
        29–62
    treatment technologies for developing
        countries, 173–193
Drinking water disinfection
    chlorine
        chemistry of, 66–67
        chloramination of drinking water,
            71–72
        disinfection by-products, 69–71
        inactivation of microorganisms,
            67–68
        on pathogens, adverse effects, 69
    chlorine dioxide, 72–74
    introduction, 65–66
    ozone
        by-products, 75–76
        inactivation mechanisms, 75
        inactivation of pathogens and
            parasites, 74–75
        introduction, 74
    physical removal/inactivation of
        microbial pathogens, 86
    membrane filtration, 87
    nanomaterials, 87–89

ultrahigh hydrostatic pressure, 87
ultrasound, 87
ultraviolet (UV) light
    categories of UV lamps, 77–78
    controlling factors, 81
    coupling with other technologies, 84
    damage repair, 78–81
    disinfection of drinking water, 84
    introduction, 76–77
    mechanism of damage, 78
    pathogen and protozoan parasites
        inactivation, 81–84, 82*t*
    use of photocatalysts in, 85–86
Drinking water filtration plant, 36*f*
    conventional filter plants, 36
    softening plants, 36
Drinking water quality
    effect of service lines and indoor
        plumbing, 55–57
    modified carbon filters
        Ag-impregnated blotting paper,
            60–61
        Ag-impregnated carbon filters, 59–60
        POU devices, 61
        reverse osmosis (RO), 61
    point-of-use devices, 58–59
Drinking water safety, 30–33
    and bioterrorism, 153–170
Drinking water treatment
    drinking water safety, worldwide
        concern, 30–33
    esthetic and other concerns, 117–138
    history of, 31*t*–32*t*
    introduction, 29
    microbiological quality of source water,
        33–35
    processes (in treatment plant), 35–36
    process microbiology and fate of
        pathogens and parasites, 36–55
    technologies for developing countries,
        173–193
    in United States, 33
DSF, *see* Dual-stage filtration
Dual-stage filtration (DSF), 47

EAggEC, *see* Enteroaggregative
Early warning systems (EWS), 165–168
Echinamoeba, 11
Echoviruses, 37

EHEC, *see* Enterohemorrhagic
EIEC, *see* Enteroinvasive
Encephalitozoon cuniculi, 24
Encephalitozoon intestinalis, 25, 84
Endotoxins, 41, 134
Entamoeba histolytica, 25, 47, 130
Enteric adenoviruses, 20
Enteric viruses, 109
Enteroaggregative (EAggEC), 9
Enterobacter, 13, 49, 99
Enterobacter cloacae, 111
Enterococci, 35
Enterococcus spp., 111
Enterocytozoon bienusi, 25
Enterohemorrhagic (EHEC), 9
Enteroinvasive (EIEC), 9
Enteropathogenic (EPEC), 9
Enterotoxigenic (ETEC), 9
Enteroviruses, 4
Environmental SEM (ESEM), 104
EPA, *see* US Environmental Protection
    Agency
EPEC, *see* Enteropathogenic
Epicoccum, 128
Erosion, 101
Escherichia coli, 3, 6, 9, 10*t*, 35, 44, 45, 68,
    75, 111
Escherichia coli O157:H7, 3, 4, 9
ESEM, *see* Environmental SEM
ETEC, *see* Enterotoxigenic
Euplotes, 130
EWS, *see* Early warning systems
Exponential model, 211

FDA, *see* Fluorescein diacetate
FESEM, *see* Field-emission scanning
    electron microscopy
Field-emission scanning electron
    microscopy (FESEM), 48
Filtration, 42
FISH, *see* Fluorescence *in situ* hybridization
FITC, *see* Fluorescein isothiocyanate
Flavobacterium, 13, 34, 37, 49, 99
Flocculation, 40, 183–184
Fluorescein diacetate (FDA), 22, 104
Fluorescein isothiocyanate (FITC), 22
Fluorescence *in situ* hybridization (FISH),
    202
Fomites, 5, 5*t*

Food, Drug and Cosmetic Act, 203
Foodborne transmission, infectious agents,
    4
  in United States, 4*t*
Francisella tularensis, 158–159
Free-living amoebas (FLA), 14, 131–132,
    132*t*
Fungi, 127–129
Fusarium, 127, 160

GAC, *see* Granular activated carbon
Gallionella, 111, 134
Gastroenteritis, 18
Gentamicin, 108
Geosmin, 118
Giardia, 3, 20–22
Giardia cysts, 21–22, 46
Giardia lamblia, 20, 23*f*, 35, 37, 68,
    130
Giardiasis, 3
Giardia spp., 21
Granular activated carbon (GAC), 48
Groundwater, 34–35

HAA, *see* Haloacetic acids
Haloacetic acids (HAA), 69–70
Hartmanella vermiformis, 57, 131
Hartmannella, 130, 131
HAV, *see* Hepatitis A virus
HBV, *see* Hepatitis B virus
HCV, *see* Hepatitis C virus
Health-based targets (for drinking water),
    207–208
Helicobacter pylori, 14, 37, 101
Hemolytic uremic syndrome (HUS), 8
Hemorrhagic fever viruses, 159
Hepatitis A virus (HAV), 4, 16–17
Hepatitis B virus (HBV), 16
Hepatitis C virus (HCV), 16
Hepatitis E virus (HEV), 16, 17
Hepatoxins, 122
Herminiimonas glaciei, 199
Heterotrophic plate count (HPC), 102
HEV, *see* Hepatitis E virus
Household water treatment (HWT), 175*f*,
    176*f*, 187*f*, 190*f*
  AQUAPOT Project, 188
  biosand filter, 177–179, 178*f*
  boiling water, 176–177

Household water treatment (HWT)
    (*Continued*)
    ceramic water filters, 179–180, 179*f*,
        180*t*
    coagulation–flocculation–sedimentation,
        183–184
    LifeStraw^R Family, 188–189, 188*f*
    safe water system (SWS), 184–187
    solar disinfection, 180–183
    UV treatment units, 189
    water purification tablets, 183
HPC, *see* Heterotrophic plate count
Human caliciviruses, 18
HUS, *see* Hemolytic uremic syndrome
HWT, *see* Household water treatment
Hydrogenophaga, 49
Hyphomicrobium, 111
Hypochlorous acid, 67, 69

Information Sharing and Analysis Centers
    (ISAC), 169
INT, *see* 2-*p*-(iodophenyl)-3-
    (*p*-nitrophenyl)-5- tetrazolium
    chloride
Invertebrates
    control of microinvertebrates, 133
    as reservoir of pathogens and parasites,
        133
    in water distribution systems, 132–133
Iron bacteria, 134
ISAC, *see* Information Sharing and
    Analysis Centers

Klebsiella, 13, 99
Klebsiella pneumoniae, 111, 145
Klebsiella terrigena, 40

LCR, *see* Lead and copper rule
Lead and copper rule (LCR), 55
Legionella, 4, 12*f*, 37, 107–108, 111,
    212–214
Legionella pneumophila, 11–12, 34, 35,
    107–108
Legionella spp., 57
Leptospira, 14–15
Leptothrix, 134
LifeStraw^R Family, 188–189, 188*f*
Lipopolysaccharides (LPS), 41
Listeria monocytogenes, 4

Long Term 2 Enhanced Surface Water
    Treatment Rule (LT2ESWTR), 82
LT2ESWTR, *see* Long Term 2 Enhanced
    Surface Water Treatment Rule
LuxS/autoinducer 2, 101
Lysobacter (gamma-proteobacteria), 99

M. avium subspecies paraturberculosis
    (MAP), 13
MAC, *see* *Mycobacterium avium* complex
MAIC, *see* *Mycobacterium avium*
    intracellulare complex
Manganese bacteria, 134
MAP, *see* *M. avium subspecies*
    paraturberculosis
MDG, *see* Millennium Development Goal
Membrane filtration, 51–53, 52*t*, 87
2-Methyl isoborneol (MIB), 118
Methylobacterium, 99
Methylophilus (betaproteobacteria), 99
Methyl tertiary butyl ether (MTBE), 118
MIB, *see* 2-methyl isoborneol
Microbial contaminants (in drinking water),
    26*t*
    pathogens and parasites, transmission
        routes to, 2*f*
        airborne transmission, 4
        fomites, 5
        foodborne transmission, 4
        person-to-person transmission, 1–3
        vector-borne transmission, 5
        waterborne transmission, 3
    pathogens and parasites of health
        concern, 6, 6*t*
        bacterial pathogens, 7–15
        protozoan parasites, 7, 20–26
        viral pathogens, 7, 15–20
Microbial pathogens and parasites (in water
    treatment plants)
    activated carbon, 48–51
    coagulation-flocculation-sedimentation,
        39–41, 40*f*
    disinfection, 55
    filtration
        diatomaceous earth filtration, 47–48
        rapid sand filtration, 46–47
        slow sand filtration, 42–46, 43*f*
    introduction, 36–38
    membrane filtration, 51–53, 52*t*

nanotechnology, 53–55
pretreatment of source water
   microstrainers, 38
   prechlorination, 39
   river bank filtration, 38–39
   roughing filters, 38
   storage of raw water, 38
   water softening, 41–42
Micrococcus, 13, 49
Microcystin, 161
Microcystis, 121, 122
Microcystis aeruginosa, 40, 79, 121, 161
Micromonospora, 130
Microsporidia, 25
Microsporidia protozoan parasites, 37
Microstrainers, 38
MID, *see* Minimal infective dose
Millennium Development Goal (MDG), 174
Mineral water, 198
Minimal infective dose (MID), 6, 6*t*
Moraxella, 13, 49, 99
MTBE, *see* Methyl tertiary butyl ether
Mucor, 127, 128
Munition, 153
Mycobacterium, 4, 34, 111
Mycobacterium avium, 14, 107, 111
Mycobacterium avium complex (MAC), 13
Mycobacterium avium intracellulare
   complex (MAIC), 37
Mycobacterium chelonae, 14, 107
Mycobacterium fortuitum, 74
Mycobacterium gordonae, 14
Mycobacterium intracellulare, 111
Mycobacterium mucogenicum, 14
Mycobacterium peregrinum, 14
Mycobacterium spp., 57
Mycotoxins, 160–161

Naegleria, 25
Naegleria fowleri, 25, 34, 83, 130
Naegleria. Hartmannella, 11
Nanomaterials, 87–89
Nanotechnology (in water treatment plants),
   53–55
NAS, *see* US National Academy of Science
Natural organic matter (NOM), 119, 143
NDMA, *see* *N*-nitrosodimethylamine
Neurotoxins, 122
Nitrifying bacteria

in chloraminated water, 135–136
   control in water distribution systems,
      136–137
Nitrobacter, 135
Nitrococcus, 135
Nitrosococcus, 135
Nitrosolobus, 135
Nitrosomonas, 135
Nitrosospira, 135
Nitrosovibrio, 135
Nitrospira, 135
NLV, *see* Norwalk-like virus
N-nitrosodimethylamine (NDMA), 72
Nocardia, 98, 130
NOM, *see* Natural organic matter
Nontubercular mycobacteria, 106
Nontubercular mycobacteria (NTM),
   13–14
Noroviruses, 4
Norwalk-like virus (NLV), 18, 19*f*, 19*t*
Nostoc, 122
Novosphingobium, 99
NTM, *see* Nontubercular mycobacteria

Ochrobactrum intermedium, 199
Opportunistic bacterial pathogens, 13–14,
   105–106
Opportunistic pathogens, 37, 199
Oscillatoria, 121, 122
Oxidation processes, 120
Ozone
   by-products, 75–76
   inactivation mechanisms, 75
   inactivation of pathogens and parasites,
      74–75
   introduction, 74

PAC, *see* Powdered activated carbon
PAHO, *see* Pan American Health
   Organization
PAME, *see* Primary meningoencephalitis
Pan American Health Organization
   (PAHO), 184
Paracoccus, 49
Payload, 153
Pedobacter, 200
Penicillium, 127, 160
Peptides, 101
Perchlorate, 197

Person-to-person transmission, infectious
agents, 1–3
PET, *see* Polyethylene terephthalate
Phialophora, 128
Phoma, 128
Phormidium, 122
Photocatalysts, 85–86
Photooxidation, 181
Photoreactivation, 78–79, 78–81, 80*f*
PI, *see* Propidium iodide
2-*p*-(Iodophenyl)-3-(*p*-nitrophenyl)-5-
tetrazolium chloride (INT), 104
Planktothrix, 122
Point-of-entry (POE) devices, 58, 58*t*
Point-of-use (POU) devices, 58–59, 58*t*, 59*f*
Polaromonas, 49
Polyelectrolytes, 41
Polyethylene terephthalate (PET), 197
Polymyxin E, 13
Polyvinyl chloride (PVC), 91
Portable water treatment systems, 189–192,
192*t*
Powdered activated carbon (PAC), 48, 119
Prechlorination, 39
Prevotella, 7
Primary meningoencephalitis (PAME), 83
Propidium iodide (PI), 22
Proteus, 13
Protozoa
FLAs as reservoirs of pathogens,
131–132
FLAs grazing role in water treatment
plants, 131
in water treatment plants, 130
Protozoan parasites, 108–109
Cryptosporidium parvum, 22–24
Cyclospora, 24
Entamoeba histolytica, 25
Giardia, 20–22
microsporidia, 25
Naegleria, 25
Toxoplasma gondii, 25–26
web resources, 27
Providencia, 13
Pseudomonas, 13, 37, 49, 98, 99
Pseudomonas aeruginosa, 13, 34, 84, 96*f*,
111, 144
Pseudomonas fluorescens, 144, 146
Pseudomonas spp., 199

Public Health Security and Prevention
Preparedness Act, 169
PVC, *see* Polyvinyl chloride

QCM, *see* Quartz crystal microbalance
QMRA, *see* Quantitative microbial risk
assessment
Quantitative microbial risk assessment
(QMRA), 208–209
dose-response assessment, 210–212
examples
Legionella, 212–214
protozoan parasites, 214–216
viruses, 214
exposure assessment, 210
hazard identification and
characterization, 210
risk characterization, 212
risk management and communication,
212
Quartz crystal microbalance (QCM), 104

Ralstonia, 99, 200
Rapid sand filtration, 46–47
Reactive oxygen species (ROS), 85, 180
Resanding, 44
Reverse osmosis (RO), 61
Reverse transcriptase-polymerase chain
reaction (RT-PCR), 17
Rhizoctonia, 128
Rhodotorula, 128
Ricin toxin, 161
Rifampin, 108
River bank filtration (RBF), 38–39
Roof-harvested rainwater, 35
ROS, *see* Reactive oxygen species
Rotaviruses, 4, 17, 18*f*, 208
Roughing filters, 38
RT-PCR, *see* Reverse
transcriptase-polymerase chain
reaction

Safe Drinking Water Act (SDWA), 21, 91
Safe water system (SWS), 184–187
Salmonella, 4, 8, 35, 46
Salmonella paratyphi, 8
Salmonella typhi, 8
Salmonella typhimurium, 8, 75
Saxitoxin, 161

Scanning confocal microscopy (SCLM), 104
Scanning electron microscopy (SEM), 104
Scenedesmus, 121
Schizothrix calcicola, 134
Schmutzdecke, 44
SCLM, *see* Scanning confocal microscopy
Scouring, 101
SDWA, *see* Safe Drinking Water Act
Selenastrum, 121
SEM, *see* Scanning electron microscopy
Serratia, 13
SFBW, *see* Spent filter backwash water
Shellfish, 4
Shiga toxin, 8
Shigella, 8–9, 46
Shigella boydii, 8
Shigella dysenteriae, 8
Shigella flexneri, 8
Shigella sonnei, 8
Sinura uvella, 121
Sloughing, 101
Slow sand filtration, 42–46, 43*f*
Smallpox virus, 159
SODIS, *see* Solar water disinfection
Solar disinfection, 180–183
Solar water disinfection (SODIS), 180–181, 183*t*
Sparkling water, 198
Spent filter backwash water (SFBW), 47
Sphaerotilus, 134
Sphingomonadaceae, 120
Sphingomonas, 49, 99
Sphingomonas sp., 199
Sphingopyxis, 99
Sphingopyxis alaskensis, 120
Spirillum, 146
Spring water, 198
Stachybotrys, 128
Staphylococcus aureus, 6, 111
Stenotrophomonas, 98
Stenotrophomonas maltophilia, 100, 199, 200
Streptococcus anginosus, 101
Streptomyces, 130
Sulfisoxazole, 108
Sulfur bacteria, 134
Surface waters, 34
Surface Water Treatment Rule (SWTR), 48

SWS, *see* Safe water system
SWTR, *see* Surface Water Treatment Rule
Synura petersenii, 121

Tap water, 198
Taste and odors problems (in water treatment plants), 117
control approaches, 119–120
sources, 118–119
TEM, *see* Transmission electron microscopy
Tetrahymena, 12
Tetrodotoxin, 161
Thiobacillus concretivorus, 134
THM, *see* Trihalomethanes
TOC, *see* Total organic carbon
Total organic carbon (TOC), 143
Total organic halogen (TOX), 71
TOX, *see* Total organic halogen
Toxoplasma gondii, 25–26
Toxothrix, 134
Trace elements, 97
Transmission electron microscopy (TEM), 104
Trichoderma, 127
Trihalomethanes (THM), 69, 70–71
Triphenyl tetrazolium chloride (TTC), 104
TTC, *see* Triphenyl tetrazolium chloride

UHP, *see* Ultrahigh hydrostatic pressure
Ultrahigh hydrostatic pressure (UHP), 87
Ultrasound, 87
Ultraviolet (UV) light
categories of UV lamps, 77–78
controlling factors, 81
coupling with other technologies, 84
damage repair, 78–81
disinfection of drinking water, 84
introduction, 76–77
mechanism of damage, 78
pathogen and protozoan parasites inactivation, 81–84
UNESCO, *see* United Nations Educational Scientific and Cultural Organization
United Nations Educational Scientific and Cultural Organization (UNESCO), 173

US Environmental Protection Agency (EPA), 203
US National Academy of Science (NAS), 207
UV treatment units, 189

Vahlkampfia, 11
Variovorax, 99, 200
VBNC, *see* Viable but not culturable
Vector-borne transmission (infectious agents), 4
Vertebrates, 5
Verticillium, 127
Viable but not culturable (VBNC), 8, 14, 34, 106, 144, 201
Vibrio cholerae, 8, 179
Vibrio sp., 4
Viral pathogens, 15–16, 16*t*
  hepatitis viruses, 16–17
  viral gastroenteritis, 17–20
  web resources, 27
VOC, *see* Volatile organic compounds
Volatile organic compounds (VOC), 197
Vorticella, 130

Waste residuals, water treatment plants, 55
Water-based diseases, 3
Waterborne diseases, 3
  caused by bacteria, 15*t*
  caused by protozoa, 21*t*
  in developing countries, 174–175
Waterborne transmission (infectious agents), 3
Water distribution system (WDS)
  advantages of biofilms, 109–110
  biofilm control and prevention, 112–114
  biofilm development in
    accumulation factors, 97–99
    detachment from surfaces, 101–102
    ecology, 99–100
    gene exchange and quorum sensing, 100–101
    introduction, 93
    methods used in biofilm study, 102–105
    processes, 94–97

disadvantages of biofilms, 110–112
esthetic and other concerns, 117–138
growh of pathogens and other microorganisms in
  antibiotic resistance genes (ARG), 108
  antibiotic-resistant bacteria (ARB), 108
  earlier studies, 105
  enteric viruses, 109
  Legionella in distribution system, 107–108
  Legionella in hot-water tanks, 107–108
  nontubercular mycobacteria, 106
  opportunistic bacterial pathogens, 105–106
  protozoan parasites, 108–109
  introduction, 91–93
  major problems
    biocorrosion, 92
    microbial growth, 92
    nitrification problems, 92
    pathogen survival and growth, 92
  nitrifying bacteria in, 135–137
Water purification tablets, 183
Water-related diseases, 3
Water softening, 41–42
Water sources, microbiological quality
  groundwater, 34–35
  roof-harvested rainwater, 35
  surface waters, 34
Water-washed diseases, 3
WDS, *see* Water distribution system
WHO, *see* World Health Organization
World Health Organization (WHO), 30

Xanthobacter, 98
Xanthomonas, 99

Yersinia, 9
Yersinia enterocolitica, 9
Yersinia pestis, 158
Yersinia spp., 9

Zoonoses, 6